# Student Solutions Manual

## J. Richard Christman
**Professor Emeritus**
*U. S. Coast Guard Academy*

# UNDERSTANDING PHYSICS

**Karen Cummings**
*Rensselaer Polytechnic Institute*

**Priscilla W. Laws**
*Dickinson College*

**Edward F. Redish**
*University of Maryland*

**Patrick J. Cooney**
*Millersville University*

WILEY

**JOHN WILEY & SONS, INC.**

Cover image:   ©Antonio M. Rosario/The Image Bank/Getty Image

To order books or for customer service, please call 1-800-CALL-WILEY (225-5945).

ISBN 0-471-46439-2

Printed in the United States of America

10 9 8 7 6 5 4 3 2 1

Printed and bound by Courier Kendalliville, Inc.

# PREFACE

This solutions manual is designed for use with the textbook *Understanding Physics* by Karen Cummings, Priscilla Laws, Edward Redish, and Patrick Cooney. Its primary purpose is to show students by example how to solve various types of problems given at the ends of chapters in the text.

Most of the solutions start from definitions or fundamental relationships and the final equation is derived. This technique highlights the fundamentals and at the same time gives students the opportunity to review the mathematical steps required to obtain a solution. The mere plugging of numbers into equations derived in the text is avoided for the most part. We hope students will learn to examine any assumptions that are made in setting up and solving each problem.

Selection of the problems included in this manual and their solutions are the responsibility of the author alone.

**Acknowledgements**   Many good people at John Wiley & Sons helped put together *A Student's Study Guide*. Thanks go to Stuart Johnson, the Acquisitions Editor, for suggesting the scope of the project. The Project Editor Geraldine Osnato oversaw the complicated logistics involved and handled a myriad of day-to-day chores associated with writing and publishing the solution manual. Rosa Bryant was instrumental in seeing the project through production. The author is also grateful for the encouragement and strong support of his wife, Mary Ellen Christman.

J. Richard Christman
Professor Emeritus
U.S. Coast Guard Academy
New London, CT 06320

# TABLE OF CONTENTS

# Chapter 1

## 1

(a) Use the conversion factors given in Appendix D and the definitions of the SI prefixes given in Table 1–2: $1\,\text{m} = 3.281\,\text{ft}$ and $1\,\text{s} = 10^9\,\text{ns}$. Thus

$$3.0 \times 10^8\,\text{m/s} = \left(\frac{3.0 \times 10^8\,\text{m}}{\text{s}}\right)\left(\frac{3.281\,\text{ft}}{\text{m}}\right)\left(\frac{\text{s}}{10^9\,\text{ns}}\right) = 0.98\,\text{ft/ns}.$$

(b) Use $1\,\text{m} = 10^3\,\text{mm}$ and $1\,\text{s} = 10^{12}\,\text{ps}$. Thus

$$3.0 \times 10^8\,\text{m/s} = \left(\frac{3.0 \times 10^8\,\text{m}}{\text{s}}\right)\left(\frac{10^3\,\text{mm}}{\text{m}}\right)\left(\frac{\text{s}}{10^{12}\,\text{ps}}\right) = 0.30\,\text{mm/ps}.$$

## 5

You need to convert meters to astronomical units and seconds to minutes. Use $1\,\text{m} = 1 \times 10^{-3}\,\text{km}$, $1\,\text{AU} = 1.50 \times 10^8\,\text{km}$, and $60\,\text{s} = 1\,\text{min}$. Thus

$$3.0 \times 10^8\,\text{m/s} = \left(\frac{3.0 \times 10^8\,\text{m}}{\text{s}}\right)\left(\frac{10^{-3}\,\text{km}}{\text{m}}\right)\left(\frac{\text{AU}}{1.50 \times 10^8\,\text{km}}\right)\left(\frac{60\,\text{s}}{\text{min}}\right) = 0.12\,\text{AU/min}.$$

## 11

Use the given conversion factors.

(a) The distance $d$ in rods is

$$d = 4.0\,\text{furlongs} = \frac{(4.0\,\text{furlongs})(201.168\,\text{m/furlong})}{5.0292\,\text{m/rod}} = 160\,\text{rods}.$$

(b) The distance in chains is

$$d = 4.0\,\text{furlongs} = \frac{(4.0\,\text{furlongs})(201.168\,\text{m/furlong})}{20.17\,\text{m/chain}} = 40\,\text{chains}.$$

## 15

The volume of ice is given by the product of the area of the semicircle and the thickness. The area is half the area of a circle: $A = \pi R^2/2$, where $R$ is the radius. Thus the volume is given by $V = \pi R^2 T/2$, where $T$ is the thickness. Since there are $10^3$ m in 1 km and $10^2$ cm in 1 m,

$$R = (2000\,\text{km})(10^3\,\text{m/km})(10^2\,\text{cm/m}) = 2000 \times 10^5\,\text{cm}.$$

Also substitute

$$T = (3000 \, \text{m})(10^2 \, \text{cm/m}) = 3000 \times 10^2 \, \text{cm} \,.$$

Thus

$$V = \frac{1}{2}\pi(2000 \times 10^5 \, \text{cm})^2(3000 \times 10^2 \, \text{cm}) = 1.9 \times 10^{22} \, \text{cm}^3 \,.$$

## 19

If $M$ is the mass of Earth, $m$ is the average mass of an atom in Earth, and $N$ is the number of atoms, then $M = Nm$ or $N = M/m$. The values for $M$ and $m$ must have the same units. Convert $m$ to kg. According to Appendix D, $1 \, \text{u} = 1.661 \times 10^{-27} \, \text{kg}$. Thus

$$N = \frac{M}{m} = \frac{5.98 \times 10^{24} \, \text{kg}}{(40 \, \text{u})(1.661 \times 10^{-27} \, \text{kg/u})} = 9.0 \times 10^{49} \,.$$

## 23

(a) Let $\rho$ be the mass per unit volume of iron. It is the same for a single atom and a large chunk. If $M$ is the mass and $V$ is the volume of an atom, then $\rho = M/V$, or $V = M/\rho$. To obtain the volume in $\text{m}^3$, first convert $\rho$ to $\text{kg/m}^3$: $\rho = (7.87 \, \text{g/cm}^3)(10^{-3} \, \text{kg/g})(10^6 \, \text{cm}^3/\text{m}^3) = 7.87 \times 10^3 \, \text{kg/m}^3$. Then

$$V = \frac{M}{\rho} = \frac{9.27 \times 10^{-26} \, \text{kg}}{7.87 \times 10^3 \, \text{kg/m}^3} = 1.18 \times 10^{-29} \, \text{m}^3 \,.$$

(b) Set $V = 4\pi R^3/3$, where $R$ is the radius of an atom, and solve for $R$:

$$R = \left(\frac{3V}{4\pi}\right)^{1/3} = \left[\frac{3(1.18 \times 10^{-29} \, \text{m}^3)}{4\pi}\right]^{1/3} = 1.41 \times 10^{-10} \, \text{m} \,.$$

The center-to-center distance between atoms is twice the radius or $2.82 \times 10^{-10} \, \text{m}$.

## 27

The volume of the box is the volume of one mole of sugar cubes. This is

$$(6.02 \times 10^{23} \, \text{sugar cubes})(1 \, \text{cm/sugar cube})^3 = 6.02 \times 10^{23} \, \text{cm}^3 \,.$$

The length of a cube edge is the cube root of this or $8.4 \times 10^7 \, \text{cm}$.

## 29

The mass of a single hydrogen atom is $m = 1.0 \, \text{u} = 1.66 \times 10^{-27} \, \text{kg}$. If the total mass of the collection is $M$ and there are $N$ hydrogen atoms in it, then $M = Nm$ and

$$N = \frac{M}{m} = \frac{1.0 \, \text{kg}}{1.66 \times 10^{-27} \, \text{kg}} = 6.0 \times 10^{26} \, \text{atoms} \,.$$

**31**

(a) According to the diagram 212 S is equivalent to 258 W, so

$$50\,S = \frac{258\,W}{212\,S}(50\,S) = 60.8\,W\,.$$

(b) According to the diagram 212 S − 32 S (= 180 S) is equivalent to 216 Z − 60 Z (= 156 Z), so

$$50\,S = \frac{156\,Z}{180\,S}(50\,S) = 43.3\,Z\,.$$

**37**

The volume of the container in pints is

$$V = \frac{(1.0\,m)(0.12\,m)(0.20\,m)(1000\,L/m^3)}{0.4732\,L/pint} = 50.7\,pints\,,$$

where the conversion factor for cubic meters and liters was taken from Appendix D. You expected to get between

$$(50.7\,pints)(26\,oysters/pint) = 1318\,oysters$$

and

$$(50.7\,pints)(38\,oysters/pint) = 1927\,oysters\,.$$

You actually got between

$$(50.7\,pints)(8\,oysters/pint) = 405\,oysters$$

and

$$(50.7\,pints)(12\,oysters/pint) = 608\,oysters\,.$$

If you expected the least possible number and got the greatest possible number you were 1318 − 608 = 710 oysters short. If you expected greatest possible number and got the least possible number you were 1927 − 405 = 1522 oysters short.

**43**

You actually sailed (24.5 mi)(1.609 km/mi) = 39.4 km.
You should have sailed (24.5 nautical miles)(1.852 m/nautical mile) = 45.4 km.
You are 45.4 km − 39.4 km = 5.97 km from the pirate ship.
Conversion factors for the conversions of miles to kilometers and nautical miles to kilometers can be found in Appendix D.

**45**

The volume of mud that slips into the valley is

$$V = (2.5 \times 10^3\,m)(0.80 \times 10^3\,m)(2.0\,m) = 4.0 \times 10^6\,m^3\,.$$

This must equal the product of the area $A$ covered after the slide and the depth $d$ then: $V = Ad$. So the depth is

$$d = \frac{V}{A} = \frac{4.0 \times 10^6 \, \text{m}^3}{(0.40 \times 10^3 \, \text{m}^2} = 250 \, \text{m}.$$

The volume $V'$ of mud covering an area $A'$ ($= 4.0 \, \text{m}^2$) is

$$V' = A'd = (4.0 \, \text{m}^2)(250 \, \text{m} = 100 \, \text{m}^3$$

and its mass is

$$m = (100 \, \text{m}^3)(1900 \, \text{kg/m}^3) = 1.9 \times 10^5 \, \text{kg}.$$

# Chapter 2

## 1

Assume the ball travels with constant speed and use $\Delta x = v\Delta t$, where $\Delta x$ is the horizontal distance traveled, $\Delta t$ is the time, and $v$ is the speed. Convert $v$ to meters per second. According to Appendix D, $1\,\text{km/h} = 0.2778\,\text{m/s}$, so $160\,\text{km/h} = (160)(0.2778\,\text{m/s}) = 44.45\,\text{m/s}$. Thus

$$\Delta t = \frac{\Delta x}{v} = \frac{18.4\,\text{m}}{44.5\,\text{m/s}} = 0.414\,\text{s}\,.$$

This may also be written $414\,\text{ms}$.

## 7

(a) Substitute, in turn, $t = 1$, 2, 3, and 4 s into the expression $x = (3\,\text{m/s})t - (4\,\text{m/s}^2)t^2 + (1\,\text{m/s}^3)t^3$:

$$x(1\,\text{s}) = (3\,\text{m/s})(1\,\text{s}) - (4\,\text{m/s}^2)(1\,\text{s})^2 + (1\,\text{m/s}^3(1\,\text{s})^3 = 0$$
$$x(2\,\text{s}) = (3\,\text{m/s})(2\,\text{s}) - (4\,\text{m/s}^2)(2\,\text{s})^2 + (1\,\text{m/s}^3)(2\,\text{s})^3 = -2\,\text{m}$$
$$x(3\,\text{s}) = (3\,\text{m/s})(3\,\text{s}) - (4\,\text{m/s}^2)(3\,\text{s})^2 + (1\,\text{m/s}^3)(3\,\text{s})^3 = 0$$
$$x(4\,\text{s}) = (3\,\text{m/s})(4\,\text{s}) - (4\,\text{m/s}^2)(4\,\text{s})^2 + (1\,\text{m/s}^3)(4\,\text{s})^3 = 12\,\text{m}\,.$$

(b) The displacement during an interval is the position at the end of the interval minus the position at the beginning. For the interval from $t = 0$ to $t = 4\,\text{s}$, the displacement is

$$\Delta\vec{x} = [x(4\,\text{s}) - x(0)]\,\hat{\imath} = (12\,\text{m} - 0)\,\hat{\imath} = (+12\,\text{m})\,\hat{\imath}\,.$$

The displacement is in the positive $x$ direction.

(c) The average velocity during an interval is defined as the displacement over the interval divided by the duration of the interval: $\langle\vec{v}\rangle = (\Delta x/\Delta t)\hat{\imath}$. For the interval from $t = 2\,\text{s}$ to $t = 4\,\text{s}$ the displacement is

$$\Delta x = x(4\,\text{s}) - x(2\,\text{s}) = 12\,\text{m} - (-2\,\text{m}) = 14\,\text{m}$$

and the time interval is

$$\Delta t = 4\,\text{s} - 2\,\text{s} = 2\,\text{s}\,.$$

Thus

$$\langle\vec{v}\rangle = \frac{\Delta x}{\Delta t}\,\hat{\imath} = \frac{14\,\text{m}}{2\,\text{s}}\,\hat{\imath} = (7\,\text{m/s})\,\hat{\imath}\,.$$

(d) The solid curve on the graph to the right shows the coordinate $x$ as a function of time. The slope of the dotted line is the average velocity between $t = 2.0\,\text{s}$ and $t = 4.0\,\text{s}$.

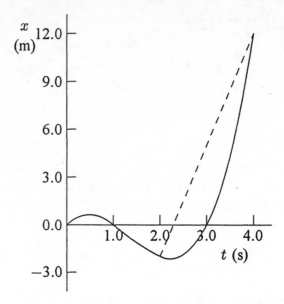

## 17

If $\vec{v}_1$ is the velocity at the beginning of a time interval (at time $t_1$) and $\vec{v}_2$ is the velocity at the end (at time $t_2$), then the average acceleration in the interval is given by $\langle\vec{a}\rangle = (\vec{v}_2 - \vec{v}_1)/(t_2 - t_1)$. Take $t_1 = 0$, $\vec{v}_1 = (18\,\text{m/s})\hat{\imath}$, $t_2 = 2.4\,\text{s}$, and $\vec{v}_2 = -(30\,\text{m/s})\hat{\imath}$. Then

$$\langle\vec{a}\rangle = \frac{-(30\,\text{m/s})\hat{\imath} - (18\,\text{m/s})\hat{\imath}}{2.4\,\text{s}} = -(20\,\text{m/s}^2)\hat{\imath}.$$

The negative sign indicates that the acceleration is opposite to the original direction of travel (the positive $x$ direction).

## 21

(a) Suppose the muon is traveling in the positive $x$ direction. Then $v_{2x} = v_{1x} + a_x(t_2 - t_1)$. The time to stop can be found by setting $v_{2x} = 0$ and solving for $t_2 - t_1$: $t_2 - t_1 = -v_{1x}/a_x$. Substitute this expression into

$$x_2 - x_1 = v_{1x}(t_2 - t_1) + \tfrac{1}{2}a_x(t_2 - t_1)^2$$

to obtain

$$x_2 - x_1 = -\frac{v_{1x}^2}{2a_x}.$$

Use $v_{1x} = 5.00 \times 10^6\,\text{m/s}$ and $a_x = -1.25 \times 10^{14}\,\text{m/s}^2$. Notice that since the muon slows the initial velocity and the acceleration must have opposite signs. The result is

$$x_2 - x_1 = -\frac{(5.00 \times 10^6\,\text{m/s})^2}{2(-1.25 \times 10^{14}\,\text{m/s}^2)} = 0.100\,\text{m}.$$

(b) Here are the graphs of the coordinate $x$ and velocity component $v_x$ of the muon from the time it enters the field to the time it stops:

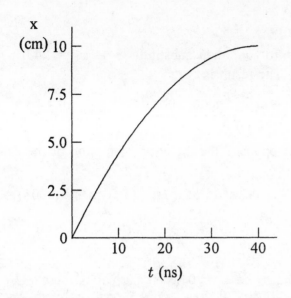

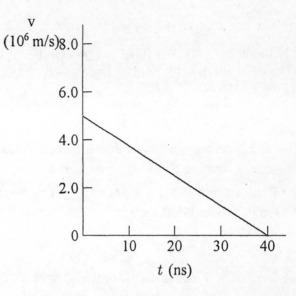

## 23

Use $v_{2x} = v_{1x} + a_x(t_2 - t_1)$, an equation that is valid for motion with constant acceleration along the $x$ axis. Take $t_1 = 0$ to be the time when the velocity component is $+9.6\,\text{m/s}$. Then $v_{1x} = 9.6\,\text{m/s}$.

(a) Since we wish to calculate the velocity for a time *before* $t_1 = 0$, the value we use for $t_2$ is negative: $t_2 = -2.5\,\text{s}$. Thus

$$v_{2x} = (9.6\,\text{m/s}) + (3.2\,\text{m/s}^2)(-2.5\,\text{s}) = 1.6\,\text{m/s}\,.$$

(b) Now $t_2 = +2.5\,\text{s}$ and

$$v_{2x} = (9.6\,\text{m/s}) + (3.2\,\text{m/s}^2)(2.5\,\text{s}) = 18\,\text{m/s}\,.$$

## 25

(a) Solve $v_{2x} = v_{1x} + a_x(t_2 - t_1$ for $t_2 - t_1$: $t_2 - t_1 = (v_{2x} - v_{1x})/a_x$. Substitute $v_{2x} = 0.1(3.0 \times 10^8\,\text{m/s}) = 3.0 \times 10^7\,\text{m/s}$, $v_{1x} = 0$, and $a_x = 9.8\,\text{m/s}^2$. The result is $t_2 - t_1 = 3.1 \times 10^6\,\text{s}$. This is 1.2 months.

(b) Evaluate $x_1 - x_1 = v_{1x}(t_2 - t_1) + \frac{1}{2}a_x(t_2 - t_1)^2$. The result is $x_2 - x_1 = \frac{1}{2}(9.8\,\text{m/s}^2)(3.1 \times 10^6\,\text{s})^2 = 4.7 \times 10^{13}\,\text{m}$.

## 27

Use $v_{2x} = v_{1x} + a_x(t_2 - t_1)$ to eliminate $t_2 - t_1$ from $x_2 - x_1 = v_{1x}(t_2 - t_1) + \frac{1}{2}a_x(t_2 - t_1)^2$ and obtain $v_{2x}^2 = v_{1x}^2 + 2a_x(x_2 - x_1)$. Solve this equation for $a_x$. Substitute $v_{1x} = 1.50 \times 10^5\,\text{m/s}$, $v_{2x} = 5.70 \times 10^6\,\text{m/s}$, and $x_2 - x_1 = 1.0\,\text{cm} = 0.010\,\text{m}$. The result is

$$a_x = \frac{v_{2x}^2 - v_{1x}^2}{2(x_2 - x_1)} = \frac{(5.70 \times 10^6\,\text{m/s})^2 - (1.50 \times 10^5\,\text{m/s})^2}{2(0.010\,\text{m})} = 1.62 \times 10^{15}\,\text{m/s}^2\,.$$

**31**

(a) Solve $x_2 - x_1 = v_{1x}(t_2 - t_1) + \frac{1}{2}a_x(t_2 - t_1)^2$ for $a_x$, then substitute $x_2 - x_1 = 24.0\,\text{m}$, $v_{1x} = 56.0\,\text{km/h} = 15.55\,\text{m/s}$, and $t_2 - t_1 = 2.00\,\text{s}$. The result is

$$a_x = \frac{2\left[(x_2 - x_1) - v_{1x}(t_2 - t_1)\right]}{(t_2 - t_1)^2} = \frac{2\left[24.0\,\text{m} - (15.55\,\text{m/s})(2.00\,\text{s})\right]}{(2.00\,\text{s})^2} = -3.56\,\text{m/s}^2\,.$$

The negative sign indicates that the acceleration is opposite the direction of motion of the car. The car is slowing down.

(b) Evaluate $v_{2x} = v_{1x} + a_x(t_2 - t_1)$. You should get $v_{2x} = 15.55\,\text{m/s} - (3.56\,\text{m/s}^2)(2.00\,\text{s}) = 8.43\,\text{m/s}$, which is $30.3\,\text{km/h}$.

**33**

(a) Let $x_1 = 0$ at the first point and take the time to be $t_1 = 0$ when the car is there. Use $v_{2x} = v_{1x} + a_x(t_2 - t_1)$ to eliminate $a_x$ from $x_2 - x_1 = v_{1x}(t_2 - t_1) + \frac{1}{2}a_x(t_2 - t_1)^2$. The first equation yields $a_x = (v_{2x} - v_{1x})/(t_2 - t_1)$, so $x_2 - x_1 = v_{1x}(t_2 - t_1) + \frac{1}{2}(v_{2x} - v_{1x})(t_2 - t_1) = \frac{1}{2}(v_{2x} + v_{1x})(t_2 - t_1)$. Solve for $v_{1x}$ and substitute $x_2 - x_1 = 60.0\,\text{m}$, $v_{2x} = 15.0\,\text{m/s}$, and $t_2 - t_1 = 6.00\,\text{s}$. Your result should be

$$v_{1x} = \frac{2\left[(x_2 - x_1) - v_{2x}(t_2 - t_1)\right]}{t_2 - t_1} = \frac{2\left[(60.0\,\text{m}) - (15.0\,\text{m/s})(6.00\,\text{s})\right]}{6.00\,\text{s}} = 5.00\,\text{m/s}\,.$$

(b) Substitute $v_{2x} = 15.0\,\text{m/s}$, $v_{1x} = 5.00\,\text{m/s}$, and $t_2 - t_1 = 6.00\,\text{s}$ into $a_x = (v_{2x} - v_{1x})/(t_2 - t_1)$. The result is $a_x = (15.0\,\text{m/s} - 5.00\,\text{m/s})/(6.00\,\text{s}) = 1.67\,\text{m/s}^2$.

(c) Use $v_{2x} = v_{1x} + a_x(t_2 - t_1)$ to eliminate $t_2 - t_1$ from $x_2 - x_1 = v_{1x}(t_2 - t_1) + \frac{1}{2}a_x(t_2 - t_1)^2$ and obtain $v_{2x}^2 = v_{1x}^2 + 2a_x(x_2 - x_1)$. Solve for $x_2 - x_1$, then substitute $v_{2x} = 0$, $v_{1x} = 5.0\,\text{m/s}$, and $a_x = 1.67\,\text{m/s}^2$:

$$x_2 - x_1 = \frac{v_{2x}^2 - v_{1x}^2}{2a_x} = -\frac{(5.0\,\text{m/s})^2}{2(1.67\,\text{m/s}^2)} = -7.50\,\text{m}\,.$$

(d) To draw the graphs you need to know the time at which the car is at rest. Solve $v_x = v_{1x} + a_x(t_2 - t_1) = 0$ for $t_2 - t_1$ to find $t_2 - t_1 = -v_{1x}/a_x = -(5.00\,\text{m/s})/(1.67\,\text{m/s}^2) = -3.0\,\text{s}$. Your graphs should look like this:

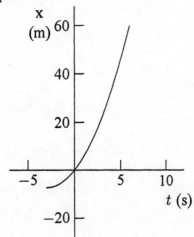

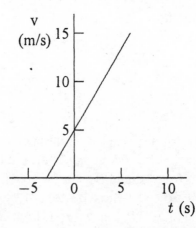

## 35

Let $t_r$ be the reaction time and $t_b$ be the braking time. Then the total distance traveled by the car is given by $x = v_{1x}t_r + v_{1x}t_b + \frac{1}{2}a_x t_b^2$, where $v_{1x}$ is the $x$ component of the velocity at time zero and $a_x$ is the acceleration. After the brakes are applied the $x$ component of the car's velocity is given by $v_{2x} = v_{1x} + a_x t_b$. Use this equation, with $v_{2x} = 0$, to eliminate $t_b$ from the first equation. According to the second equation, $t_b = -v_{1x}/a_x$, so

$$x = v_{1x}t_r - \frac{v_{1x}^2}{a_x} + \frac{1}{2}\frac{v_{1x}^2}{a_x} = v_{1x}t_r - \frac{1}{2}\frac{v_{1x}^2}{a_x} \,.$$

Write this equation twice, once for each of the two different initial velocities ($\vec{v}_{A1}$ and $\vec{v}_{B1}$):

$$x_A = v_{A1x}t_r - \frac{1}{2}\frac{v_{A1x}^2}{a_x}$$

and

$$x_B = v_{B1x}t_r - \frac{1}{2}\frac{v_{B2x}^2}{a_x} \,.$$

Solve these equations simultaneously for $t_r$ and $a$. You should get

$$t_r = \frac{v_{B1x}^2 x_A - v_{A1x}^2 x_B}{v_{A1x}v_{B1x}(v_{B1x} - v_{A1x})}$$

and

$$a_x = -\frac{1}{2}\frac{v_{B1x}v_{A1x}^2 - v_{A1x}v_{B1x}^2}{v_{B1x}x_A - v_{A1x}x_B} \,.$$

Substitute $x_A = 56.7\,\text{m}$, $v_{A1x} = 80.5\,\text{km/h} = 22.36\,\text{m/s}$, $x_B = 24.4\,\text{m}$, and $v_{B1x} = 48.3\,\text{km/h} = 13.42\,\text{m/s}$. The results are

$$t_r = \frac{(13.42\,\text{m/s})^2(56.7\,\text{m}) - (22.36\,\text{m/s})^2(24.4\,\text{m})}{(22.36\,\text{m/s})(13.42\,\text{m/s})(13.42\,\text{m/s} - 22.36\,\text{m/s})} = 0.74\,\text{s}$$

and

$$a_x = -\frac{1}{2}\frac{(13.42\,\text{m/s})(22.36\,\text{m/s})^2 - (22.36\,\text{m/s})(13.42\,\text{m/s})^2}{(13.42\,\text{m/s})(56.7\,\text{m}) - (22.36\,\text{m/s})(24.4\,\text{m})} = -6.2\,\text{m/s}^2 \,.$$

## 37

(a) Take the $x$ axis to be vertical with the upward direction positive. Use $v_{2x} = v_{1x} + a_x\,\Delta t$, where $v_{1x}\ (= 0)$ is the $x$ component of its velocity at the bottom of its run, $v_{2x}\ (= 305\,\text{m/min} = 5.08\,\text{m/s})$ is the $x$ component of its velocity when it has reached its top speed, $\Delta t$ is the time required to reach top speed, and $a_x$ is the $x$ component of its acceleration. The solution for $\Delta t$ is $\Delta t = v_{2x}/a_x = (5.08\,\text{m/s})/1.22\,\text{m/s}^2) = 4.17\,\text{s}$. During this time it goes the distance

$$x_2 - x_1 = v_{1x}\,\Delta t + \tfrac{1}{2}a_x(\Delta t)^2 = \tfrac{1}{2}(1.22\,\text{m/s}^2)(4.17\,\text{s})^2 = 10.6\,\text{m} \,.$$

(b) The motion is symmetric so the elevator takes 4.17 s to slow to a stop at the top of its run and its travels 10.6 m while slowing. It travels $190\,\text{m} - 10.6\,\text{m} - 10.6\,\text{m} = 168.8\,\text{m}$ at a constant speed of 5.08 m/s and this takes $(168.8\,\text{m})/(5.08\,\text{m/s}) = 33.2\,\text{s}$. The total time of the trip is $4.17\,\text{s} + 4.17\,\text{s} + 33.2\,\text{s} = 41.5\,\text{s}$.

## 43

Suppose the car is traveling along the $x$ axis. Let $v_{1x}$ ($= 42\,\text{mph} = 18.8\,\text{m/s}$) be the $x$ component of the car's velocity when the driver first applies the brakes and $v_{2x}$ ($= 0$) be the $x$ component of the velocity when the car comes to a stop. The $x$ component of the car's acceleration is $a_x = -8.9\,\text{m/s}^2$. If $\Delta t$ is the time required to stop, then $v_{2x} = v_{1x} + a_x\,\Delta t$. The stopping time is

$$\Delta t = -\frac{v_{1x}}{a_x} = -\frac{18.7\,\text{m/s}}{-8.9\,\text{m/s}^2} = 2.11\,\text{s}.$$

The distance traveled by the car in this time is

$$x_2 - x_1 = v_{1x}\,\Delta t + \tfrac{1}{2}a_x(\Delta t)^2 = (18.8\,\text{m/s})(2.11\,\text{s}) + \tfrac{1}{2}(-8.9\,\text{m/s}^2)(2.11\,\text{s})^2 = 20\,\text{m}.$$

The car stops in time.

## 47

Take the $x$ axis to be vertical and the downward direction to be positive. The coordinate of the flare obeys $x_2 - x_1 = v_{1x}(t_2 - t_1) + \tfrac{1}{2}a_x(t_2 - t_1)^2$. Suppose the flare leaves the shuttle at time $t_1 = 0$, when its coordinate is $x_1 = 0$. It has an $x$ component of velocity $v_{1x}$ then. It hits the ground at time $t_2 = 8.5$ s, when its coordinate is $x_2 = 100$ m. The $x$ component of its acceleration is $a_x = 9.8\,\text{m/s}^2$. The solution for $v_{1x}$ is

$$v_{2x} = \frac{x_2 - \tfrac{1}{2}a_x t_2^2}{t_2} = \frac{(100\,\text{m}) - \tfrac{1}{2}(9.8\,\text{m/s}^2)(8.5\,\text{s})^2}{8.5\,\text{s}} = -30\,\text{m/s}.$$

The shuttle is going upward at 30 m/s.

# Chapter 3

## 1

Take the positive $x$ direction to be the direction of motion. According to Newton's second law, the magnitude of the force is given by $F = ma$, where $a$ is the magnitude of the neutron's acceleration. Use kinematics to find the acceleration that brings the neutron to rest in a distance $d$. Assume the acceleration is constant and solve $v_{2x}^2 = v_{1x}^2 + 2a_x d$ for the acceleration component $a_x$:

$$a_x = \frac{v_{2x}^2 - v_{1x}^2}{2d} = \frac{0 - (1.4 \times 10^7 \, \text{m/s})^2}{2(1.0 \times 10^{-14} \, \text{m})} = -9.8 \times 10^{27} \, \text{m/s}^2 \,.$$

The magnitude of the force is $F = ma = (1.67 \times 10^{-27} \, \text{kg})(9.8 \times 10^{27} \, \text{m/s}^2) = 16 \, \text{N}$.

## 5

Take the $x$ axis to be along the direction of travel. According to Newton's second law $F_x^{\text{net}} = ma_x$, $F = ma$, where $F_x$ is the $x$ component of the net force, $a_x$ is the $x$ component of the acceleration, and $m$ is the mass. The acceleration component can be found using constant acceleration kinematics. Solve $v_{2x} = v_{1x} + a_x \Delta t$ for $a_x$: $a_x = (v_{2x} - v_{1x})/\Delta t$. Here $\vec{v}_1$ (= 0) is the velocity when the sled starts accelerating, $\vec{v}_2$ is the final velocity. The final velocity component is $v_{2x} = (1600 \, \text{km/h})(1000 \, \text{m/km})/(3600 \, \text{s/h}) = 444 \, \text{m/s}$, so the acceleration component is $a_x = (444 \, \text{m/s})/(1.8 \, \text{s}) = 247 \, \text{m/s}^2$ and the force component is $F_x = (500 \, \text{kg})(247 \, \text{m/s}^2) = 1.2 \times 10^5 \, \text{N}$. This is also the magnitude of the net force.

## 15

The free-body diagram is shown on the right. $\vec{T}$ is the tension force of the cable and $m\vec{g}$ is the force of gravity. If the positive $x$ direction is upward, then Newton's second law gives $T - mg = ma_x$, where $a_x$ is the $x$ component of the acceleration. Solve for the magnitude of the tension: $T = m(g + a_x)$. Use constant acceleration kinematics to find the acceleration. If $\vec{v}_2$ (= 0) is the final velocity, $\vec{v}_1 = -(12 \, \text{m/s})\hat{\imath}$ is the initial velocity, and $\Delta \vec{x} = (-42 \, \text{m})\hat{\imath}$ is the magnitude of the displacement from the starting to the stopping points, then $v_{2x}^2 = v_{1x}^2 + 2a_x \Delta x$. Solve for $a_x$: $a_x = (v_{2x} - v_{1x}^2/2\Delta x = (0 - (-12 \, \text{m/s})^2)/2(-42 \, \text{m}) = 1.71 \, \text{m/s}^2$. Now go back to calculate the tension: $T = m(g + a_x) = (1600 \, \text{kg})(9.8 \, \text{m/s}^2 + 1.71 \, \text{m/s}^2) = 1.8 \times 10^4 \, \text{N}$.

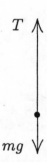

## 19

The forces on the balloon are the force of gravity $m\vec{g}$, down, and the force of the air $\vec{F}^{\text{air}}$, up. Take the positive direction to be upward. When the mass is $M$ (before the ballast is thrown out) the acceleration is downward and Newton's second law, in terms of magnitudes, gives $F^{air} - Mg = -Ma$. After the ballast is thrown out the mass is $M - m$, where $m$ is the mass of the ballast, and the acceleration is upward. Newton's second law then gives $F^{air} - (M - m)g = (M - m)a$. The first equation yields $F_{air} = M(g - a)$ and the second yields $M(g - a) - (M - m)g = (M - m)a$. Solve for $m$: $m = 2Ma/(g + a)$.

## 23

(a) The links are numbered from bottom to top. The forces on the bottom link are the force of gravity $m\vec{g}$, downward, and the force $\vec{F}_{2 \text{ on } 1}$ of link 2, upward. Take the positive direction to be upward. Then Newton's second law for this link, in terms of magnitudes, gives $F_{2 \text{ on } 1} - mg = ma$. Thus $F_{2 \text{ on } 1} = m(a + g) = (0.100 \text{ kg})(2.50 \text{ m/s}^2 + 9.8 \text{ m/s}^2) = 1.23 \text{ N}$. If the positive $x$ direction is upward this force is $\vec{F}_{2 \text{ on } 1} = (1.23 \text{ N})\hat{\imath}$.

(b) The forces on the second link are the force of gravity $m\vec{g}$, downward, the force $\vec{F}_{1 \text{ on } 2}$ of link 1, downward, and the force $\vec{F}_{3 \text{ on } 2}$ of link 3, upward. According to Newton's third law $\vec{F}_{1 \text{ on } 2}$ has the same magnitude as $\vec{F}_{2 \text{ on } 1}$. Newton's second law for the second link gives $F_{3 \text{ on } 2} - F_{1 \text{ on } 2} - mg = ma$, so $F_{3 \text{ on } 2} = m(a + g) + F_{1 \text{ on } 2} = (0.100 \text{ kg})(2.50 \text{ m/s}^2 + 9.8 \text{ m/s}^2) + 1.23 \text{ N} = 2.46 \text{ N}$. This force is $\vec{F}_{3 \text{ on } 2} = (2.46 \text{ N})\hat{\imath}$.

(c) Newton's second for link 3 is $F_{4 \text{ on } 3} - F_{2 \text{ on } 3} - mg = ma$, so $F_{4 \text{ on } 3} = m(a + g) + F_{2 \text{ on } 3} = (0.100 \text{ N})(2.50 \text{ m/s}^2 + 9.8 \text{ m/s}^2) + 2.46 \text{ N} = 3.69 \text{ N}$, where Newton's third law was used to write $F_{2 \text{ on } 3} = F_{3 \text{ on } 2}$. This force is $\vec{F}_{4 \text{ on } 3} = (3.69 \text{ N})\hat{\imath}$.

(d) Newton's second law for link 4 is $F_{5 \text{ on } 4} - F_{3 \text{ on } 4} - mg = ma$, so $F_{5 \text{ on } 4} = m(a + g) + F_{3 \text{ on } 4} = (0.100 \text{ kg})(2.50 \text{ m/s}^2 + 9.8 \text{ m/s}^2) + 3.69 \text{ N} = 4.92 \text{ N}$, where Newton's third law was used to write $F_{3 \text{ on } 4} = F_{4 \text{ on } 3}$. This force is $\vec{F}_{3 \text{ on } 4} = 4.92 \text{ N})\hat{\imath}$.

(e) Newton's second law for the top link gives $F - F_{4 \text{ on } 5} - mg = ma$, so $F = m(a + g) + F_{4 \text{ on } 5} = (0.100 \text{ kg})(2.50 \text{ m/s}^2 + 9.8 \text{ m/s}^2) + 4.92 \text{ N} = 6.15 \text{ N}$, where Newton's third law as used to write $F_{4 \text{ on } 5} = F_{5 \text{ on } 4}$. This force is $\vec{F} = 6.15 \text{ N})\hat{\imath}$.

(f) Each link has the same mass and the same acceleration, so the same net force acts on each of them: $\vec{F}^{net} = m\vec{a} = (0.100 \text{ kg})(2.50 \text{ m/s}^2)\hat{\imath} = (0.25 \text{ N})\hat{\imath}$.

## 27

(a) Take the positive $y$ direction to be upward and use $y_2 - y_1 = v_{1y}t - \frac{1}{2}g(t_2 - t_1)^2$, with $y_1$ (= 0) the coordinate at time $t_1 = 0$ and $y_{2y}$ (= 0.544 m) the coordinate at time $t_2 = 0.200$ s. Solve for $v_{1y}$:

$$v_{1y} = \frac{y_{2y} + \frac{1}{2}g(t_2 - t_1)^2}{t_2 - t_1} = \frac{0.544 \text{ m} + \frac{1}{2}(9.8 \text{ m/s}^2)(0.200 \text{ s})^2}{0.200 \text{ s}} = 3.70 \text{ m/s}.$$

(b) Use $v_{2y} = v_{1y} - gt = 3.70 \text{ m/s} - (9.8 \text{ m/s}^2)(0.200 \text{ s}) = 1.74 \text{ m/s}$.

(c) Use $v_{2y}^2 = v_{1y}^2 - 2g(y_2 - y_1)$, with $v_{2y} = 0$. Solve for $y_2$:

$$y_2 = \frac{v_{1y}^2}{2g} = \frac{(3.7 \text{ m/s})^2}{2(9.8 \text{ m/s}^2)} = 0.698 \text{ m}.$$

It goes $0.698 \text{ m} - 0.544 \text{ m} = 0.154 \text{ m}$ higher.

## 33

(a) Take the $y$ axis to be upward and place the origin on the ground, under the balloon. Since the package is dropped, its initial velocity ($\vec{v}_1$) is the same as the velocity of the balloon, $+12\,\text{m/s}$, up. The initial coordinate of the package is $y_1 = 80\,\text{m}$; when it hits the ground its coordinate is $y_2 = 0$. Solve $y_2 - y_1 = v_{1y}(t_2 - t_1)t - \frac{1}{2}g(t_2 - t_1)^2$ for the time of flight $t_2 - t_1$:

$$t_2 - t_1 = \frac{v_{1y}}{g} \pm \sqrt{\frac{v_{1y}^2}{g^2} + \frac{2y_1}{g}} = \frac{12\,\text{m/s}}{9.8\,\text{m/s}^2} + \sqrt{\frac{(12\,\text{m/s})^2}{(9.8\,\text{m/s}^2)^2} + \frac{2(80\,\text{m})}{9.8\,\text{m/s}^2}} = 5.4\,\text{s},$$

where the positive solution was used. A negative value for $t_2 - t_1$ corresponds to a time before the package was dropped and cannot be accepted.

(b) Use $v_{2y} = v_{1y} - g(t_2 - t_1)$ to find that $v_{2y} = 12\,\text{m/s} - (9.8\,\text{m/s}^2)(5.4\,\text{s}) = -41\,\text{m/s}$. Its speed is $41\,\text{m/s}$.

## 39

Let $g$ be the gravitational strength near the surface of Callisto, $m$ be the mass of the landing craft, $a$ be the magnitude of the acceleration of the landing craft, and $F$ be the magnitude of the rocket thrust. Take the downward direction to be positive. Then Newton's second law, in terms of magnitudes, gives $mg - F = ma$. If the magnitude of the thrust is $F_1$ ($= 3260\,\text{N}$), then the acceleration is zero, so $mg - F_1 = 0$. If the magnitude of the thrust is $F_2$ ($= 2200\,\text{N}$), then the magnitude of the acceleration is $a_2$ ($= 0.39\,\text{m/s}^2$), so $mg - F_2 = ma_2$. (a) The first equation gives the weight of the landing craft: $mg = F_1 = 3260\,\text{N}$.

(b) The second equation gives the mass:

$$m = \frac{mg - F_2}{a_2} = \frac{3260\,\text{N} - 2200\,\text{N}}{0.39\,\text{m/s}^2} = 2.7 \times 10^3\,\text{kg}.$$

(c) The weight divided by the mass gives the gravitational strength:

$$g = \frac{3260\,\text{N}}{2.7 \times 10^3\,\text{kg}} = 1.2\,\text{N/kg}^2.$$

The free-fall acceleration is $1.2\,\text{m/s}^2$.

## 43

(a) and (b) Let $W$ be the weight of particle. Its mass is then $m = W/g = (22\,\text{N})/(9.8\,\text{N/kg}) = 2.2\,\text{kg}$. At a place where $g = 4.9\,\text{m/s}^2$ its mass is still $2.2\,\text{kg}$ but weight is

$$W = mg = (2.2\,\text{kg})(4.9\,\text{N/kg}) = 11\,\text{N}.$$

(c) and (d) At a place where $g = 0$ its mass is still $2.2\,\text{kg}$ but its weight is zero.

## 47

(a) The free-body diagrams are shown on the right. $\vec{F}$ is the applied force and $\vec{f}$ is the force of block 1 on block 2. Note that $\vec{F}$ is applied only to block 1 and that block 2 exerts the force $-\vec{f}$ on block 1. Newton's third law has thereby been taken into account.

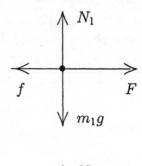

Newton's second law for block 1, in terms of magnitudes, gives $F - f = m_1 a$, where $a$ is the magnitude of the acceleration. The second law for block 2 gives $f = m_2 a$. Since the blocks move together they have the same acceleration and the same symbol is used in both equations. Use the second equation to obtain an expression for $a$: $a = f/m_2$. Substitute into the first equation to get $F - f = m_1 f/m_2$. Solve for $f$:

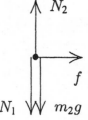

$$f = \frac{Fm_2}{m_1 + m_2} = \frac{(3.2\,\text{N})(1.2\,\text{kg})}{2.3\,\text{kg} + 1.2\,\text{kg}} = 1.1\,\text{N}.$$

(b) If $\vec{F}$ is applied to block 2 instead of block 1, the force of contact is

$$f = \frac{Fm_1}{m_1 + m_2} = \frac{(3.2\,\text{N})(2.3\,\text{kg})}{2.3\,\text{kg} + 1.2\,\text{kg}} = 2.1\,\text{N}.$$

(c) The acceleration of the blocks is the same in the two cases. Since the contact force $\vec{f}$ is the only horizontal force on one of the blocks it must be just right to give that block the same acceleration as the block to which $\vec{F}$ is applied. In the second case the contact force accelerates a more massive block than in the first case, so it must be larger.

## 63

(a) The carts have the same acceleration. Consider them to be a single object with a mass of $M = 7.5\,\text{kg}$. Since $\vec{F}_B$ is the only horizontal force on this object, Newton's second law gives the magnitude of the acceleration as $a = F_B/M = (20.0\,\text{N})/(7.5\,\text{kg}) = 2.67\,\text{m/s}^2$. The only horizontal force on cart A is the force of cart B, so $F_{B \to A} = m_A a = (2.5\,\text{kg})(2.67\,\text{m/s}^2) = 6.7\,\text{N}$. The force of B on A has a magnitude of $6.7\,\text{N}$ and is to the left. According to Newton's third law the force of A on B has the same magnitude and is to the right.

(b) The acceleration of the carts has the same magnitude as in part (a) but now the force of A on B must give B that acceleration, so $F_{A \to B} = m_B a = (5.0\,\text{kg})(2.67\,\text{m/s}^2) = 13\,\text{N}$. The force of A on B has a magnitude of $1.3\,\text{N}$ and is to the right. According to Newton's third law the force of B on A has the same magnitude and is to the left. (c) The magnitude of the acceleration of the carts is the same for the two cases. In part (b) the contact force accelerates a greater mass than in part (a) so it must greater.

# Chapter 4

## 7

(a) Use $A = \sqrt{A_x^2 + A_y^2}$ to obtain $A = \sqrt{(-25.0\,\text{m})^2 + (+40.0\,\text{m})^2} = 47.2\,\text{m}$.

(b) The tangent of the angle between the vector and the positive $x$ direction is

$$\tan\theta = \frac{A_y}{A_x} = \frac{-40.0\,\text{m}}{25.0\,\text{m}} = -1.6.$$

The inverse tangent is $-58.0°$ or $-58.0° + 180° = 122°$. The first angle has a positive cosine and a negative sine. It is not correct. The second angle has a negative cosine and a positive sine. It is correct for a vector with a negative $x$ component and a positive $y$ component.

## 9

The point $P$ is displaced vertically by $2R$, where $R$ is the radius of the wheel. It is displaced horizontally by half the circumference of the wheel, or $\pi R$. Since $R = 45.0\,\text{cm}$, the horizontal component of the displacement is $1.41\,\text{m}$ and the vertical component of the displacement is $0.900\,\text{m}$. If the $x$ axis is horizontal and the $y$ axis is vertical, the vector displacement is $\vec{r} = (1.41\,\text{m})\hat{\imath} + (0.900\,\text{m})\hat{\jmath}$. The displacement has a magnitude of $r = \sqrt{(1.41\,\text{m})^2 + (0.900\,\text{m})^2} = 1.68\,\text{m}$ and it is $\tan^{-1}[(0.900\,\text{m}/(1.41\,\text{m})] = 32.5°$ above the floor.

## 13

The diagram shows the displacement vectors for the two segments of her walk, labeled $\vec{r}_1$ and $\vec{r}_2$, and the final displacement vector, labeled $\vec{r}$. Take the $x$ axis to run from west to east and the $y$ axis to run from south to north. Then the components of $\vec{r}_1$ are $r_{1x} = (250\,\text{m})\sin 30° = 125\,\text{m}$ and $r_{1y} = (250\,\text{m})\cos 30° = 216.5\,\text{m}$. The components of $\vec{r}_2$ are $r_{2x} = 175\,\text{m}$ and $r_{2y} = 0$. The components of the resultant displacement are $r_x = r_{1x} + r_{2x} = 125\,\text{m} + 175\,\text{m} = 300\,\text{m}$ and $r_y = r_{1y} + r_{2y} = 216.5\,\text{m} + 0 = 216.5\,\text{m}$.

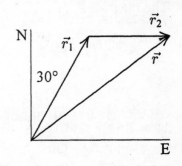

(a) The magnitude of the resultant displacement is

$$r = \sqrt{r_x^2 + r_y^2} = \sqrt{(300\,\text{m})^2 + (216.5\,\text{m})^2} = 370\,\text{m}.$$

(b) The angle $\theta$ that the resultant displacement makes with the positive $x$ axis is

$$\theta = \tan^{-1}\frac{r_y}{r_x} = \tan^{-1}\frac{216.5\,\text{m}}{300\,\text{m}} = 36°.$$

The second solution ($\theta = 36° + 180° = 216°$) is rejected because it is associated with a vector that is opposite in direction to $\vec{r}$.

(c) The total distance walked is $d = 250\,\text{m} + 175\,\text{m} = 425\,\text{m}$.

(d) The total distance walked is greater than the magnitude of the final displacement. A glance at the diagram should show you why: $\vec{r}_1$ and $\vec{r}_2$ are not along the same line.

## 17

The vectors are shown on the diagram to the right. Let the $x$ axis run from west to east and the $y$ axis run from south to north. Then $a_x = 5.0\,\text{m}$, $a_y = 0$, $b_x = -(4.0\,\text{m})\sin 35° = -2.29\,\text{m}$, and $b_y = (4.0\,\text{m})\cos 35° = 3.28\,\text{m}$.

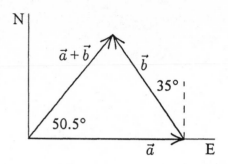

(a) Let $\vec{c} = \vec{a} + \vec{b}$. Then $c_x = a_x + b_x = 5.0\,\text{m} - 2.29\,\text{m} = 2.71\,\text{m}$ and $c_y = a_y + b_y = 0 + 3.28\,\text{m} = 3.28\,\text{m}$. The magnitude of $\vec{c}$ is

$$c = \sqrt{c_x^2 + c_y^2} = \sqrt{(2.71\,\text{m})^2 + (3.28\,\text{m})^2} = 4.3\,\text{m}.$$

(b) The angle $\theta$ that $\vec{c}$ makes with the positive $x$ axis is

$$\theta = \tan^{-1}\frac{c_y}{c_x} = \tan^{-1}\frac{3.28\,\text{m}}{2.71\,\text{m}} = 50.4°.$$

The second solution ($\theta = 50.4° + 180° = 126°$) is rejected because it is associated with a vector with a direction opposite to that of $\vec{c}$.

(c) The vector $\vec{b} - \vec{a}$ is found by adding $-\vec{a}$ to $\vec{b}$. The result is shown on the diagram to the right. Let $\vec{c} = \vec{b} - \vec{a}$. Then $c_x = b_x - a_x = -2.29\,\text{m} - 5.0\,\text{m} = -7.29\,\text{m}$ and $c_y = b_y - a_y = 3.28\,\text{m} - 0 = 3.28\,\text{m}$. The magnitude of $c$ is

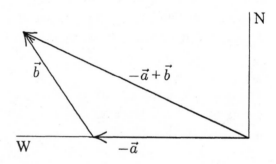

$$c = \sqrt{c_x^2 + c_y^2} = \sqrt{(-7.29\,\text{m})^2 + (3.28\,\text{m})^2}$$

$$= 8.0\,\text{m}.$$

(d) The tangent of the angle $\theta$ that $\vec{c}$ makes with the positive $x$ axis is

$$\tan\theta = \frac{c_y}{c_x} = \frac{3.28\,\text{m}}{-7.29\,\text{m}} = -4.50,.$$

There are two solutions: $-24.2°$ and $155.8°$. As the diagram shows, the second solution is correct. The vector $\vec{c}$ is 24° north of west.

## 25

Let $\vec{a}$ and $\vec{b}$ be two vectors such their sum is perpendicular to their difference. Put the $x$ axis along the direction of their sum and the $y$ axis along the direction of their difference. Since the $y$ component of their sum vanishes their individual $y$ components must have the same magnitude and opposite signs. Since the $x$ component of their difference vanishes their individual $x$ components must have the same value. Let $\vec{r}_1 = x\hat{\imath} + y\hat{\jmath}$ be one of the vectors. Then the other is $\vec{r}_2 = x\hat{\imath} - y\hat{\jmath}$. The two vectors have the same magnitude.

## 31

According to the problem statement $\vec{A} + \vec{B} = 6.0\hat{i} + 1.0\hat{j}$ and $\vec{A} - \vec{B} = -4.0\hat{i} + 7.0\hat{j}$. Add the two equations to obtain $2\vec{A} = 2.0\hat{i} + 48.0\hat{j}$. Divide by 2 to obtain $\vec{A} = 1.0\hat{i} + 4.0\hat{j}$. The magnitude of $\vec{A}$ is $\sqrt{(1.0)^2 + (4.0)^2} = 4.1$.

## 37

Take the $x$ axis to run from west to east and the $y$ axis to run from south to north. Let $\vec{r}$ $(= (5.6\,\text{km})\hat{j})$ be the position of the base camp relative the starting point of the explorer, let $\vec{r_1}$ be the position of the explorer when the snow clears, again relative to his starting point, and let $\vec{r_2}$ be the vector from the explorer's position to the base camp. Then $\vec{r} = \vec{r_1} + \vec{r_2}$. Thus $\vec{r_2} = \vec{r} - \vec{r_1}$. Use the component method to carry out the vector subtraction. The $x$ component of $\vec{r_1}$ is

$$r_{1x} = (7.8\,\text{km})\cos 50° = 5.0\,\text{km}$$

and the $y$ component is

$$r_{2y} = (7.8\,\text{km})\sin 50° = 6.0\,\text{km}.$$

Thus

$$r_{2x} = r_x - r_{1x} = 0 - 5.0\,\text{km} = -5.0\,\text{km}$$

and

$$r_{2y} = r_y - r_{1y} = 5.6\,\text{km} - 6.0\,\text{km} = -0.4\,\text{km}.$$

The distance he must travel is equal to the magnitude of $\vec{r_2}$, which is

$$r_2 = \sqrt{r_{2x}^2 + r_{2y}^2} = \sqrt{(-5.0\,\text{km})^2 + (0.4\,\text{km})^2} = 5.0\,\text{km}.$$

The angle $\phi$ his path should make with the positive $x$ direction is given by

$$\tan\phi = \frac{r_{2y}}{r_{2x}} = \frac{0.4\,\text{km}}{5.0\,\text{km}} = 0.08.$$

The angle is either 44.5° or 4.5° + 180° = 184.5°. Only the second angle gives a negative $x$ component and a positive $y$ component. You can also give the direction as 4.5° south of west.

## 43

(a) and (b) Add the four displacement vectors to obtain $\vec{r} = (-60\,\text{m} + b_x)\hat{i} + (-80\,\text{m} + c_y)\hat{j}$. Thus $-60\,\text{m} + b_x = -140\,\text{m}$, so $b_x = -80\,\text{m}$, and $-80\,\text{m} + c_y = 30\,\text{m}$, so $c_y = 110\,\text{m}$.

(c) The magnitude of the overall displacement is the square root of the sum of the squares of its components:

$$r = \sqrt{x^2 + y^2} = \sqrt{(-140\,\text{m})^2 + (30\,\text{m})^2} = 143\,\text{m}.$$

(d) The angle $\phi$ between $\vec{r}$ and the positive $x$ direction is given by

$$\tan\phi = y/x = (30\,\text{m})/(-140\,\text{m}) = -0.214.$$

The angle is either $-12.1°$ or $-12.1° + 180° = 168°$. The first of these leads to a vector with a positive $x$ component and a negative $y$ component and so cannot be correct. The second angle leads to a vector with a negative $x$ component and a positive $y$ component and is correct.

## 45

Because $\vec{A}$ is along the $x$ axis $\vec{A} = A\,\hat{\imath}$. Let $\vec{C} = \vec{A} + \vec{B}$. Because $\vec{C}$ is along the $y$ axis $\vec{C} = C\,\hat{\jmath}$. The $x$ component of the vector addition equation gives $A + B_x = 0$ and the $y$ component gives $C = B_y$. Now $C = 2A$ so the last equation becomes $2A = B_y$. The square of the magnitude of $\vec{B}$ is $B^2 = B_x^2 + B_y^2 = (-A)^2 + 4A^2 = 5A^2$. This means that

$$A = \frac{B}{\sqrt{5}} = \frac{8.0\,\text{m}}{\sqrt{5}} = 3.6\,\text{m}.$$

## 49

(a) and (b) The $x$ component is

$$a_x = (17.0\,\text{m})\cos 56.0° = 9.51\,\text{m}$$

and the $y$ component is

$$a_y = (17.0\,\text{m})\sin 56.0° = 14.1\,\text{m}.$$

(c) and (d) The vector $\vec{a}$ makes the angle $56.0° - 18.0° = 38.0°$ with the $x'$ axis, so its $x'$ component is

$$a_x' = (17.0\,\text{m})(\cos 38° = 13.4\,\text{m}$$

and its $y'$ component is

$$a_x' = (17.0\,\text{m})\sin 38° = 10.5\,\text{m}.$$

# Chapter 5

## 1

(a) Take the positive $y$ direction to be downward. Then the $y$ coordinate of the bullet at time $t_2$ is given by $y_2 = y_1 + \frac{1}{2}g(t_2 - t_1)^2$, where $y_1$ is its coordinate at time $t_1$. If $t_2 - t_1$ is the time of flight and $y_2 - y_1$ is the distance the bullet hits below the target, then

$$t_2 - t_1 = \sqrt{\frac{2(y_2 - y_1)}{g}} = \sqrt{\frac{2(0.019\,\text{m})}{9.8\,\text{m/s}^2}} = 6.2 \times 10^{-2}\,\text{s}.$$

(b) The speed of the bullet as it emerges from the rifle is the $x$ component of its initial velocity. If $x_1$ is the horizontal coordinate of the bullet at time $t_1$ and $x_2$ is its horizontal coordinate at time $t_2$, then $x_2 - x_1 = v_{1x}(t_2 - t_1)$, where $v_{1x}$ is the $x$ component of the initial velocity. The solution is

$$v_{1x} = \frac{x_2 - x_1}{t_2 - t_1} = \frac{30\,\text{m}}{6.2 \times 10^{-2}\,\text{s}} = 480\,\text{m/s}.$$

## 5

(a) The horizontal component of the velocity is constant. If $\ell$ is the length of a plate and $\Delta t$ is the time an electron is between the plates, then $\ell = v_{1x}\Delta t$, where $v_{1x}$ is the initial speed. Thus

$$\Delta t = \frac{\ell}{v_{1x}} = \frac{2.0\,\text{cm}}{1.0 \times 10^9\,\text{cm/s}} = 2.0 \times 10^{-9}\,\text{s}.$$

(b) The vertical displacement of the electron is

$$\Delta y = \frac{1}{2}a_y(\Delta t)^2 = \frac{1}{2}(1.0 \times 10^{17}\,\text{cm/s}^2)(2.0 \times 10^{-9}\,\text{s})^2 = 0.20\,\text{cm},$$

where down was taken to be positive.

(c) and (d) The $x$ component of velocity is $v_{2x} = v_{1x} = 1.0 \times 10^9$ cm/s and the $y$ component is $v_{2y} = a_y\,\Delta t = (1.0 \times 10^{17}\,\text{cm/s}^2)(2.0 \times 10^{-9}\,\text{s}) = 2.0 \times 10^8\,\text{cm/s}$.

## 9

Take the $y$ axis to be vertical with the positive direction upward and take the $x$ axis to be horizontal with the direction of firing positive. Let $\vec{v}_1$ be the firing velocity and $\theta_1$ be the firing angle. Suppose the target is a horizontal distance $d$ away. The kinematic equations are $d = (v_1\Delta t)\cos\theta_1$ and $0 = (v_1\Delta t)\sin\theta_1 - \frac{1}{2}g(\Delta t)^2$, where $\Delta t$ is the time of flight. Eliminate $\Delta t$ and solve for $\theta_1$. The first equation gives $\Delta t = d/v_1\cos\theta_1$. This expression is substituted into the second equation to obtain $2v_1^2\sin\theta_1\cos\theta_1 - gd = 0$. Use the trigonometric identity $\sin\theta_1\cos\theta_1 = \frac{1}{2}\sin(2\theta_1)$ to obtain $v_1^2\sin(2\theta_1) = gd$. This means

$$\sin(2\theta_1) = \frac{gd}{v_1^2} = \frac{(9.8\,\text{N/kg})(45.7\,\text{m})}{(460\,\text{m/s})^2} = 2.12 \times 10^{-3},$$

so $\theta_1 = 0.1215°$ and $\theta_1 = 0.0607°$. If the gun is aimed at a point a distance $\ell$ above the target, then $\tan\theta_1 = \ell/d$ or $\ell = d\tan\theta_1 = (45.7\,\text{m})\tan 0.0607° = 0.0484\,\text{m} = 4.84\,\text{cm}$.

**11**

Take the $y$ axis to be upward. Then the height of the projectile after it has been traveling for a time $\Delta t$ is $\Delta y = (v_1 \Delta t) \sin \theta_1 - \frac{1}{2} g(\Delta t)^2$ and the $y$ component of the velocity then is given by $v_{2y} = v_1 \sin \theta_1 - g \Delta t$. The maximum height occurs when $v_{2y} = 0$. This means $\Delta t = (v_1/g) \sin \theta_1$ and

$$\Delta y = v_1 \left(\frac{v_1}{g}\right) \sin \theta_1 \sin \theta_1 - \frac{1}{2} \frac{g(v_1 \sin \theta_1)^2}{g^2} = \frac{(v_1 \sin \theta_1)^2}{g} - \frac{1}{2} \frac{(v_1 \sin \theta_1)^2}{g} = \frac{(v_1 \sin \theta_1)^2}{2g} .$$

**15**

Take the $y$ axis to be vertical and upward to be the positive direction. Take the $x$ axis to be horizontal and positive in the direction of the kick. Let $\vec{v}_1$ be the initial velocity of the ball. The $x$ component of the ball's displacement is $\Delta x = 46$ m and the $y$ component is $\Delta y = -1.5$ m. The time of flight is $\Delta t = 4.5$ s. Since $\Delta x = v_{1x} \Delta t$,

$$v_{1x} = \frac{\Delta x}{\Delta t} = \frac{46 \text{ m}}{4.5 \text{ s}} = 10.2 \text{ m/s} .$$

Since $\Delta y = v_{1y} \Delta t - \frac{1}{2} g(\Delta t)^2$,

$$v_{1y} = \frac{\Delta y + \frac{1}{2} g(\Delta t)^2}{\Delta t} = \frac{(-1.5 \text{ m}) + \frac{1}{2}(9.8 \text{ N/kg})(4.5 \text{ s})^2}{4.5 \text{ s}} = 21.7 \text{ m/s} .$$

The magnitude of the initial velocity is $v_1 = \sqrt{v_{1x}^2 + v_{1y}^2} = \sqrt{(10.2 \text{ m/s})^2 + (21.7 \text{ m/s})^2} = 24 \text{ m/s}$. The kicking angle satisfies $\tan \theta_1 = v_{1y}/v_{1x} = (21.7 \text{ m/s})/(10.2 \text{ m/s}) = 2.13$. The angle is $\theta_1 = 64.8°$.

**23**

You want to know how high the ball is from the ground when its horizontal distance from home plate is 97.5 m. To calculate this quantity you need to know the components of the initial velocity of the ball. Use the range information. Take the $y$ axis to be vertical with the positive direction to be upward. Take the $x$ axis to be horizontal with the positive direction the direction the ball is hit. If $\Delta x = 107$ m and $\Delta y = 0$ are the components of the ball's displacement, then $\Delta x = v_{1x} \Delta t$ and $0 = v_{1y} \Delta t - \frac{1}{2} g(\Delta t)^2$, where $\Delta t$ is the time of flight and $\vec{v}_1$ is the initial velocity of the ball of the ball. The second equation gives $\Delta t = 2v_{1y}/g$ and this is substituted into the first equation. Use $v_{1x} = v_{1y}$, which is true since the initial angle is $\theta_1 = 45°$. The result is $\Delta x = 2v_{1y}^2/g$. Thus

$$v_{1y} = \sqrt{\frac{g \Delta x}{2}} = \sqrt{\frac{(9.8 \text{ N/kg})(107 \text{ m})}{2}} = 22.9 \text{ m/s} .$$

Now take $\Delta x$ and $\Delta y$ to be the components of the displacement when the ball is at the fence. Again $\Delta x = v_{1x} t$ and $\Delta y = v_{1y} \Delta t - \frac{1}{2} g(\Delta t)^2$. The time to reach the fence is given by $\Delta t = \Delta x/v_{1x} = (97.5 \text{ m})/(22.9 \text{ m/s}) = 4.26$ s. When this is substituted into the second equation the result is

$$\Delta y = v_{1y} \Delta t - \frac{1}{2} g(\Delta t)^2 = (22.9 \text{ m/s})(4.26 \text{ s}) - \frac{1}{2}(9.8 \text{ N/kg})(4.26 \text{ s})^2 = 8.63 \text{ m} .$$

Since the ball started 1.22 m above the ground, it is 8.63 m + 1.22 m = 9.85 m above the ground when it gets to the fence and it is 9.85 m − 7.32 m = 2.53 m above the top of the fence. It goes over the fence.

## 29

Let $\vec{r}_1 = (5.0\,\text{m})\hat{\imath} - (6.0\,\text{m})\hat{\jmath}$ and $\vec{r}_2 = -(2.0\,\text{m})\hat{\imath} + (6.0\,\text{m})\hat{\jmath}$. The displacement is

$$\vec{r}_2 - \vec{r}_1 = [-2.0\,\text{m}) - (5.0\,\text{m})]\,\hat{\imath} + [(6.0\,\text{m}) - (-6.0\,\text{m})]\,\hat{\jmath} = (-7.0\,\text{m})\hat{\imath} + (12\,\text{m})\hat{\jmath}.$$

The displacement is in the $xy$ plane.

## 33

The average velocity is the total displacement divided by the time interval. The total displacement $\vec{r}$ is the sum of three displacements, each calculated as the product of a velocity and a time interval. The first has a magnitude of $(60.0\,\text{km/h})(40.0\,\text{min})/(60.0\,\text{min/h}) = 40.0\,\text{km}$. Its direction is east. If we take the $x$ axis to be toward the east and the $y$ axis to be toward the north, then this displacement is $\vec{r}_1 = (40.0\,\text{km})\hat{\imath}$.

The second has a magnitude of $(60.0\,\text{km/h})(20.0\,\text{min})/(60.0\,\text{min/h}) = 20.0\,\text{km}$. Its direction is $50.0°$ east of north, so it may be written $\vec{r}_2 = (20.0\,\text{km})\sin 50.0°\,\hat{\imath} + (20.0\,\text{km})\cos 50.0°\,\hat{\jmath} = (15.3\,\text{km})\hat{\imath} + (12.9\,\text{km})\hat{\jmath}$.

The third has a magnitude of $(60.0\,\text{km/h})(50.0\,\text{min})/(60.0\,\text{min/h}) = 50.0\,\text{km}$. Its direction is west, so the displacement may be written $\vec{r}_3 = (-50\,\text{km})\hat{\imath}$. The total displacement is

$$\vec{r} = \vec{r}_1 + \vec{r}_2 + \vec{r}_3 = (40.0\,\text{km})\hat{\imath} + (15.3\,\text{km})\hat{\imath} + (12.9\,\text{km})\hat{\jmath} - (50\,\text{km})\hat{\imath}$$
$$= (5.3\,\text{km})\hat{\imath} + (12.9\,\text{km})\hat{\jmath}.$$

The total time for the trip is $40\,\text{min} + 20\,\text{min} + 50\,\text{min} = 110\,\text{min} = 1.83\,\text{h}$. Divide $\vec{r}$ by this interval to obtain an average velocity of $\langle\vec{v}\rangle = (2.90\,\text{km/h})\hat{\imath} + (7.05\,\text{km/h})\hat{\jmath}$. The average velocity has a magnitude of $7.62\,\text{km/h}$ and is directed $67.6°$ north of east.

## 41

(a) The velocity of the particle at any time $t_2$ is given by $\vec{v}_2 = \vec{v}_1 + \vec{a}(t_2 - t_1)$, where $\vec{v}_1$ is the velocity at time $t_1$ and $\vec{a}$ is the acceleration. The $x$ component is $v_{2x} = v_{1x} + a_x(t_2 - t_1) = 3.00\,\text{m/s} - (1.00\,\text{m/s}^2)(t_2 - t_1)$ and the $y$ component is $v_{2y} = v_{1y} + a_y(t_2 - t_1) = -(0.500\,\text{m/s}^2)(t_2 - t_1)$. When the particle reaches its maximum $x$ coordinate $v_{2x} = 0$. This means $3.00\,\text{m/s} - (1.00\,\text{m/s}^2)(t_2 - t_1) = 0$ or $t_2 - t_1 = 3.00\,\text{s}$. The $y$ component of the velocity at this time is $v_{2y} = (-0.500\,\text{m/s}^2)(3.00\,\text{s}) = -1.50\,\text{m/s}$. The velocity is $\vec{v}_2 = (-1.50\,\text{m/s})\hat{\jmath}$.

(b) Since the particle starts from the origin its coordinates at time $t_2$ are $x_2 = v_{1x}(t_2 - t_1) + \frac{1}{2}a_x(t_2 - t_1)^2$ and $y_2 = v_{1y}t + \frac{1}{2}a_y(t_2 - t_1)^2$. For $t_2 - t_1 = 3.00\,\text{s}$ their values are

$$x_2 = (3.00\,\text{m/s})(3.00\,\text{s}) - \frac{1}{2}(1.00\,\text{m/s}^2)(3.00\,\text{s})^2 = 4.50\,\text{m}$$

and

$$y_2 = -\frac{1}{2}(0.500\,\text{m/s}^2)(3.00\,\text{s})^2 = -2.25\,\text{m}.$$

The position vector is $\vec{r} = (4.50\,\text{m})\hat{\imath} - (2.25\,\text{m})\hat{\jmath}$.

## 45

(a) The velocity is the derivative of the position vector with respect to time:

$$\vec{v} = \frac{d}{dt}\left((1\,\text{m})\hat{i} + (4\,\text{m/s}^2)t^2\,\hat{j}\right) = (8\,\text{m/s}^2)t\,\hat{j}.$$

(b) The acceleration is the derivative of the velocity with respect to time:

$$\vec{a} = \frac{d}{dt}\left[(8\,\text{m/s}^2)t\right]\hat{j} = (8\,\text{m/s}^2)\hat{j}.$$

## 51

(a) Let $r$ be the radius of the orbit and $a$ be the magnitude of the centripetal acceleration, then solve $a = v^2/r$ for the speed $v$:

$$v = \sqrt{ra} = \sqrt{(5.0\,\text{m})(7.0)(9.8\,\text{N/kg})} = 19\,\text{m/s}.$$

(b) If the astronaut goes around $N$ times in time $\Delta t$, then his speed is $v = 2\pi Nr/\Delta t$. Solve for $N/\Delta t$, then substitute $v = 19\,\text{m/s}$ and $r = 5.0\,\text{m}$:

$$\frac{N}{\Delta t} = \frac{v}{2\pi r} = \frac{(19\,\text{m/s})(60\,\text{s/min})}{2\pi(5.0\,\text{m})} = 35\,\text{rev/min}.$$

(c) If the astronaut rotates through 35 rev each minute, then the time for one revolution is $T = 1/(35\,\text{min}^{-1}) = 2.86 \times 10^{-2}\,\text{min} = 1.71\,\text{s}$.

## 57

To calculate the centripetal acceleration of the stone you need to know its speed while it is being whirled around. This the same as its initial speed when it flies off. Use the projectile motion equations to find that speed. Take the $y$ axis to be upward and the $x$ axis to be horizontal. Suppose the stone leaves the string with velocity $\vec{v}_i$ and its time of flight is $\Delta t$ Then the components of the displacement of the stone when it is a projectile are given by $\Delta x = v_i \Delta t$ and $\Delta y = -\frac{1}{2}g(\Delta t)^2$. It hits the ground when $\Delta x = 10\,\text{m}$ and $\Delta y = -2.0\,\text{m}$. Note that the initial velocity is horizontal. Solve the second equation for the time of flight: $\Delta t = \sqrt{-2\,\Delta y/g}$. Substitute this expression into the first equation and solve for $v_i$:

$$v_i = \Delta x\sqrt{-\frac{g}{2\,\Delta y}} = (10\,\text{m})\sqrt{-\frac{9.8\,\text{N/kg}}{2(-2.0\,\text{m})}} = 15.7\,\text{m/s}.$$

The magnitude of the centripetal acceleration is $a = v_i^2/r = (15.7\,\text{m/s})^2/(1.5\,\text{m}) = 160\,\text{m/s}^2$.

# Chapter 6

## 7

Label the two forces $\vec{F}_1$ and $\vec{F}_2$. According to Newton's second law, $\vec{F}_1 + \vec{F}_2 = m\vec{a}$, so $\vec{F}_2 = m\vec{a} - \vec{F}_1$. In unit vector notation $\vec{F}_1 = (20.0\,\text{N})\,\hat{\imath}$ and

$$\vec{a} = -(12\sin 30°\,\text{m/s}^2)\,\hat{\imath} - (12\cos 30°\,\text{m/s}^2)\,\hat{\jmath} = -(6.0\,\text{m/s}^2)\,\hat{\imath} - (10.4\,\text{m/s}^2)\,\hat{\jmath}.$$

Thus

$$\vec{F}_2 = (2.0\,\text{kg})(-6.0\,\text{m/s}^2)\,\hat{\imath} + (2.0\,\text{kg})(-10.4\,\text{m/s}^2)\,\hat{\jmath} - (20.0\,\text{N})\,\hat{\imath} = (-32\,\text{N})\,\hat{\imath} - (21\,\text{N})\,\hat{\jmath}.$$

(b) and (c) The magnitude of $\vec{F}_2$ is $F_2 = \sqrt{F_{2x}^2 + F_{2y}^2} = \sqrt{(-32\,\text{N})^2 + (-21\,\text{N})^2} = 38\,\text{N}$. The angle that $\vec{F}_2$ makes with the positive $x$ axis is given by $\tan\theta = F_{2y}/F_{2x} = (21\,\text{N})/(32\,\text{N}) = 0.656$. The angle is either 33° or 33° + 180° = 213°. Since both the $x$ and $y$ components are negative the correct result is 213°.

## 9

In all three cases the scale is not accelerating, which means that the two cords exert forces of equal magnitude on it. The scale reads the magnitude of either of these forces. In each case since the salami is not accelerating you know that the magnitude of the tension force of the cord attached to the salami must be the same as the weight of the salami. Thus the scale reading is $mg$, where $m$ is the mass of the salami. Its value is $(11.0\,\text{kg})(9.8\,\text{N/kg}) = 108\,\text{N}$.

## 11

(a) The free-body diagram for the bureau is shown on the right. $\vec{F}$ is the applied force, $\vec{f}$ is the force of friction, $\vec{N}$ is the normal force of the floor, and $\vec{F}^{\text{grav}}$ is the force of gravity. Take the $x$ axis to be horizontal and the $y$ axis to be vertical. Assume the bureau does not move and write Newton's second law. The $x$ component is $F - f = 0$ and the $y$ component is $N - F^{\text{grav}} = 0$. The force of friction is then equal in magnitude to the applied force: $f = F$. The normal force is equal in magnitude to the force of gravity: $N = F^{\text{grav}}$. As $F$ increases, $f$ increases until $f = \mu^{\text{static}} N$. Then the bureau starts to move. The minimum force that must be applied to start the bureau moving is $F = \mu^{\text{static}} N = \mu^{\text{static}} F^{\text{grav}} = \mu^{\text{static}} mg = (0.45)(45\,\text{kg})(9.8\,\text{N/kg}) = 200\,\text{N}$, where $mg$ was substituted for $F^{\text{grav}}$.

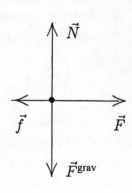

(b) The equation for $F$ is the same but the mass is now $45\,\text{kg} - 17\,\text{kg} = 28\,\text{kg}$. Thus $F = \mu^{\text{static}}mg = (0.45)(28\,\text{kg})(9.8\,\text{N/kg}) = 120\,\text{N}$.

## 15

(a) The free-body diagram for the crate is shown on the right. $\vec{F}$ is the force of the person on the crate, $\vec{f}$ is the force of friction, $\vec{N}$ is the normal force of the floor, and $\vec{F}^{\text{grav}}$ is the force of gravity. The magnitude of the force of friction is given by $f = \mu^{\text{kin}}N$, where $\mu^{\text{kin}}$ is the coefficient of kinetic friction. The vertical component of Newton's second law is used to find the normal force. Since the vertical component of the acceleration is zero, $N - F^{\text{grav}} = 0$ and $N = F^{\text{grav}}$. Thus $f = \mu^{\text{kin}}N = \mu^{\text{kin}}F^{\text{grav}} = \mu^{\text{kin}}mg = (0.35)(55\,\text{kg})(9.8\,\text{N/kg}) = 190\,\text{N}$, where $mg$ was substituted for $F^{\text{grav}}$.

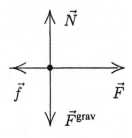

(b) Use the horizontal component of Newton's second law to find the acceleration. Since $F - f = ma$, $a = (F - f)/m = (220\,\text{N} - 189\,\text{N})/(55\,\text{kg}) = 0.56\,\text{m/s}^2$.

## 17

(a) The free-body diagram for the puck is shown on the right. $\vec{N}$ is the normal force of the ice on the puck, $\vec{f}$ is the force of friction, and $\vec{F}^{\text{grav}}$ is the force of gravity. The horizontal component of Newton's second law is $-f = ma_x$, where the positive direction is taken to be the direction of motion of the puck (to the right in the diagram). Constant acceleration kinematics can be used to find the acceleration. Solve $v_2^2 = v_1^2 + 2a_x\,\Delta x$ for $a_x$. Since the final velocity is zero, $a_x = -v_1^2/2\,\Delta x$. This result is substituted into the second law equation to obtain

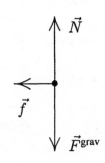

$$f = \frac{mv_1^2}{2\,\Delta x} = \frac{(0.110\,\text{kg})(6.0\,\text{m/s})^2}{2(15\,\text{m})} = 0.13\,\text{N}.$$

(b) Use $f = \mu^{\text{kin}}N$. The vertical component of Newton's second law is $N - F^{\text{grav}} = 0$, so $N = F^{\text{grav}} = mg$ and $f = \mu^{\text{kin}}mg$. Solve for $\mu^{\text{kin}}$:

$$\mu^{\text{kin}} = \frac{f}{mg} = \frac{0.13\,\text{N}}{(0.110\,\text{kg})(9.8\,\text{m/s}^2)} = 0.12.$$

## 21

A cross section of the cone of sand is shown on the right. To pile the most sand without extending the radius, sand is added to make the height $h$ as great as possible. Eventually, however, the sides become so steep that sand at the surface begins to slide. You want to find the greatest height (or greatest slope) for which the sand does not slide.

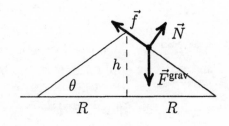

A grain of sand is shown on the diagram and the forces on it are labeled. $\vec{N}$ is the normal force of the surface, $\vec{F}^{\text{grav}}$ is the force of gravity, and $\vec{f}$ is the force of friction. Take the $x$ axis to be parallel to the plane with the positive direction down the plane and the $y$ axis to be in the direction of the normal force. Assume the grain does not slide, so its acceleration is zero. Then the $x$ component of Newton's second law is $F^{\text{grav}} \sin\theta - f = 0$ and the $y$ component is $N - F^{\text{grav}} \cos\theta = 0$. The first equation gives $f = F^{\text{grav}} \sin\theta$ and the second gives $N = F^{\text{grav}} \cos\theta$. If the grain does not slide, the condition $f < \mu^{\text{static}} N$ must hold. This means $F^{\text{grav}} \sin\theta < \mu^{\text{static}} F^{\text{grav}} \cos\theta$ or $\tan\theta < \mu^{\text{static}}$. The surface of the cone has the greatest slope (and the height of the cone is the greatest) if $\tan\theta = \mu^{\text{static}}$.

Since $R$ and $h$ are two sides of a right triangle, $h = R\tan\theta$. Replace $\tan\theta$ with $\mu^{\text{static}}$ to obtain $h = \mu^{\text{static}} R$ and use this to substitute for $h$ in the equation $V = Ah/3$ for the volume of the cone. Also replace the area $A$ of the base with $\pi R^2$. The result is $V = \pi\mu^{\text{static}} R^3/3$.

## 23

(a) The free-body diagram for the crate is shown on the right. $\vec{T}$ is the tension force of the rope on the crate, $\vec{N}$ is the normal force of the floor on the crate, $\vec{F}^{\text{grav}}$ is the force of gravity, and $\vec{f}$ is the force of friction. Take the $x$ axis to be horizontal to the right and the $y$ axis to be vertically upward. Assume the crate is motionless. The $x$ component of Newton's second law is then $T\cos\theta - f = 0$ and the $y$ component is $T\sin\theta + N - F^{\text{grav}} = 0$, where $\theta \, (= 15°)$ is the angle between the rope and the horizontal. The first equation gives $f = T\cos\theta$ and the second gives $N = F^{\text{grav}} - T\sin\theta$. If the crate is to remain at rest, $f$ must be less than $\mu^{\text{static}} N$, or $T\cos\theta < \mu^{\text{static}}(F^{\text{grav}} - T\sin\theta)$. When

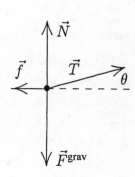

the tension force is sufficient to just start the crate moving $T\cos\theta = \mu^{\text{static}}(F^{\text{grav}} - T\sin\theta)$. When $mg$ is substituted for $F^{\text{grav}}$ this becomes $T\cos\theta = \mu^{\text{static}}(mg - T\sin\theta)$. Solve for $T$:

$$T = \frac{\mu^{\text{static}} mg}{\cos\theta + \mu^{\text{static}}\sin\theta} = \frac{(0.50)(68\,\text{kg})(9.8\,\text{N/kg})}{\cos 15° + 0.50\sin 15°} = 300\,\text{N}.$$

(b) The second law equations for the moving crate are $T\cos\theta - f = ma_x$ and $N + T\sin\theta - mg = 0$. Now $f = \mu^{\text{kin}} N$. The second equation gives $N = mg - T\sin\theta$, as before, so $f = \mu^{\text{kin}}(mg - T\sin\theta)$. This expression is substituted for $f$ in the first equation to obtain $T\cos\theta - \mu^{\text{kin}}(mg - T\sin\theta) = ma_x$, so the acceleration component is

$$a_x = \frac{T(\cos\theta + \mu^{\text{kin}}\sin\theta)}{m} - \mu^{\text{kin}} g.$$

Its numerical value is

$$a_x = \frac{(304\,\text{N})(\cos 15° + 0.35\sin 15°)}{68\,\text{kg}} - (0.35)(9.8\,\text{m/s}^2) = 1.3\,\text{m/s}^2.$$

(a) Free-body diagrams for the blocks $A$ and $C$, considered as a single object, and for the block $B$ are shown on the right. $\vec{T}_A$ is the tension force of the rope on block A, $\vec{N}$ is the normal force of the table on block $A$, $\vec{f}$ is the force of friction, $\vec{F}_{AC}^{\text{grav}}$ is gravitational force on blocks $A$ and $C$, and $\vec{F}_B^{\text{grav}}$ is the gravitational force on block $B$. Assume the blocks are not moving. For the blocks on the table take the $x$ axis to be to the right and the $y$ axis to be upward. The $x$ component of Newton's second

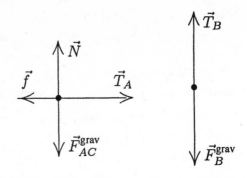

law is then $T_A - f = 0$ and the $y$ component is $N - F_{AC}^{\text{grav}} = 0$. For block $B$ take the downward direction to be positive. Then Newton's second law for that block is $F_B^{\text{grav}} - T_B = 0$. This yields $T_B = F_B^{\text{grav}}$. The magnitudes of the tension forces are the same, so $T_A = F_B^{\text{grav}}$. The first of the Newton's second law equations gives $f = T_A$, so $f = F_B^{\text{grav}}$. The second of the Newton's second law equations gives $N = F_{AC}^{\text{grav}}$. If sliding is not to occur, $f$ must be less than $\mu^{\text{static}} N$, or $F_B^{\text{grav}} < \mu^{\text{static}} F_{AC}^{\text{grav}}$. The smallest that $F_{AC}^{\text{grav}}$ can be with the blocks still at rest is $F_{AC}^{\text{grav}} = F_B^{\text{grav}}/\mu^{\text{static}} = (22\,\text{N})/(0.20) = 110\,\text{N}$. Since the weight of block $A$ is 44 N, the least weight for $C$ is $110\,\text{N} - 44\,\text{N} = 66\,\text{N}$.

(b) The second law equations become $T_A - f = m_A a_{Ax}$, $N - F_A^{\text{grav}} = 0$, and $F_B^{\text{grav}} - T_B = m_B a_{By}$. In addition, $f = \mu^{\text{kin}} N$. The second equation gives $N = F_A^{\text{grav}}$, so $f = \mu^{\text{kin}} F_A^{\text{grav}}$. The two blocks move together, so $a_{By} = a_{Ax}$. Furthermore, $T_A = T_B$. Thus the third equation becomes $F_B^{\text{grav}} - T_A = m_B a_{Ax}$. This gives $T_A = F_B^{\text{grav}} - m_B a_{Ax}$. Substitute the expressions for $T_A$ and $f$ into the first Newtons's law equation to obtain $F_B^{\text{grav}} - m_B a_{Ax} - \mu^{\text{kin}} F_A^{\text{grav}} = m_A a_{Ax}$. Solve for $a_{Ax}$:

$$a_{Ax} = \frac{F_B^{\text{grav}} - \mu^{\text{kin}} F_A^{\text{grav}}}{m_A + m_B} = \frac{22\,\text{N} - (0.15)(44\,\text{N})}{(44\,\text{N} + 22\,\text{N})(9.8\,\text{N/kg})} = 2.3\,\text{m/s}^2 \,.$$

The free-body diagrams for block $B$ and for the knot just above block $A$ are shown on the right. $\vec{T}_B$ is the tension force of the rope on block $B$, $\vec{T}_{1K}$ is the tension force of the same rope on the knot, $\vec{T}_{2K}$ is the tension force of the other rope on knot, $\vec{f}$ is the force of friction of the horizontal surface on block $B$, $\vec{N}$ is the normal force of the surface on block $B$, $\vec{F}_A^{\text{grav}}$ is the gravitational force on block A, and $\vec{F}_B^{\text{grav}}$ is the gravitational force on block B. $\theta$ ($= 30°$) is the angle between the second rope and the horizontal.

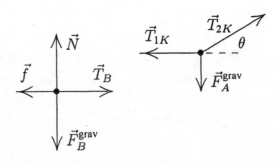

For each object take the $x$ axis to be horizontal and the $y$ axis to be vertical. The $x$ component of Newton's second law for block $B$ is then $T_B - f = 0$ and the $y$ component is $N - F_B^{\text{grav}} = 0$. The $x$ component of Newton's second law for the knot is $T_{2K} \cos\theta - T_{1K} = 0$ and the $y$ component is $T_{2K} \sin\theta - F_A^{\text{grav}} = 0$. Use $T_B = T_{1K}$ and eliminate the tension forces to find expressions for

$f$ and $N$ in terms of $F_A^{\text{grav}}$ and $F_B^{\text{grav}}$, then select $F_A^{\text{grav}}$ so $f = \mu^{\text{static}} N$. The second Newton's law equation gives $N = F_B^{\text{grav}}$ immediately. The third gives $T_{2K} = T_{1K}/\cos\theta$. Substitute this expression into the fourth equation to obtain $T_B = F_A^{\text{grav}}/\tan\theta$. Substitute $F_A^{\text{grav}}/\tan\theta$ for $T_B$ in the first equation to obtain $f = F_A^{\text{grav}}/\tan\theta$. For the blocks to remain stationary $f$ must be less than $\mu^{\text{static}} N$ or $F_A^{\text{grav}}/\tan\theta < \mu^{\text{static}} F_B^{\text{grav}}$. The greatest that $F_A^{\text{grav}}$ can be is the value for which $F_A^{\text{grav}}/\tan\theta = \mu^{\text{static}} F_B^{\text{grav}}$. Solve for $F_A^{\text{grav}}$:

$$F_A^{\text{grav}} = \mu^{\text{static}} F_B^{\text{grav}} \tan\theta = (0.25)(711\,\text{N}) \tan 30° = 100\,\text{N}.$$

## 31

(a) First check to see if the bodies start to move. Assume they remain at rest, compute the force of friction that holds them at rest, and compare its magnitude with $\mu^{\text{static}} N$.

The free-body diagrams are shown on the right. $\vec{T}_B$ is the tension force of the string on B, $\vec{T}_A$ is the tension force of the string on A, $\vec{f}$ is the force of friction on B, $\vec{N}$ is the magnitude of the normal force of the plane on B, $\vec{F}_A^{\text{grav}}$ is the force of gravity on A, and $\vec{F}_B^{\text{grav}}$ is the force of gravity on body B. $\theta$ is the angle of incline of the plane (40°).

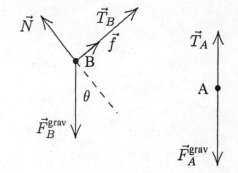

For B take the $x$ axis to be up the plane and the $y$ axis to be in the direction of the normal force. The $x$ component of Newton's second law is then $T_B + f_x - F_B^{\text{grav}} \sin\theta = 0$ and the $y$ component is $N - F_B^{\text{grav}} \cos\theta = 0$.

For $A$ take the positive direction to be downward. The second law equation for this object is $F_A^{\text{grav}} - T_A = 0$. The third equation gives $T_A = F_A^{\text{grav}}$. Since the tension forces on A and B have the same magnitude, $T_B = F_A^{\text{grav}}$. Substitute this into the first equation to obtain $F_A^{\text{grav}} + f_x - F_B^{\text{grav}} \sin\theta = 0$. Thus $f_x = F_B^{\text{grav}} \sin\theta - F_A^{\text{grav}} = (102\,\text{N}) \sin 40° - 32\,\text{N} = 34\,\text{N}$. The positive result indicates that the force of friction is up the plane.

The second equation gives $N = F_B^{\text{grav}} \cos\theta = (102\,\text{N}) \cos 40° = 78\,\text{N}$, so $\mu^{\text{static}} N = (0.56)(78\,\text{N}) = 44\,\text{N}$. Since the magnitude of the force of friction that holds the bodies motionless is less than $\mu^{\text{static}} N$ the bodies remain at rest. Their accelerations are zero.

(b) Since $B$ is moving up the plane the force of friction is down the plane and has the magnitude $f = \mu^{\text{kin}} N$. The second law equations become $T_B - \mu^{\text{kin}} N - F_B^{\text{grav}} \sin\theta = m_B a_{Bx}$, $N - F_B^{\text{grav}} \cos\theta = 0$, and $F_A^{\text{grav}} - T_A = m_A a_{Ax}$, where $a_{Ax}$ and $a_{Bx}$ are acceleration components. Because the objects move together and because the coordinate axis is chosen to be up the plane for $B$ and downward for $A$, $a_{Ax} = a_{Bx}$. Furthermore, the tension forces on the two objects have the same magnitude, so $T_A = T_B$. Substitute $N = F_B^{\text{grav}} \cos\theta$, from the second equation, and $T_B = T_A = F_A^{\text{grav}} - m_A a$, from the third, into the first to obtain

$$F_A^{\text{grav}} - m_A a_x - \mu^{\text{kin}} F_B^{\text{grav}} \cos\theta - F_B^{\text{grav}} \sin\theta = m_B a_x.$$

Solve for $a_x$:

$$a_x = \frac{F_A^{\text{grav}} - F_B^{\text{grav}} \sin\theta - \mu^{\text{kin}} F_B^{\text{grav}} \cos\theta}{m_A + m_B}$$

$$= \frac{32\,\text{N} - (102\,\text{N})\sin 40° - (0.25)(102\,\text{N})\cos 40°}{(32\,\text{N} + 102\,\text{N})/(9.8\,\text{N/kg})} = -3.9\,\text{m/s}^2.$$

The acceleration is down the plane. The objects are slowing down. Notice that $m = F^{\text{grav}}/g$ was used to calculate the masses in the denominator.

(c) Now $B$ is moving down the plane, so the force of friction is up the plane and has a magnitude that is given by $\mu^{\text{kin}} N$. The second law equations become $T_B + \mu^{\text{kin}} N - F_B^{\text{grav}} \sin\theta = m_B a_x$, $N - F_B^{\text{grav}} \cos\theta$, and $F_A^{\text{grav}} - T_A = m_A a_x$. Substitute $N = F_B^{\text{grav}} \cos\theta$, from the second equation, and $T_B = T_A = F_A^{\text{grav}} - m_A a_x$, from the third, into the first to obtain

$$F_A^{\text{grav}} - m_A a_x + \mu^{\text{kin}} F_B^{\text{grav}} \cos\theta - F_B^{\text{grav}} \sin\theta = m_B a_x.$$

Solve for $a_x$:

$$a_x = \frac{F_A^{\text{grav}} - F_B^{\text{grav}} \sin\theta + \mu^{\text{kin}} F_B^{\text{grav}} \cos\theta}{m_A + m_B}$$

$$= \frac{32\,\text{N} - (102\,\text{N})\sin 40° + (0.25)(102\,\text{N})\cos 40°}{(32\,\text{N} + 102\,\text{N})/(9.8\,\text{m/s}^2)} = -0.98\,\text{m/s}^2.$$

The acceleration is again down the plane. The objects are speeding up.

## 37

The free-body diagrams for the slab and block are shown on the right. $\vec{F}$ is the force applied to the block, $\vec{N}_f$ is the normal force of the floor on the slab, $\vec{N}_s$ is the normal force of the slab on the block, $\vec{N}_s$ is the normal force of the slab on the block, $\vec{f}_s$ is the frictional force of the slab on the block, $\vec{f}_b$ is the frictional force of the block on the slab, $\vec{F}_s^{\text{grav}}$ is the gravitational force on the slab, and $\vec{F}_b^{\text{grav}}$ is the gravitational force on the block. Newton's third law tells us that $\vec{f}_s = -\vec{f}_b$: the forces of friction on the two objects have the same magnitude but are in opposite directions. Similarly the normal forces of the blocks on each other have the same magnitude and are in opposite directions: $\vec{N}_b = -\vec{N}_s$.

Take positive $x$ to be to the left and positive $y$ to be upward. The $x$ component of Newton's second law for the slab is then $f_b = m_s a_{sx}$ and the $y$ component is $N_f - N_b - F_s^{\text{grav}} = 0$, where $\vec{a}_s$ is the acceleration of the slab. The $x$ component of Newton's second law for the block is $F - f_s = m_b a_{bx}$ and the $y$ component is $N_s - F_b^{\text{grav}} = 0$, where $\vec{a}_b$ is the acceleration of the block.

First check to see if the block slides on the slab. Assume it does not. Then $\vec{a}_s = \vec{a}_b$. Use $a_x$ to denote the $x$ components of both these accelerations. Also use $f$ to denote the magnitudes of the frictional forces between the slab and the block and $N$ to denote the magnitudes of their normal forces. Use $f = m_s a_x$ to eliminate $a_x$ from $F - f = m_b a$, then solve for $f$. The result is

$$f = \frac{m_s F}{m_s + m_b} = \frac{(40\,\text{kg})(100\,\text{N})}{40\,\text{kg} + 10\,\text{kg}} = 80\,\text{N}\,.$$

According to the last of the second law equations, $N = F_b^{\text{grav}}$, so

$$\mu^{\text{static}} N = \mu^{\text{static}} F_b^{\text{grav}} = (0.60)(10\,\text{kg})(9.8\,\text{N/kg}) = 59\,\text{N}\,.$$

Since $f > \mu^{\text{static}} N$ the block slides on the slab and their accelerations are different. Carefully note that $N$, not $N_f$, was used to compute the upper limit of the static frictional force. The forces of friction are exerted by the slab and block on each other, so the normal force these objects exert on each other is used to compute the upper limit.

Since the block slides on the slab, the magnitude of the frictional force is given by $f = \mu^{\text{kin}} N$. The $x$ component of Newton's second law for the slab is now $\mu^{\text{kin}} N = m_s a_{sx}$, the $x$ component of Newton's second law for the block is $F - \mu^{\text{kin}} N = m_b a_{bx}$, and the $y$ component is $N - F_b^{\text{grav}} = 0$. The last of these gives $N = F_b^{\text{grav}}$, as before. Substitute into the first equation and solve for $a_{sx}$:

$$a_{sx} = \frac{\mu^{\text{kin}} F_b^{\text{grav}}}{m_s} = \frac{(0.40)(10\,\text{kg})(9.8\,\text{N/kg})}{40\,\text{kg}} = 0.98\,\text{m/s}^2\,.$$

Substitute $N = F_b^{\text{grav}}$ into the second equation and solve for $a_{bx}$:

$$a_{bx} = \frac{F - \mu^{\text{kin}} F_b^{\text{grav}}}{m_b} = \frac{100\,\text{N} - (0.40)(10\,\text{kg})(9.8\,\text{m/s}^2)}{10\,\text{kg}} = 6.1\,\text{m/s}^2\,.$$

Both these accelerations are leftward.

## 41

Let the magnitude of the frictional force be $\alpha v$, where $\alpha = 70\,\text{N} \cdot \text{s/m}$. Take the direction of the boat's motion to be positive. Newton's second law is then $-\alpha v = m\,dv/dt$. Thus

$$\int_{v_0}^{v} \frac{dv}{v} = -\frac{\alpha}{m} \int_0^t dt\,,$$

where $v_0$ is the velocity at time zero and $v$ is the velocity at time $t$. The integrals can be evaluated, with the result

$$\ln \frac{v}{v_0} = -\frac{\alpha t}{m}\,.$$

Take $v = v_0/2$ and solve for $t$:

$$t = \frac{m}{\alpha} \ln 2 = \frac{1000\,\text{kg}}{70\,\text{N} \cdot \text{s/m}} \ln 2 = 9.9\,\text{s}\,.$$

## 43

Use Eq. 6–23 of the text: $D = \frac{1}{2}C\rho Av^2$, where $\rho$ is the air density, $A$ is the cross-sectional area of the missile, $v$ is the speed of the missile, and $C$ is the drag coefficient. The area is given by $A = \pi R^2$, where $R$ (= 26.5 cm = 0.265 m) is the radius of the missile. Thus $D = \frac{1}{2}(0.75)(1.2\,\text{kg/m}^3)\pi(0.265\,\text{m})^2(250\,\text{m/s})^2 = 6.2 \times 10^3\,\text{N}$.

## 47

The acceleration of the electron is vertical and for all practical purposes the only force acting on it is the electric force. The force of gravity is much smaller. Take the $x$ axis to be in the direction of the initial velocity and the $y$ axis to be in the direction of the electrical force. Place the origin at the initial position of the electron. Since the force and acceleration are constant the kinematic equations are $\Delta x = v_0\,\Delta t$ and $\Delta y = \frac{1}{2}a_y(\Delta t)^2 = \frac{1}{2}(F/m)(\Delta t)^2$, where $\vec{F} = m\vec{a}$ was used to substitute for the acceleration $\vec{a}$. The time taken by the electron to travel a distance $x$ (= 30 mm) horizontally is $t = x/v_0$ and its deflection in the direction of the force is

$$y = \frac{1}{2}\frac{F}{m}\left(\frac{x}{v_0}\right)^2 = \frac{1}{2}\left(\frac{4.5 \times 10^{-16}\,\text{N}}{9.11 \times 10^{-31}\,\text{kg}}\right)\left(\frac{30 \times 10^{-3}\,\text{m}}{1.2 \times 10^7\,\text{m/s}}\right)^2 = 1.5 \times 10^{-3}\,\text{m}\,.$$

## 51

The free-body diagram is shown on the right, with the tension force $\vec{T}$ of the string, the force of gravity $\vec{F}^{\text{grav}}$, and the force $\vec{F}$ of the air labeled. Take the coordinate system to be as shown. The $x$ component of the net force is $T\sin\theta - F$ and the $y$ component is $T\cos\theta - F^{\text{grav}}$, where $\theta = 37°$. Since the sphere is motionless the net force on it is zero. Answer the questions in the reverse order to that given. (b) Solve $T\cos\theta - F^{\text{grav}} = 0$ for the tension force of the string: $T = F^{\text{grav}}/\cos\theta = (3.0 \times 10^{-4}\,\text{kg})(9.8\,\text{N/kg})/\cos 37° = 3.7 \times 10^{-3}\,\text{N}$. (a) Solve $T\sin\theta - F = 0$ for the force of the air: $F = T\sin\theta = (3.7 \times 10^{-3}\,\text{N})\sin 37° = 2.2 \times 10^{-3}\,\text{N}$.

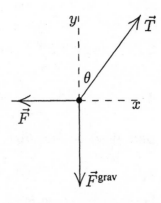

## 61

(a) Consider a small segment of the rope. It has mass and is pulled down by the gravitational force of the Earth. Equilibrium is reached because neighboring portions of the rope pull up on it. Since the tension is a force along the rope at least one of the neighboring portions must slope up away from the segment we are considering. Then the tension has an upward component. This means the rope sags.

(b) The only force acting with a horizontal component is the applied force $\vec{F}$. Consider the block and rope as a single object and write Newton's second law for it: $F_x = (M + m)a_x$, where $a_x$ is the horizontal component of the acceleration and the positive $x$ direction is taken to be to the right. The acceleration component is given by $a_x = F/(M + m)$.

(c) The force of the rope is the only force with a horizontal component acting on the block. Let this force be $\vec{F}_r$. Then Newton's second law for the block gives

$$F_{rx} = Ma_x = \frac{MF}{M+m},$$

where the expression found above for $a_x$ has been substituted.

(d) Consider the block and half the rope to be a single object, with mass $M+\frac{1}{2}m$. The horizontal force on it is the tension at the midpoint of the rope. Let this force be $\vec{T}_m$ and use Newton's second law to find its value:

$$T_{mx} = (M + \frac{1}{2}m)a_x = \frac{(M + \frac{1}{2}m)F_x}{(M+m)} = \frac{(2M+m)F_x}{2(M+m)}.$$

## 65

The magnitude of the acceleration of the car as it rounds the curve is given by $v^2/R$, where $v$ is the speed of the car and $R$ is the radius of the curve. Since the road is horizontal, only the frictional force of the road on the tires provides the force to produce this acceleration. The horizontal component of Newton's second law is $f = mv^2/R$. If $N$ is the normal force of the road on the car and $m$ is the mass of the car, the vertical component of the second law is $N - mg = 0$. Thus $N = mg$ and $\mu^{\text{static}} N = \mu^{\text{static}} mg$. If the car does not slip, $f \leq \mu^{\text{static}} mg$. This means $v^2/R \leq \mu^{\text{static}} g$, or $v \leq \sqrt{\mu^{\text{static}} Rg}$. The maximum speed with which the car can round the curve without slipping is

$$v_{\max} = \sqrt{\mu^{\text{static}} Rg} = \sqrt{(0.60)(30.5 \text{ m})(9.8 \text{ m/s}^2)} = 13 \text{ m/s}.$$

## 69

For the puck to remain at rest the magnitude of the tension force $T$ of the cord must equal the magnitude of the gravitational force $F^{\text{grav}}$ on the cylinder. The tension force supplies the centripetal force that keeps the puck in its circular orbit, so Newton's second law gives $T = mv^2/r$. Thus $F^{\text{grav}} = mv^2/r$ and since $F^{\text{grav}} = Mg$, this becomes $Mg = mv^2/r$. The solution for $v$ is $v = \sqrt{Mgr/m}$.

## 71

(a) At the highest point the seat pushes up on the student with a force of magnitude $N$ (= 556 N). Earth pulls down with a force of magnitude $W$ (= 667 N). The seat is pushing up with a force that is smaller than the student's weight in magnitude. The student feels light at the highest point.

(b)When the student is at the highest point, the net force toward the center of the circular orbit is $W - N$ and, according to Newton's second law, this must be $mv^2/R$, where $v$ is the speed of the student and $R$ is the radius of the orbit. Thus $mv^2/R = W - N = 667 \text{ N} - 556 \text{ N} = 111 \text{ N}$.

The force of the seat when the student is at the lowest point is upward, so the net force toward the center of the circle is $N - W$ and $N - W = mv^2/R$. Solve for $N$:

$$N = \frac{mv^2}{R} + W = 111\,\text{N} + 667\,\text{N} = 778\,\text{N}.$$

(c) At the highest point $W - N = mv^2/R$, so $N = W - mv^2/R$. If the speed is doubled, $mv^2/R$ increases by a factor of 4, to 444 lb. Then $N = 667\,\text{lb} - 444\,\text{lb} = 223\,\text{N}$.

## 75

(a) The free-body diagram for the ball is shown on the right. $\vec{T}_u$ is the tension force of the upper string, $\vec{T}_\ell$ is the tension force of the lower string, and $m$ is the mass of the ball. Note that the tension force of the upper string is greater than the tension force of the lower string. It must balance the downward pull of gravity and the force of the lower string.

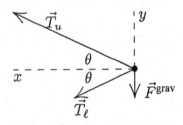

(b) Take the $x$ axis to be to the left, toward the center of the circular orbit, and the $y$ axis to be upward. Since the magnitude of the acceleration is $a = v^2/R$, the $x$ component of Newton's second law is

$$T_u \cos\theta + T_\ell \cos\theta = \frac{mv^2}{R},$$

where $v$ is the speed of the ball and $R$ is the radius of its orbit. Since $F^{\text{grav}} = mg$, the $y$ component is

$$T_u \sin\theta - T_\ell \sin\theta - mg = 0.$$

The second equation gives the tension force of the lower string: $T_\ell = T_u - mg/\sin\theta$. Since the triangle is equilateral $\theta = 30°$. Thus

$$T_\ell = 35\,\text{N} - \frac{(1.34\,\text{kg})(9.8\,\text{M/kg})}{\sin 30°} = 8.74\,\text{N}.$$

(c) The net force is radially inward and has magnitude $F_{\text{net}} = (T_u + T_\ell)\cos\theta = (35\,\text{N} + 8.74\,\text{N})\cos 30° = 37.9\,\text{N}$.

(d) Use $F^{\text{net}} = mv^2/R$. The radius of the orbit is $[(1.70\,\text{m})/2)]\tan 30° = 1.47\,\text{m}$. Thus

$$v = \sqrt{\frac{RF_{\text{net}}}{m}} = \sqrt{\frac{(1.47\,\text{m})(37.9\,\text{N})}{1.34\,\text{kg}}} = 6.45\,\text{m/s}.$$

# Chapter 7

## 7

If $\langle F \rangle$ is the magnitude of the average force, then the magnitude of the impulse is $J = \langle F \rangle \Delta t$, where $\Delta t$ is the time interval over which the force is exerted. This equals the magnitude of the change in the momentum of the ball and since the ball is initially at rest it equals the magnitude of the final momentum $mv$. When $\langle F \rangle \Delta t = mv$ is solved for $v$ the result is

$$v = \frac{\langle F \rangle \Delta t}{m} = \frac{(50\,\mathrm{N})(10 \times 10^{-3}\,\mathrm{s})}{0.20\,\mathrm{kg}} = 2.5\,\mathrm{m/s}.$$

## 17

Take the magnitude of the force to be $F = At$, where $A$ is a constant of proportionality. The condition that $F = 50\,\mathrm{N}$ when $t = 4.0\,\mathrm{s}$ leads to

$$A = \frac{50\,\mathrm{N}}{4.0\,\mathrm{s}} = 12.5\,\mathrm{N/s}.$$

The magnitude of the impulse exerted on the object is

$$J = \int_0^{4.0\,\mathrm{s}} F\,dt = \int_0^{4.0\,\mathrm{s}} At\,dt = \frac{1}{2}At^2 \Big|_0^{4.0\,\mathrm{s}} = \frac{1}{2}(12.5\,\mathrm{N/s})(4.0\,\mathrm{s})^2 = 100\,\mathrm{N\cdot s}.$$

This equals the magnitude of the change in the momentum of the ball or since the ball started from rest it equals the magnitude of the final momentum: $J = mv_2$. Thus

$$v_2 = \frac{J}{m} = \frac{100\,\mathrm{N\cdot s}}{10\,\mathrm{kg}} = 10\,\mathrm{m/s}.$$

## 19

(a) Suppose the pellets are initially traveling in the positive $x$ direction. If $m$ is the mass of a pellet and $\vec{v}$ is its velocity as it hits the wall, then its momentum is

$$\vec{p} = m\vec{v} = (2.0 \times 10^{-3}\,\mathrm{kg})(500\,\mathrm{m/s})\hat{\imath} = (1.00\,\mathrm{kg\cdot m/s})\hat{\imath}.$$

(b) The force on the wall is given by the rate at which momentum is transferred from the pellets to the wall. Since the pellets do not rebound, each pellet that hits transfers $\vec{p} = (1.00\,\mathrm{kg\cdot m/s})\hat{\imath}$. If $\Delta N$ pellets hit in time $\Delta t$, then the average rate at which momentum is transferred is

$$\frac{d\vec{p}_{\mathrm{sys}}}{dt} = -\frac{\vec{p}\,\Delta N}{\Delta t} = -(1.0\,\mathrm{kg\cdot m/s})(10\,\mathrm{s}^{-1})\hat{\imath} = -(10\,\mathrm{N})\hat{\imath}.$$

This is the average force of the wall on the pellets. According to Newton's third law the average force of the pellets on the wall has the same magnitude but is in the opposite direction. It is $\langle \vec{F}^{\mathrm{on\ wall}} \rangle = +(10\,\mathrm{N})\hat{\imath}.$

(c) If $\Delta t$ is the time interval for a pellet to be brought to rest by the wall, then the average force exerted on the wall by a pellet is negative of the rate of change of the pellet's momentum. That is, it is

$$\langle \vec{F} \rangle = \frac{\vec{p}}{\Delta t} = \frac{1.0\,\text{kg} \cdot \text{m/s}}{0.6 \times 10^{-3}\,\text{s}} \hat{\imath} = (1.7 \times 10^3\,\text{N}) \hat{\imath}.$$

The force is in the direction of the initial velocity of the pellet.

(d) In part (c) the force is averaged over the time a pellet is in contact with the wall, while in part (b) it is averaged over the time for many pellets to hit the wall. Most of this time no pellet is in contact with the wall, so the average force in part (b) is much less than the average force in part (c).

## 21

Consider first the lighter part. Suppose the impulse has magnitude $J$ and is in the positive direction. Let $m_A$ be the mass of the part and $v_A$ be its speed after the bolts are exploded. Assume both parts are at rest before the explosion. Then $J = m_A v_A$, so

$$v_A = \frac{J}{m_A} = \frac{300\,\text{N} \cdot \text{s}}{1200\,\text{kg}} = 0.25\,\text{m/s}.$$

The impulse on the heavier part has the same magnitude, so $J = m_B v_B$, where $m_B$ is the mass and $v_B$ is the final speed of the part. Thus

$$v_B = \frac{J}{m_B} = \frac{300\,\text{N} \cdot \text{s}}{1800\,\text{kg}} = 0.167\,\text{m/s}.$$

The impulses on the two parts are in opposite directions so the final velocities are also in opposite directions and the relative speed of the parts after the explosion is $0.25\,\text{m/s} + 0.167\,\text{m/s} = 0.417\,\text{m/s}$.

## 25

Let $m_m$ be the mass of the meteor and $m_E$ be the mass of Earth. Earth's mass can be found in Appendix C. Let $v_m$ be the speed of the meteor just before the collision and let $v$ be the speed of Earth (with the meteor embedded) just after the collision. Take the speed of Earth before the collision to be zero. The total momentum of the Earth-meteor system is conserved during the collision, so $m_m v_m = (m_m + m_E)v$. This means

$$v = \frac{v_m m_m}{m_m + m_E} = \frac{(7200\,\text{m/s})(5 \times 10^{10}\,\text{kg})}{5.98 \times 10^{24}\,\text{kg} + 5 \times 10^{10}\,\text{kg}} = 6 \times 10^{-11}\,\text{m/s}.$$

This is about $2\,\text{mm/y}$.

## 31

No external forces with horizontal components act on the cart-man system and the vertical forces sum to zero, so the total momentum of the system is conserved. Let $m_c$ be the mass of the cart, $\vec{v}$ be its initial velocity, and $\vec{v}_c$ be its final velocity (after the man jumps off). Let $m_m$ be the mass of the man. His initial velocity is the same as that of the cart and his final velocity is zero. Conservation of momentum yields

$$(m_m + m_c)\vec{v} = m_c\vec{v}_c.$$

The final speed of the cart is

$$v_c = \frac{v(m_m + m_c)}{m_c} = \frac{(2.3 \text{ m/s})(75 \text{ kg} + 39 \text{ kg})}{39 \text{ kg}} = 6.7 \text{ m/s}.$$

The cart speeds up by $6.7 \text{ m/s} - 2.3 \text{ m/s} = 4.4 \text{ m/s}$. In order to slow himself, the man gets the cart to push backward on him by pushing forward on it, so the cart speeds up.

## 37

(a) Assume no external forces act on the system composed of the two parts of the last stage. The total momentum of the system is conserved. Let $m_c$ be the mass of the rocket case and $m_p$ be the mass of the payload. At first they are traveling together with velocity $\vec{v}$. After the clamp is released the case has velocity $\vec{v}_c$ and the payload has velocity $\vec{v}_p$. Conservation of momentum yields

$$(m_c + m_p)\vec{v} = m_c\vec{v}_c + m_p\vec{v}_p.$$

After the clamp is released the payload, having the lesser mass, will be traveling at the greater speed. Write $\vec{v}_p = \vec{v}_c + \vec{v}^{\text{rel}}$, where $\vec{v}^{\text{rel}}$ is the relative velocity of the payload. When this expression is substituted into the conservation of momentum equation the result is

$$(m_c + m_p)\vec{v} = m_c\vec{v}_c + m_p\vec{v}_c + m_p\vec{v}^{\text{rel}}.$$

Put the $x$ axis along the original direction of travel and suppose the relative velocity of the payload is in the positive $x$ direction. This means that

$$v_c = \frac{(m_c + m_p)v - m_p v^{\text{rel}}}{m_c + m_p}$$

$$= \frac{(290.0 \text{ kg} + 150.0 \text{ kg})(7600 \text{ m/s}) - (150.0 \text{ kg})(910.0 \text{ m/s})}{290.0 \text{ kg} + 150.0 \text{ kg}} = 7290 \text{ m/s}.$$

(b) The final speed of the payload is

$$v_p = v_c + v_{\text{rel}} = 7290 \text{ m/s} + 910.0 \text{ m/s} = 8200 \text{ m/s}.$$

## 41

The total momentum of the child-man system is conserved. Let $\vec{v}_m(t_1)$ be the initial velocity of the man, $\vec{v}_c(t_1)$ be the initial velocity of the child, and $\vec{v}(t_2)$ be the final velocity of both of them. Then

$$m_m\vec{v}_m(t_1) + m_c\vec{v}_c(t_1) = (m_m + m_c)\vec{v}(t_2).$$

The solution for the initial velocity of the child is

$$\vec{v}_c(t_1) = \frac{(m_m + m_c)\vec{v}(t_2) - m_m\vec{v}_m(t_1)}{m_c}.$$

Take the $x$ axis to run from west to east and the $y$ axis to run from south to north. The the $x$ component of the child's initial velocity is

$$v_{cx}(t_1) = \frac{(m_m + m_c)v_x(t_2) - m_m v_{mx}(t_1)}{m_c} = \frac{(60\,\text{kg} + 38\,\text{kg})(3.0\,\text{m/s})\cos 35°}{38\,\text{kg}} = 6.35\,\text{m/s}$$

and the $y$ component is

$$v_{cy}(t_1) = \frac{(m_m + m_c)v_y(t_2) - m_m v_{my}(t_1)}{m_c} = \frac{(60\,\text{kg} + 38\,\text{kg})(3.0\,\text{m/s})\sin 35° - (60\,\text{kg})(6.0\,\text{m/s})}{38\,\text{kg}}$$
$$= -5.04\,\text{m/s}.$$

The magnitude is

$$v_c(t_1) = \sqrt{v_{cx}^2(t_1) + v_{cy}^2(t_1)} = \sqrt{(6.35\,\text{m/s})^2 + (-5.04\,\text{m/s})^2} = 8.1\,\text{m/s}.$$

If $\theta$ is the angle between the child's initial velocity and the positive $x$ direction then

$$\tan\theta = \frac{v_{cy}(t_1)}{v_{cx}(t_1)} = \frac{-5.04\,\text{m/s}}{6.35\,\text{m/s}} = -7.94$$

and the angle is $\theta = -38°$. That is, the child is initially going in the direction $38°$ south of east.

## 49

The total momentum of the package is conserved. Let $m$ be the mass of each piece, $\vec{v}(t)$ be the velocity of the package before the explosion, $\vec{v}_A(t_2)$ be the velocity of the first piece after the explosion, $\vec{v}_B(y_2)$ be the velocity of the second piece after the explosion, and $\vec{v}_C(t_2)$ be the velocity of the third piece after the explosion. Conservation of momentum yields

$$3m\vec{v}(t_1) = m\vec{v}_A(t_2) + m\vec{v}_B(t_2) + m\vec{v}_C(t_2).$$

Notice that $m$ cancels from this equation. Take the $x$ axis to run from east to west and the $y$ axis to run from south to north. Then

$$\vec{v}_A(t_2) = (7.0\,\text{m/s})\hat{j},$$

$$\vec{v}_B(t_2) = (4.0\,\text{m/s})(\cos 210°)\hat{i} + (4.0\,\text{m/s})(\sin 210°)\hat{i} = -(3.46\,\text{m/s})\hat{i} - (2.0\,\text{m/s})\hat{j},$$

and

$$\vec{v}_C(t_2) = (4.0\,\text{m/s})(\cos 30°)\hat{i} - (4.0\,\text{m/s})(\sin 30°)\hat{j} = (3.46\,\text{m/s})\hat{i} - (2.0\,\text{m/s})\hat{j}.$$

The initial velocity of the package is

$$\vec{v}(t_1) = \frac{(7.0\,\text{m/s})\hat{j} - (3.46\,\text{m/s})\hat{i} - (2.0\,\text{m/s})\hat{j} + (3.46\,\text{m/s})\hat{i} - (2.0\,\text{m/s})\hat{j}}{3} = (1.0\,\text{m/s})\hat{j}.$$

Its magnitude is 1.0 m/s and its direction is north.

## 53

Assume no external forces act, so the total momentum
of the three-piece system is conserved. Since the total
momentum before the break-up is zero, it is also zero
after the break-up. This means the velocity vectors of
the three pieces lie in the same plane, as shown in the
diagram to the right. Conservation of the $y$ component
of momentum leads to $mv \sin\theta_1 - mv \sin\theta_2 = 0$, where
$v$ is the speed of each of the smaller pieces. Thus
$\theta_1 = \theta_2$ and, since $\theta_1 + \theta_2 = 90°$, both $\theta_1$ and $\theta_2$ are 45°.
Conservation of the $x$ component of momentum leads to $3mV = 2mv \cos\theta_1$. Thus

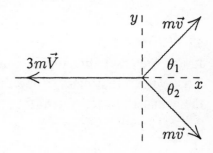

$$V = \tfrac{2}{3} v \cos\theta_1 = (2/3)(30\,\text{m/s}) \cos 45° = 14\,\text{m/s}.$$

The angle between the velocity vector of the large piece and either of the smaller pieces is
$180° - 45° = 135°$.

## 55

We need to consider only the horizontal components of the momenta of the package and sled.
Let $m_s$ be the mass of the sled and $v_{sx}$ be its initial velocity component. Let $m_p$ be the mass
of the package and let $v_x$ be the final velocity component of the sled and package together.
The horizontal component of the total momentum of the sled-package system is conserved, so
$m_s v_{sx} = (m_s + m_p) v_x$ and

$$v_x = \frac{v_{sx} m_s}{m_s + m_p} = \frac{(9.0\,\text{m/s})(6.0\,\text{kg})}{6.0\,\text{kg} + 12\,\text{kg}} = 3.0\,\text{m/s}.$$

This is also the final speed of the sled.

## 61

(a) Use Fig. 7–15 of the text. Take the cue ball to be body A and the other ball to be body B.
Conservation of the $x$ component of the total momentum of the two-ball system leads to

$$mv_A(t_1) = mv_A(t_2) \cos\theta_A + mv_B(t_2) \cos\theta_B$$

and conservation of the $y$ component leads to

$$0 = -mv_A(t_2) \sin\theta_A + mv_B(t_2) \sin\theta_B.$$

Notice that the masses are the same and cancel from the equations. Solve the second equation
for $\sin\theta_B$:

$$\sin\theta_B = \frac{v_A(t_2)}{v_B(t_2)} \sin\theta_A = \left(\frac{3.50\,\text{m/s}}{2.00\,\text{m/s}}\right) \sin 22.0° = 0.656.$$

The angle is $\theta_2 = 41.0°$.

(b) Solve the first momentum conservation equation for $v_A(t_1)$:

$$v_A(t_1) = v_A(t_2) \cos \theta_A + v_B(t_2) \cos \theta_B = (3.50 \, \text{m/s}) \cos 22.0° + (2.00 \, \text{m/s}) \cos 41.0°$$
$$= 4.75 \, \text{m/s}.$$

## 65

Ignore the gravitational pull of Jupiter and use Eq. 7–34 of the text. Let $\vec{v_1}$ be the initial velocity, $M_1$ be the initial mass, $\vec{v_2}$ be the final velocity, $M_2$ be the final mass, and $\vec{v}^{\text{rel}}$ be the speed of the exhaust gas relative to the rocket. Then

$$\vec{v_2} = \vec{v_1} + \vec{v}^{\text{rel}} \ln \frac{M_1}{M_2}.$$

For this problem $M_1 = 6090 \, \text{kg}$ and $M_2 = 6090 - 80.0 = 6010 \, \text{kg}$. Thus

$$v_2 = 105 \, \text{m/s} + (253 \, \text{m/s}) \ln \frac{6090 \, \text{kg}}{6010 \, \text{kg}} = 108 \, \text{m/s}.$$

The velocity is toward Jupiter.

# Chapter 8

## 3

(a) Put the origin at the center of Earth. Then the distance $R_{com}$ of the center of mass of the Earth-Moon system is given by

$$R_{com} = \frac{m_M r_M}{m_M + m_E},$$

where $m_M$ is the mass of the Moon, $m_E$ is the mass of Earth, and $r_M$ is their separation. These values are given in Appendix C. The numerical result is

$$R_{com} = \frac{(7.36 \times 10^{22}\,\text{kg})(3.82 \times 10^8\,\text{m})}{7.36 \times 10^{22}\,\text{kg} + 5.98 \times 10^{24}\,\text{kg}} = 4.64 \times 10^6\,\text{m}.$$

(b) The radius of Earth is $R_E = 6.37 \times 10^6$ m, so $R_{com}/R_E = 0.73$.

## 5

(a) Let $x_A$ (= 0) and $y_A$ (= 0) be the coordinates of one particle, $x_B$ (= 1.0 m) and $y_B$ (= 2.0 m) be the coordinates of the second particle, and $x_C$ (= 2.0 m) and $y_C$ (= 1.0 m) be the coordinates of the third. Designate the corresponding masses by $m_A$ (= 3.0 kg), $m_B$ (= 8.0 kg), and $m_C$ (= 4.0 kg). Then the $x$ coordinate of the center of mass is

$$X_{com} = \frac{m_A x_A + m_B x_B + m_C x_C}{m_A + m_B + m_B}$$
$$= \frac{0 + (8.0\,\text{kg})(1.0\,\text{m}) + (4.0\,\text{kg})(2.0\,\text{m})}{3.0\,\text{kg} + 8.0\,\text{kg} + 4.0\,\text{kg}} = 1.1\,\text{m}.$$

(b) The $y$ coordinate is

$$Y_{com} = \frac{m_A y_A + m_B y_B + m_C y_C}{m_A + m_B + m_C}$$
$$= \frac{0 + (8.0\,\text{kg})(2.0\,\text{m}) + (4.0\,\text{kg})(1.0\,\text{m})}{3.0\,\text{kg} + 8.0\,\text{kg} + 4.0\,\text{kg}} = 1.3\,\text{m}.$$

(c) As the mass of the topmost particle is increased the center of mass shifts toward that particle. In the limit as the topmost particle is much more massive than the others, the center of mass is nearly at the position of that particle.

## 11

(a) Since the can is uniform its center of mass is at its geometrical center, a distance $H/2$ above its base. The center of mass of the soda alone is at its geometrical center, a distance $x/2$ above the base of the can. When the can is full this is $H/2$. Thus the center of mass of the can and the soda it contains is a distance

$$h = \frac{M(H/2) + m(H/2)}{M + m} = \frac{H}{2}$$

above the base, on the cylinder axis.

(b) We now consider the can alone. The center of mass is $H/2$ above the base, on the cylinder axis.

(c) As $x$ decreases the center of mass of the can and the soda in it at first drops, then rises to $H/2$ again.

(d) When the top surface of the soda is a distance $x$ above the base of the can the mass of the soda in the can is $m_p = m(x/H)$, where $m$ is the mass when the can is full $(x = H)$. The center of mass of the soda alone is a distance $x/2$ above the base of the can. Hence

$$h = \frac{M(H/2) + m_p(x/2)}{M + m_p} = \frac{M(H/2) + m(x/H)(x/2)}{M + (mx/H)} = \frac{MH^2 + mx^2}{2(MH + mx)}.$$

Find the lowest position of the center of mass of the can and soda by setting the derivative of $h$ with respect to $x$ equal to 0 and solving for $x$. The derivative is

$$\frac{dh}{dx} = \frac{2mx}{2(MH + mx)} - \frac{(MH^2 + mx^2)m}{2(MH + mx)^2} = \frac{m^2x^2 + 2MmHx - MmH^2}{2(MH + mx)^2}.$$

The solution to $m^2x^2 + 2MmHx - MmH^2 = 0$ is

$$x = \frac{MH}{m}\left[-1 + \sqrt{1 + \frac{m}{M}}\right].$$

The positive root is used since $x$ must be positive.

Now substitute the expression you found for $x$ into $h = (MH^2 + mx^2)/2(MH + mx)$. After some algebraic manipulation you should obtain

$$h = \frac{HM}{m}\left(\sqrt{1 + \frac{m}{M}} - 1\right).$$

## 19

The $x$ coordinate of the center of mass of the three-particle system is given by

$$X_{\text{com}} = \frac{m_A x_A + m_B x_B + m_C x_C}{m_A + m_B + m_C}.$$

Take particle C to be the particle whose position is unknown. Then

$$
\begin{aligned}
x_C &= \frac{(m_A + m_B + m_C)X_{\text{com}} - m_A x_A - m_B x_B}{m_C} \\
&= \frac{(2.00\,\text{kg} + 4.00\,\text{kg} + 3.00\,\text{kg})(-0.500\,\text{m}) - (2.00\,\text{kg})(-1.20\,\text{m}) - (4.00\,\text{kg})(0.600\,\text{m})}{3.00\,\text{kg}} \\
&= -1.50\,\text{m}.
\end{aligned}
$$

The $y$ coordinate of the center of mass is given by

$$Y_{\text{com}} = \frac{m_A y_A + m_B y_B + m_C y_C}{m_A + m_B + m_C},$$

so

$$y_C = \frac{(m_A + m_B + m_C)Y_{com} - m_A y_A - m_B y_B}{m_C}$$

$$= \frac{(2.00\,kg + 4.00\,kg + 3.00\,kg)(-0.700\,m) - (2.00\,kg)(0.500\,m) - (4.00\,kg)(-0.750\,m)}{3.00\,kg}$$

$$= -1.43\,m\,.$$

## 23

Let $m_c$ be the mass of the Chrysler and $\vec{v}_c$ be its velocity. Let $m_f$ be the mass of the Ford and $\vec{v}_f$ be its velocity. Then the velocity of the center of mass is

$$\vec{v}_{com} = \frac{m_c \vec{v}_c + m_f \vec{v}_f}{m_c + m_f}\,.$$

Place the $x$ axis along the path of travel of the cars. Then the $x$ component of the velocity of the center of mass is

$$v_{com\,x} = \frac{m_c v_{cx} + m_f v_{fx}}{m_c + m_f} = \frac{(2400\,kg)(80\,km/h) + (1600\,kg)(60\,km/h)}{2400\,kg + 1600\,kg} = 72\,km/h\,.$$

Notice that the velocities of the two cars are in the same direction, so the two terms in the numerator have the same sign.

## 27

You need to find the coordinates of the point where the shell explodes and the velocity of the fragment that does not fall straight down. These become the initial conditions for a projectile motion problem to determine where it lands.

Consider first the motion of the shell from firing to the time of the explosion. Place the origin at the firing point, take the $x$ axis to be horizontal, and take the $y$ axis to be vertically upward. The $y$ component of the velocity just before the explosion is given by $v_y = v_{1y} - gt$ and this is zero at time $t = v_{1y}/g = (v_1/g)\sin\theta_1$, where $v_1$ is the initial speed and $\theta_1$ is the firing angle. The coordinates of the highest point on the trajectory are

$$x_2 = v_{1x}t = v_1 t \cos\theta_1 = \frac{v_1^2}{g}\sin\theta_1\cos\theta_1 = \frac{(20\,m/s)^2}{9.8\,m/s^2}\sin 60°\cos 60° = 17.7\,m$$

and

$$y = v_{01}t - \frac{1}{2}gt^2 = \frac{1}{2}\frac{v_1^2}{g}\sin^2\theta_1 = \frac{1}{2}\frac{(20\,m/s)^2}{9.8\,m/s^2}\sin^2 60° = 15.3\,m\,.$$

Consider the system consisting of the shell before the explosion and the two pieces after the explosion. It follows the trajectory of a projectile and since the trajectory is symmetric about the highest point it lands a distance $2x_2 = 2(17.2\,m$ from the firing point. Both pieces land at the same time and they land with their center of mass this distance from the firing point. Since their masses are equal the center of mass is halfway between them. Since the piece that falls straight down is 17.1 m from the center of mass the other piece is also 17.7 m from the center of mass, but on the far side. Thus it is $3x_2 = 3(17.7\,\text{m} = 53.1\,\text{m}$ from the firing point.

## 29

(a) Place the origin of a coordinate system at the center of the pulley, with the $x$ axis horizontal and to the right and with the $y$ axis downward. The center of mass is halfway between the containers, at $x = 0$ and $y = \ell$, where $\ell$ is the vertical distance from the pulley center to either of the containers. Since the diameter of the pulley is 50 mm, the center of mass is 25 mm from each container.

(b) Suppose 20 g is transferred from the container on the left to the container on the right. The container on the left then has mass $m_A = 480\,\text{g}$ and is at $x_A = -25\,\text{mm}$. The container on the right has mass $m_B = 520\,\text{g}$ and is at $x_B = +25\,\text{mm}$. The $x$ coordinate of the center of mass is

$$X_{\text{com}} = \frac{m_A x_A + m_B x_B}{m_A + m_B} = \frac{(480\,\text{g})(-25\,\text{mm}) + (520\,\text{g})(25\,\text{mm})}{480\,\text{g} + 520\,\text{g}} = 1.0\,\text{mm}.$$

The $y$ coordinate is still $\ell$. The center of mass is 26 mm from the lighter container, along the line that joins the center of the bodies.

(c) When they are released, the heavier container moves downward and the lighter container moves upward, so the center of mass, which must remain closer to the heavier container, moves downward.

(d) Because the containers are connected by the string that runs over the pulley, their accelerations have the same magnitude but are in opposite directions. If $a_y$ is the vertical component of the acceleration of container B, then $-a_y$ is the vertical component of the acceleration of container A. The vertical component of the acceleration of the center of mass is

$$a_{\text{com } y} = \frac{m_A(-a_y) + m_B a_y}{m_A + m_B} = a_y \frac{m_B - m_A}{m_A + m_B}.$$

We must resort to Newton's second law to find the acceleration of each container. The force of gravity $m_A g$, down, and the tension force of the string $T$, up, act on the lighter container. The second law for it is $m_A g - T = -m_A a_y$. The same forces act on the heavier container and for it the second law is $m_B g - T = m_B a_y$. The first equation gives $T = m_A g + m_A a$. This is substituted into the second equation to obtain $m_B g - m_A g - m_A a_y = m_B a_y$, so $a_y = (m_B - m_A)g/(m_A + m_B)$. Thus

$$a_{\text{com } y} = \frac{g(m_B - m_A)^2}{(m_A + m_B)^2} = \frac{(9.8\,\text{m/s}^2)(520\,\text{g} - 480\,\text{g})^2}{(480\,\text{g} + 520\,\text{g})^2} = 1.6 \times 10^{-2}\,\text{m/s}^2.$$

The acceleration is downward.

## 31

Take the $x$ axis to be horizontal with the positive direction to the right in the figure. Let $m_b$ be the mass of the boat and $x_{b1}$ be its initial coordinate. Let $m_d$ be the mass of the dog and $x_{d1}$ be its initial coordinate. The coordinate of the center of mass is

$$X_{\text{com}} = \frac{m_b x_{b1} + m_d x_{d1}}{m_b + m_d}.$$

Now the dog walks a distance $d$ to the left on the boat. The new coordinates $x_{b2}$ and $x_{d2}$ are related by $x_{b2} = x_{d2} + d$, so the coordinate of the center of mass can be written

$$X_{\text{com}} = \frac{m_b x_{b2} + m_d x_{d2}}{m_b + m_d} = \frac{m_b x_{d2} + m_b d + m_d x_{d2}}{m_b + m_d}.$$

Since the net external force on the boat-dog system is zero the velocity of the center of mass does not change. Since the boat and dog were initially at rest the velocity of the center of mass is zero. The center of mass remains at the same place and the two expressions we have written for $x_{\text{com}}$ must equal each other. This means $m_b x_{b1} + m_d x_{d1} = m_b x_{d2} + m_b d + m_d x_{d2}$. Solve for $x_{d2}$:

$$x_{d2} = \frac{m_b x_{b1} + m_d x_{d1} - m_b d}{m_b + m_d}$$

$$= \frac{(18\,\text{kg})(6.1\,\text{m}) + (4.5\,\text{kg})(6.1\,\text{m}) - (18\,\text{kg})(2.4\,\text{m})}{18\,\text{kg} + 4.5\,\text{kg}} = 4.2\,\text{m}.$$

## 39

(a) No external forces act horizontally on the cannon-shell system, so the horizontal component of the velocity of its center of mass is constant. The cannon and shell are initially at rest so the horizontal component of the center of mass remains at rest. Let $m_c$ be the mass of the cannon and $\vec{v}_c$ be its velocity after firing. Let $m_s$ be the mass of the shell and $\vec{v}_s$ be its velocity after firing. Take the $x$ axis to be horizontal. Then $m_c v_{cx} + m_s v_{sx} = 0$. The velocities here are relative to the ground and are related by $v_{cx} = v_{sx} - v_x^{\text{rel}}$, where $v_x^{\text{rel}}$ is the horizontal component of the shell velocity relative to the cannon. Use this expression to substitute for $v_{cx}$ in the previous equation and obtain $m_c v_{sx} - m_c v_x^{\text{rel}} + m_s v_{sx} = 0$. The solution for the horizontal component of the shell velocity is

$$v_{sx} = \frac{m_c v_x^{\text{rel}}}{m_c + m_s} = \frac{m_c v^{\text{rel}} \cos\theta}{m_c + m_s} = \frac{(1400\,\text{kg})(556\,\text{m/s}) \cos 39°}{1400\,\text{kg} + 70\,\text{kg}} = 412\,\text{m/s}.$$

Here $\theta$ is the elevation angle of the canon.

The vertical component of the shell velocity is $v_{sy} = v^{\text{rel}} \sin\theta = (556\,\text{m/s}) \sin 39° = 350\,\text{m/s}$ and the speed of the shell relative to the ground is

$$v_s = \sqrt{v_{sx}^2 + v_{sy}^2} = \sqrt{(412\,\text{m/s})^2 + (350\,\text{m/s})^2} = 541\,\text{m/s}.$$

(b) The firing angle $\phi$ relative to the ground is given by

$$\tan\phi = \frac{v_{sy}}{v_{sx}} = \frac{350\,\text{m/s}}{412\,\text{m/s}} = 0.849$$

and the angle is $\phi = 40.3°$.

## 43

(a) The velocity of the package before it explodes is the same as the velocity of the center of mass of the system consisting of the package before the explosion and the three pieces after the explosion. The explosion does not change the velocity of the center of mass so we calculate its value after the explosion. Take the $x$ axis to run from west to east and the $y$ axis to run from south to north. Let $m$ be the mass of each piece and let $v_A$, $v_B$, and $v_C$ be the speeds of the pieces. Then the velocity of the center of mass of the three-piece system is

$$\vec{v}_{\text{com}} = \frac{m\left[(7.0\,\text{m/s})\hat{\jmath} + (4.0\,\text{m/s})(-\cos 30°\,\hat{\imath} - \sin 30°\,\hat{\jmath}) + (4.0\,\text{m/s})(\cos 30°\,\hat{\imath} - \sin 30°\,\hat{\jmath})\right]}{3m}$$

$$= (1.0\,\text{m/s})\hat{\jmath}.$$

The velocity of the package before the explosion is $1.0\,\text{m/s}$, north.

## 47

Place the origin at the heavy cart and take the positive $x$ direction to be to the right. Let $x_A$ be the mass of the heavy cart, $x_B$ be the mass of the light cart, and $x_B$ be the position of the light cart. Set $m_A$ equal to $2m$ and $x_B$ equal to $m$. Then the $x$ coordinate of the center of mass is

$$X_{\text{com}} = \frac{m_B x_B}{m_A + m_B} = \frac{m x_B}{2m + m} = \frac{x_B}{s}.$$

The distance from the heavy cart to the center of mass is exactly one-third the separation of the carts.

# Chapter 9

## 1

The kinetic energy is given by $K = \frac{1}{2}mv^2$, where $m$ is the mass and $v$ is the speed of the electron. The speed is therefore

$$v = \sqrt{\frac{2K}{m}} = \sqrt{\frac{2(6.7 \times 10^{-19}\,\text{J})}{9.11 \times 10^{-31}\,\text{kg}}} = 1.2 \times 10^6\,\text{m/s}.$$

## 5

(a) Suppose the proton travels along the $x$ axis and use the kinematic equation $v_{2x}^2 = v_{1x}^2 + 2a_x\Delta x$, where $\vec{v}_1$ is the initial velocity, $\vec{v}_2$ is the final velocity, $\Delta x$ is the $x$ component of the displacement, and $\vec{a}$ is the acceleration. This equation yields

$$v_{2x} = \sqrt{v_{1x}^2 + 2a_x\Delta x} = \sqrt{(2.4 \times 10^7\,\text{m/s})^2 + 2(3.6 \times 10^{15}\,\text{m/s}^2)(0.035\,\text{m})} = 2.9 \times 10^7\,\text{m/s}.$$

(b) The initial kinetic energy is

$$K_1 = \frac{1}{2}mv_1^2 = \frac{1}{2}(1.67 \times 10^{-27}\,\text{kg})(2.4 \times 10^7\,\text{m/s})^2 = 4.8 \times 10^{-13}\,\text{J}.$$

The final kinetic energy is

$$K_2 = \frac{1}{2}mv_2^2 = \frac{1}{2}(1.67 \times 10^{-27}\,\text{kg})(2.9 \times 10^7\,\text{m/s})^2 = 6.9 \times 10^{-13}\,\text{J}.$$

The change in kinetic energy is $\Delta K = 6.9 \times 10^{-13}\,\text{J} - 4.8 \times 10^{-13}\,\text{J} = 2.1 \times 10^{-13}\,\text{J}.$

## 11

(a) The initial kinetic energy is $K_1 = \frac{1}{2}mv_1^2$, where $m$ is the mass of the ball and $v_1$ is its speed before it hits the wall. The final kinetic energy is $K_2 = \frac{1}{2}mv_2^2$, where $v_2$ is the speed of the ball after it hits the wall. Since $K_2 - K_1/2$, $v_2 = v_1/\sqrt{2} = (5.2\,\text{m/s})/\sqrt{2} = 3.7\,\text{m/s}.$

(b) Assume the direction of travel reverses, so the ball rebounds along the line of incidence. Then the magnitude of the change in the momentum of the ball is $|\Delta\vec{p}| = m[(v_1 - (-v_2)] = m(v_1 + v_2) = (0.150\,\text{kg})(5.2\,\text{m/s} + 3.7\,\text{m/s}) = 1.3\,\text{kg}\cdot\text{m/s}.$ According to the impulse-momentum theorem this is the magnitude of the impulse on the ball.

(c) If $\Delta t$ is the time of contact between the ball and the wall, the magnitude of the average force of the wall on the ball is $J/\Delta t = (1.3\,\text{kg}\cdot\text{m/s})/(7.6 \times 10^{-3}\,\text{s}) = 1.8 \times 10^2\,\text{N}.$

## 15

Symmetry tells use that the center of mass must lie on the $y$ axis, so its $x$ coordinate is $X_{\text{com}} = R/2$. Use $Y_{\text{com}} = (1/M) \int y\,dm$ to find the $y$ coordinate. Divide the semicircle into infinitesimal strips of with $dy$ parallel to the $x$ axis. The length of the strip with coordinate $y$ is $2\sqrt{R^2 - y^2}$ and if $rho$ is the density of the material and $T$ is the thickness of the plate, the mass in the strip is $dm = 2\rho T\sqrt{R^2 - y^2}\, dy$. The coordinate is

$$Y_{\text{com}} = \frac{1}{M}\int_0^R 2\rho T\sqrt{R^2 - y^2}\, y\, dy = \frac{2\rho T}{3M}R^3\,.$$

Now the total area of the semicircle is $\pi R^2/2$ and the total mass is $M = \pi R^2\rho T/2$, so

$$Y_{\text{com}} = \frac{2\rho T}{3\pi R^2\rho T/2}R^3 = \frac{4R}{3\pi}\,.$$

## 19

Take $a$ on the graph to be the $x$ component of the acceleration. According to the graph $a$ varies linearly with the coordinate $x$ and we may write $a = \alpha x$, where $\alpha$ is the slope of the graph. Numerically, $\alpha = (20\,\text{m/s}^2)/(8.0\,\text{m}) = 2.5\,\text{s}^{-2}$. The force on the brick is in the positive $x$ direction and, according to Newton's second law, its $x$ component is given by $F_x = a/m = (\alpha/m)x$. If $x_2$ is the final coordinate, the work done by the force is

$$W = \int_0^{x_2} F_x\, dx = \frac{\alpha}{m}\int_0^{x_2} x\, dx = \frac{\alpha}{2m}x_2^2 = \frac{2.5\,\text{s}^{-2}}{2(10\,\text{kg})}(8.0\,\text{m})^2 = 800\,\text{J}\,.$$

## 23

(a) Let $F$ be the $x$ component of the force. The graph shows $F$ as a function of $x$ if $x_0$ and $F_0$ are positive. The work is negative as the object moves from $x = 0$ to $x = x_0$ and positive as it moves from $x = x_0$ to $x = 2x_0$. Since the area of a triangle is $\frac{1}{2}(\text{base})(\text{altitude})$, the work done from $x = 0$ to $x = x_0$ is $-\frac{1}{2}(x_0)(F_0)$ and the work done from $x = x_0$ to $x = 2x_0$ is $\frac{1}{2}(2x_0 - x_0)(F_0) = \frac{1}{2}(x_0)(F_0)$. The net work is the sum, which is zero.

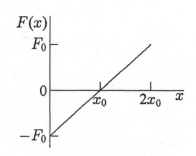

(b) The integral for the work is

$$W = \int_0^{2x_0} F_0\left(\frac{x}{x_0} - 1\right) dx = F_0\left(\frac{x^2}{2x_0} - x\right)\Big|_0^{2x_0} = 0\,.$$

## 25

(a) As the cage moves from $x = x_1$ to $x = x_2$ the work done by the spring is given by

$$W^{\text{spring}} = \int_{x_1}^{x_2}(-kx)\, dx = -\frac{1}{2}kx^2\Big|_{x_1}^{x_2} = -\frac{1}{2}k(x_2^2 - x_1^2)\,,$$

where $k$ is the spring constant for the spring. Substitute $x_1 = 0$ and $x_2 = 7.6 \times 10^{-3}\,\text{m}$. The result is

$$W^{\text{spring}} = -\frac{1}{2}(1500\,\text{N/m})(7.6 \times 10^{-3}\,\text{m})^2 = -0.043\,\text{J}\,.$$

To use SI units consistently throughout, $15\,\text{N/cm}$ is converted to $1500\,\text{N/m}$.

(b) Now substitute $x_1 = 7.6 \times 10^{-3}$ m and $x_2 = 15.2 \times 10^{-3}$ m. The result is now

$$W^{\text{spring}} = -\frac{1}{2}(1500\,\text{N/m})\left[(15.2 \times 10^{-3}\,\text{m})^2 - (7.6 \times 10^{-3}\,\text{m})^2\right] = -0.13\,\text{J}.$$

Notice this is more than twice the work done in the first interval. Although the displacements have the same magnitude the force is larger throughout the second interval.

## 29

The forces are all constant and since the canister started from rest its displacement is in the same direction as the net force. This means net work done by the forces is given by $W^{\text{net}} = F^{\text{net}}d$, where $F^{\text{net}}$ is the magnitude of the net force and $d$ is magnitude of the displacement. To find the net force, we vectorially add the three vectors. The $x$ component is

$$F_x^{\text{net}} = -F_A - F_B \sin 50° + F_C \cos 35° = -3.00\,\text{N} - (4.00\,\text{N})\sin 35° + (10.0\,\text{N})\cos 35°$$
$$= 2.127\,\text{N}$$

and the $y$ component is

$$F_y^{\text{net}} = -F_B \cos 50° + F_C \sin 35° = -(4.00\,\text{N})\cos 50° + (10.0\,\text{N})\sin 35° = 3.165\,\text{N}.$$

The magnitude of the net force is

$$F^{\text{net}} = \sqrt{(F_x^{\text{net}})^2 + (F_y^{\text{net}})^2} = \sqrt{(2.127\,\text{N})^2 + (3.165\,\text{N})^2} = 3.813\,\text{N}.$$

The work done by the net force is

$$W = F_{\text{net}}d = (3.813\,\text{N})(4.00\,\text{m}) = 15.3\,\text{J}.$$

## 31

(a) The force of the worker on the crate is constant, so the work it does is given by $W^{\text{worker}} = \vec{F}^{\text{worker}} \cdot \vec{d} = Fd \cos\phi$, where $\vec{F}^{\text{worker}}$ is the force, $\vec{d}$ is the displacement of the crate, and $\phi$ is the angle between the force and the displacement. Here $F^{\text{worker}} = 210\,\text{N}$, $d = 3.0\,\text{m}$, and $\phi = 20°$. Thus $W^{\text{worker}} = (210\,\text{N})(3.0\,\text{m})\cos 20° = 590\,\text{J}$.

(b) The force of gravity is downward, perpendicular to the displacement of the crate. The angle between this force and the displacement is 90° and $\cos 90° = 0$, so the work done by the force of gravity is zero.

(c) The normal force of the floor on the crate is also perpendicular to the displacement, so the work done by this force is also zero.

(d) These are the only forces acting on the crate, so the net work done on it is 590 J.

## 39

(a) Let $F^{\text{cable}}$ be the magnitude of the force of the cable on the astronaut. This force is upward and the force of gravity, with magnitude $mg$, is downward. Furthermore, the acceleration of the astronaut is $g/10$, upward. According to Newton's second law, $F^{\text{cable}} - mg = mg/10$, so $F^{\text{cable}} = 11mg/10$. Since the force $\vec{F}^{\text{cable}}$ and the displacement $\vec{d}$ are in the same direction the work done by $\vec{F}^{\text{cable}}$ is

$$W^{\text{cable}} = F^{\text{cable}}d = \frac{11mgd}{10} = \frac{11(72\,\text{kg})(9.8\,\text{m/s}^2)(15\,\text{m})}{10} = 1.16 \times 10^4\,\text{J}.$$

(b) The force of gravity has magnitude $mg$ and is opposite in direction to the displacement. Since $\cos 180° = -1$, it does work

$$W_g = -mgd = -(72\,\text{kg})(9.8\,\text{m/s}^2)(15\,\text{m}) = -1.1 \times 10^4\,\text{J}.$$

(c) The net work done is $W^{\text{net}} = 1.16 \times 10^4\,\text{J} - 1.06 \times 10^4\,\text{J} = 1.1 \times 10^3\,\text{J}$. Since the astronaut started from rest the work-kinetic energy theorem tells us that this must be her final kinetic energy.

(d) Since $K = \frac{1}{2}mv^2$ her final speed is

$$v = \sqrt{\frac{2K}{m}} = \sqrt{\frac{2(1.0 \times 10^3\,\text{J})}{72\,\text{kg}}} = 5.3\,\text{m/s}.$$

## 41

(a) Let $F^{\text{cord}}$ be the magnitude of the force of the cord on the block. This force is upward, while the force of gravity, with magnitude $Mg$, is downward. The acceleration is $g/4$, down. Take the downward direction to be positive. Then Newton's second law is $Mg - F^{\text{cord}} = Mg/4$, so $F^{\text{cord}} = 3Mg/4$. The force is directed opposite to the displacement, so the work it does is $W^{\text{cord}} = -F^{\text{cord}}d = -3Mgd/4$.

(b) The force of gravity is in the same direction as the displacement, so it does work $W^{\text{grav}} = Mgd$.

(c) The net work done on the block is $W^{\text{net}} = -3Mgd/4 + Mgd = Mgd/4$. Since the block starts from rest this is its kinetic energy $K$ after it is lowered a distance $d$.

(d) Since $K = \frac{1}{2}Mv^2$, where $v$ is the speed,

$$v = \sqrt{\frac{2K}{M}} = \sqrt{\frac{gd}{2}}$$

after the block is lowered a distance $d$. The result found in (c) was used.

## 45

(a) The net work done by the lifting and gravitational forces together is equal to the change in the kinetic energy of the rescuee and is therefore $W^{\text{net}} = \frac{1}{2}mv_f^2$, where $m$ is the mass of the rescuee and $v_f$ is his final speed. Numerically, $W^{\text{net}} = \frac{1}{2}(80.0\,\text{kg})(5.00\,\text{m/s})^2 = 1.00 \times 10^3\,\text{J}$. If $h$ is the distance the rescuee is raised during this phase, the work done by the gravitational force is $W^{\text{grav}} = -mgd = -(80.0\,\text{kg})(9.8\,\text{N/kg})(10.0\,\text{m}) = -7.84 \times 10^3\,\text{J}$. The work done by the lifting force is $W^{\text{lift}} = W^{\text{net}} - W^{\text{grav}} = 1.00 \times 10^3\,\text{J} - (-7.84 \times 10^3\,\text{J}) = 8.84 \times 10^3\,\text{J}$.

(b) For this phase the net work is zero and the work done by the lifting force is the negative of the work done by the gravitational force. Thus $W^{\text{lift}} = +mgh = (80.0\,\text{kg})(9.8\,\text{N/kg})(10.0\,\text{m}) = +7.84 \times 10^3\,\text{J}$.

(c) The net work is the change in the kinetic energy and is now negative since the speed of rescuee is reduced to zero. It is $W^{\text{net}} = -1.00 \times 10^3\,\text{J}$. The work done by the gravitational force is again $W^{\text{grav}} = -7.84 \times 10^3$ [J, so the work done by the lifting force is $W^{\text{lift}} = W^{\text{net}} - W^{\text{grav}} = -1.00 \times 10^3\,\text{J} - (-7.84 \times 10^3\text{J}) = 6.84 \times 10^3\,\text{J}$.

## 49

The power associated with the force $\vec{F}$ is given by $P = \vec{F} \cdot \vec{v}$, where $\vec{v}$ is the velocity of the object on which the force acts. Let $\phi\ (= 37°)$ be the angle between the force and the horizontal. Then

$$P = \vec{F} \cdot \vec{v} = Fv \cos\phi = (122\,\text{N})(5.0\,\text{m/s}) \cos 37° = 490\,\text{W}\,.$$

## 51

(a) Since the net force and the velocity are in the same direction, the power delivered by the net force is given by $P = F^{\text{net}}v$ and the work it does from time $t_1$ to time $t_2$ is given by

$$W^{\text{net}} = \int_{t_1}^{t_2} P\,dt = \int_{t_1}^{t_2} F^{\text{net}}v\,dt\,.$$

Here $v$ is the speed of the body. Since $\vec{F}^{\text{net}}$ is the net force the magnitude of the acceleration is $a = F^{\text{net}}/m$ and, since the initial velocity is $v_1 = 0$, the speed as a function of time is given by $v = v_1 + at = (F^{\text{net}}/m)t$. Thus

$$W^{\text{net}} = \int_{t_1}^{t_2} \frac{(F^{\text{net}})^2}{m}t\,dt = \frac{1}{2}\frac{(F^{\text{net}})^2}{m}(t_2^2 - t_1^2)\,.$$

For $t_1 = 0$ and $t_2 = 1.0\,\text{s}$,

$$W^{\text{net}} = \frac{1}{2}\left[\frac{(5.0\,\text{N})^2}{15\,\text{kg}}\right](1.0\,\text{s})^2 = 0.83\,\text{J}\,.$$

(b) For $t_1 = 1.0\,\text{s}$ and $t_2 = 2.0\,\text{s}$,

$$W^{\text{net}} = \frac{1}{2}\left[\frac{(5.0\,\text{N})^2}{15\,\text{kg}}\right]\left[(2.0\,\text{s})^2 - (1.0\,\text{s})^2\right] = 2.5\,\text{J}\,.$$

(c) For $t_1 = 2.0\,\text{s}$ and $t_2 = 3.0\,\text{s}$,

$$W^{\text{net}} = \frac{1}{2}\left[\frac{(5.0\,\text{N})^2}{15\,\text{kg}}\right]\left[(3.0\,\text{s})^2 - (2.0\,\text{s})^2\right] = 4.2\,\text{J}\,.$$

(d) Substitute $v = (F^{\text{net}}/m)t$ into $P = F^{\text{net}}v$ to obtain $P = (F^{\text{net}})^2 t/m$ for the power at any time $t$. At the end of the third second

$$P = \frac{(5.0\,\text{N})^2(3.0\,\text{s})}{15\,\text{kg}} = 5.0\,\text{W}\,.$$

## 55

Let the magnitude of the force be $F = \alpha v$, where $v$ is the speed and $\alpha$ is a constant of proportionality. The power required is $P = Fv = \alpha v^2$. Let $P_1$ be the power required for speed $v_1$ and $P_2$ be the power required for speed $v_2$. Divide $P_2 = \alpha v_2^2$ by $P_1 = \alpha v_1^2$, then solve for $P_2$. You should obtain $P_2 = (v_2/v_1)^2 P_1$. Since $P_1 = 7.5\,\mathrm{kW}$ and $v_2 = 3v_1$, $P_2 = (3)^2(7.5\,\mathrm{kW}) = 68\,\mathrm{kW}$.

## 70

(a) Cart A has momentum in the positive $x$ direction before the collision. During the collision cart B exerts a force on A in the negative $x$ direction and thereby slows it. Cart A continues in the positive $x$ direction but with reduced momentum. Graph B might depict this.

(b) The total momentum of the two-cart system is conserved. It is in the positive $x$ direction, the direction of motion of cart A before the collision. Graph C might depict this.

(c) The kinetic energy of cart A is reduced by the collision since the speed of that cart is reduced. Graph B might depict this.

(d) The force of cart B on cart A is zero before and after the collision and is in the negative $x$ direction during the collision. Graph G might depict this.

(e) The force of cart A on cart B is zero before and after the collision and is in the positive $x$ direction during the collision. None of the graphs depict this.

# Chapter 10

## 3

(a) The force of gravity is constant, so the work it does is given by $W^{\text{grav}} = \vec{F}^{\text{grav}} \cdot \vec{d}$, where $\vec{F}^{\text{grav}}$ is the force and $\vec{d}$ is the displacement. The force is vertically downward and has magnitude $mg$, where $m$ is the mass of the flake. The displacement is also downward and it has magnitude $r$, the radius of the bowl. Thus $W^{\text{grav}} = mgr = (2.00 \times 10^{-3}\,\text{kg})(9.8\,\text{m/s}^2)(22.0 \times 10^{-2}\,\text{m}) = 4.31 \times 10^{-3}\,\text{J}$.

(b) The force of gravity is conservative, so the change in gravitational potential energy of the flake-Earth system is the negative of the work done: $\Delta U^{\text{grav}} = -W^{\text{grav}} = -4.31 \times 10^{-3}\,\text{J}$.

(c) The potential energy when the flake is at the top is greater than when it is at the bottom by $|\Delta U^{\text{grav}}|$. If $U^{\text{grav}} = 0$ at the bottom, then $U^{\text{grav}} = +4.31 \times 10^{-3}\,\text{J}$ at the top.

(d) If $U^{\text{grav}} = 0$ at the top, then $U^{\text{grav}} = -4.31 \times 10^{-3}\,\text{J}$ at the bottom.

(e) All the answers are proportional to the mass of the flake. If the mass is doubled, all answers are doubled.

## 7

(a) The force of gravity is constant, so the work it does is given by $W^{\text{grav}} = \vec{F}^{\text{grav}} \cdot \vec{d}$, where $\vec{F}^{\text{grav}}$ is the force and $\vec{d}$ is the displacement. The force is vertically downward and has magnitude $mg$, where $m$ is the mass of the snowball. If $h$ is the height through which the snowball drops $W^{\text{grav}} = mgh = (1.50\,\text{kg})(9.8\,\text{m/s}^2)(12.5\,\text{m}) = 184\,\text{J}$.

(b) The force of gravity is conservative, so the change in the potential energy of the snowball-Earth system is the negative of the work it does: $\Delta U^{\text{grav}} = -W^{\text{grav}} = -184\,\text{J}$.

(c) The potential energy when it reaches the ground is less than the potential energy when it is fired by $|\Delta U^{\text{grav}}|$, so $U^{\text{grav}} = -184\,\text{J}$ when the snowball hits the ground.

## 13

(a) The force of the rod on the ball is perpendicular to the path, so that force does no work on the ball. The force of gravity is conservative, so the mechanical energy of the ball-Earth system is conserved: $K_1 + U_1^{\text{grav}} = K_2 + U_2^{\text{grav}}$, where $U_1^{\text{grav}}$ is the initial potential energy of the system, $K_1$ is the initial kinetic energy, $U_2^{\text{grav}}$ is the potential energy when the ball is at the top of its path, and $K_2$ is its kinetic energy when it is there. Take the potential energy to be zero when the ball is at its initial position. At the top of its swing it is a vertical distance $L$ above this position and the potential energy is $mgL$. The final kinetic energy is zero since the ball comes to rest at the top of its path. Write $\frac{1}{2}mv_1^2$ for the initial kinetic energy. Here $m$ is the mass of the ball and $v_1$ is its initial speed. Thus $\frac{1}{2}mv_1^2 = mgL$ and $v_1 = \sqrt{2gL}$.

(b) Again $K_1 + U_1^{grav} = K_2 + U_2^{grav}$, where the final position of the ball is now at the bottom of its swing. The ball is then a distance $L$ below its initial position and the potential energy is $-mgL$. Set $K_1 = mgL$, $U_1 = 0$, $K_2 = \frac{1}{2}mv_2^2$, and $U_2^{grav} = -mgL$. The energy conservation equation becomes $mgL = \frac{1}{2}mv_2^2 - mgL$, so $v_2 = \sqrt{4gL}$.

(c) Again $K_1 + U_1^{grav} = K_2 + U_2^{grav}$. The final position of the ball is at the same height as the initial position, so $U_2^{grav} = U_1^{grav}$. This means $K_2 = K_1 = 2gL$. Thus $v_2 = \sqrt{2gL}$.

(d) None of the answers depend on the mass of the ball. They remain the same if the mass is doubled.

## 17

(a) Let $K_1$ be the kinetic energy of the snowball just after it is fired, $K_2$ be its kinetic energy just before it hits the ground, $U_1^{grav}$ be the potential energy of the snowball-Earth system at firing, and $U_2^{grav}$ be the potential energy of the system when the snowball hits the ground. The only force acting is the force of gravity and it is conservative, so $K_1 + U_1^{grav} = K_2 + U_2^{grav}$. Substitute $K_1 = \frac{1}{2}mv_1^2$ and $K_2 = \frac{1}{2}mv_2^2$, where $v_1$ is the firing speed and $v_2$ is the landing speed. Take the potential energy to be zero at firing. Then, according to the result of Problem 7, $U_2 = -184\,\text{J}$. Solve for $v_2$:

$$v_2 = \sqrt{v_1^2 - \frac{2U_2^{grav}}{m}} = \sqrt{(14.0\,\text{m/s})^2 - \frac{2(-184\,\text{J})}{1.50\,\text{kg}}} = 21.0\,\text{m/s}.$$

(b) The landing speed depends only the firing speed and the distance the snowball falls, not on the firing angle, so the speed is again $21.0\,\text{m/s}$.

(c) The final speed does not depend on the mass of the snowball. The final potential energy is proportional to the mass, so $U_2^{grav}/m$ is independent of the mass. The final speed is again $21.0\,\text{m/s}$.

## 25

Let $m$ be the mass of the block, $h$ the height from which it dropped, and $x$ the compression of the spring. Take the potential energy of the block-spring-Earth system to be zero when the block is at its initial position. The block drops a distance $h + x$ and the final gravitational potential energy is $-mg(h+x)$. Here $x$ is taken to be positive for a compression of the spring. The spring potential energy is initially zero and finally $\frac{1}{2}kx^2$. The kinetic energy is zero at both the beginning and end. Since energy is conserved $0 = -mg(h + x) + \frac{1}{2}kx^2$. This is a quadratic equation for $x$. Its solution is

$$x = \frac{mg \pm \sqrt{(mg)^2 + 2mghk}}{k}.$$

Now $mg = (2.0\,\text{kg})(9.8\,\text{m/s}^2) = 19.6\,\text{N}$ and $hk = (0.40\,\text{m})(1960\,\text{N/m}) = 784\,\text{N}$, so

$$x = \frac{19.6\,\text{N} \pm \sqrt{(19.6\,\text{N})^2 + 2(19.6\,\text{N})(784\,\text{N})}}{1960\,\text{N/m}}$$

$= 0.10\,\text{m}$ or $-0.080\,\text{m}$. Since $x$ must be positive (a compression) we accept the positive solution and reject the negative solution.

**29**

Use conservation of mechanical energy: the mechanical energy must be the same at the top of the swing as it is initially. Use Newton's second law to find the speed, and hence the kinetic energy, at the top. There the tension force $\vec{T}$ of the string and the force of gravity are both downward, toward the center of the circle. Notice that the radius of the circle is $r = L - d$, so the law can be written $T + mg = mv^2/(L - d)$, where $v$ is the speed and $m$ is the mass of the ball. When the ball passes the highest point with the least possible speed the tension is zero. Then $mg = mv^2/(L - d)$ and $v = \sqrt{g(L - d)}$.

Take the gravitational potential energy of the ball-Earth system to be zero when the ball is at the bottom of its swing. Then the initial potential energy is $mgL$. The initial kinetic energy is zero since the ball starts from rest. The final potential energy, at the top of the swing, is $2mg(L - d)$ and the final kinetic energy is $\frac{1}{2}mv^2 = \frac{1}{2}mg(L - d)$. Conservation of energy yields $mgL = 2mg(L - d) + \frac{1}{2}mg(L - d)$. The solution for $d$ is $d = 3L/5$. If $d$ is greater than this value, so the highest point is lower, then the speed of the ball is greater as it reaches that point and the ball passes the point. If $d$ is less, the ball cannot go around. Thus the value you found for $d$ is a lower limit.

**35**

The free-body diagram for the boy is shown on the right. $\vec{N}$ is the normal force of the ice on him and $\vec{F}^{\text{grav}}$ is the force of gravity, which has a magnitude $mg$, where $m$ is the mass of the boy. The net inward force has magnitude $mg\cos\theta - N$ and, according to Newton's second law, this must be equal to $mv^2/R$, where $v$ is the speed of the boy. At the point where the boy leaves the ice $N = 0$, so $g\cos\theta = v^2/R$. We wish to find his speed. If the gravitational potential energy is taken to be zero when he is at the top of the ice mound, then his potential energy at the time shown is $U^{\text{grav}} = -mgR(1 - \cos\theta)$. He starts from

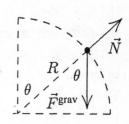

rest and his kinetic energy at the time shown is $\frac{1}{2}mv^2$. Thus conservation of energy gives $0 = \frac{1}{2}mv^2 - mgR(1 - \cos\theta)$, or $v^2 = 2gR(1 - \cos\theta)$. Substitute this expression into the equation developed from the second law to obtain $g\cos\theta = 2g(1 - \cos\theta)$. This gives $\cos\theta = 2/3$. The height of the boy above the bottom of the mound is $R\cos\theta = 2R/3$.

**41**

(a) The force exerted by the rope is constant, so the work it does is $W^{\text{rope}} = \vec{F}^{\text{rope}} \cdot \vec{d}$, where $\vec{F}^{\text{rope}}$ is the force of the rope and $\vec{d}$ is the displacement of the block. Thus the work done by the force of the rope is $W^{\text{rope}} = F^{\text{rope}}d\cos\theta = (7.68\,\text{N})(4.06\,\text{m})\cos 15.0° = 30.1\,\text{J}$.

(b) The increase in thermal energy is $\Delta E^{\text{thermal}} = f^{\text{friction}}d$, where $f^{\text{friction}}$ is the magnitude of the frictional force. Since the velocity of the block is constant you know that the net force on it is zero and that the magnitude of the frictional force must equal the magnitude of the horizontal component of the force of the rope: $f^{\text{friction}} = (7.68\,\text{N})\cos 15° = 7.42\,\text{N}$. Thus $\Delta E^{\text{thermal}} = (7.42\,\text{N})(4.06\,\text{m}) = 30.1\,\text{J}$.

(c) We can use Newton's second law of motion to obtain the magnitude of the normal force, then use $\mu^{\text{kin}} = f^{\text{friction}}/N$ to obtain the coefficient of friction. The vertical component of the second law gives $N + F^{\text{rope}} \sin\theta - mg = 0$, where $m$ is the mass of the block and $\theta$ is the angle between that force and the horizontal. Thus $N = mg - F^{\text{rope}} \sin\theta = (3.57\,\text{kg})(9.8\,\text{m/s}^2) - (7.68\,\text{N}) \sin 15.0^\circ = 33.0\,\text{N}$ and $\mu^{\text{kin}} = f^{\text{friction}}/N = (7.42\,\text{N})/(33.0\,\text{N}) = 0.225$.

## 53

(a) To stretch the spring without a change in kinetic energy, an external force, equal in magnitude to the force of the spring but opposite to its direction, is applied. Since a spring stretched in the positive $x$ direction exerts a force in the negative $x$ direction, the applied force must have magnitude $F = (52.8\,\text{N/m})x + (38.4\,\text{N/m}^2)x^2$ and be in the positive $x$ direction. The work it does is

$$W = \int_{0.50\,\text{m}}^{1.00\,\text{m}} \left[ (52.8\,\text{N/m})x + (38.4\,\text{N/m}^2)x^2 \right]\, dx$$

$$= \left[ \frac{52.8\,\text{N/m}}{2}x^2 + \frac{38.4\,\text{N/m}^2}{3}x^3 \right]_{0.50\,\text{m}}^{1.00\,\text{m}} = 31.0\,\text{J}.$$

(b) The spring does $31.0\,\text{J}$ of work and this must be the increase in the kinetic energy of the particle. Its speed is then

$$v = \sqrt{\frac{2K}{m}} = \sqrt{\frac{2(31.0\,\text{J})}{2.17\,\text{kg}}} = 5.35\,\text{m/s}.$$

(c) The force is conservative since the work it does as the particle goes from any point $x_1$ to any other point $x_2$ depends only on $x_1$ and $x_2$, not on details of the motion between $x_1$ and $x_2$.

## 61

(a) Let $h$ be the maximum height reached. The thermal energy generated by air resistance as the stone rises to this height is $\Delta E^{\text{thermal}} = fh$. Use $K_2 + U_2^{\text{grav}} + \Delta E^{\text{thermal}} = K_1 + U_1^{\text{grav}}$, where $K_1$ and $K_2$ are the initial and final kinetic energies and $U_1^{\text{grav}}$ and $U_2^{\text{grav}}$ are the initial and final gravitational potential energies. Take the potential energy to be zero at the throwing point. The initial kinetic energy is $K_1 = \frac{1}{2}mv_1^2$, the initial potential energy is $U_1^{\text{grav}} = 0$, the final kinetic energy is $K_2 = 0$, and the final potential energy is $U_2^{\text{grav}} = wh$. Thus $wh + fh = \frac{1}{2}mv_1^2$. Solve for $h$:

$$h = \frac{mv_1^2}{2(w+f)} = \frac{wv_1^2}{2g(w+f)} = \frac{v_1^2}{2g(1+f/w)}.$$

Here $w/g$ was substituted for $m$ and both the numerator and denominator were divided by $w$.

(b) The force of the air is downward on the trip up and upward on the trip down. It is always opposite to the direction of the velocity. Over the entire trip the increase in thermal energy is $\Delta E_{\text{th}} = 2fh$. The final kinetic energy is $K_2 = \frac{1}{2}mv_2^2$, where $v_2$ is the speed of the stone just before it hits the ground. The final potential energy is $U_2^{\text{grav}} = 0$. Thus $\frac{1}{2}mv_2^2 + 2fh = \frac{1}{2}mv_1^2$. Substitute the expression you found for $h$ to obtain

$$-\frac{2fv_1^2}{2g(1+f/w)} = \frac{1}{2}mv_2^2 - \frac{1}{2}mv_1^2.$$

This leads to

$$v_2^2 = v_1^2 - \frac{2fv_1^2}{mg(1 + f/w)} = v_1^2 - \frac{2fv_1^2}{w(1 + f/w)} = v_1^2 \left[1 - \frac{2f}{w + f}\right] = v_1^2 \frac{w - f}{w + f}.$$

Here $w$ was substituted for $mg$ and some algebraic manipulations were carried out. Thus

$$v = v_1 \left(\frac{w - f}{w + f}\right)^{1/2}.$$

## 73

(a) Let $\vec{v}_1$ be the velocity of the ball before it enters the gun and $\vec{v}_2$ be the final velocity of the ball-gun system. Since the net momentum of the system is conserved $mv_1 = (m + M)v_2$ and $v_2 = mv_1/(m + M)$.

(b) The initial kinetic energy of the system is $K_1 = \frac{1}{2}mv_1^2$ and the final kinetic energy is $K_2 = \frac{1}{2}(m + M)v_2^2 = \frac{1}{2}m^2v_1^2/(m + M)$. The difference is the energy $U^{\text{spring}}$ stored in the spring:

$$U^{\text{spring}} = \frac{1}{2}mv_1^2 - \frac{1}{2}\frac{m^2v_1^2}{(m + M)} = \frac{1}{2}mv_1^2\left(1 - \frac{m}{m + M}\right) = \frac{1}{2}mv_1^2\frac{M}{m + M}.$$

The fraction of the original energy that is stored in the spring is $U^{\text{spring}}/K_1 = M/(m + M)$.

## 83

(a) Time $t_1$ is just before the collision and time $t_2$ is just after. Let $m_A$ be the mass of the cart that is originally moving, $\vec{v}_A(t_1)$ be its velocity before the collision, and $\vec{v}_A(t_2)$ be its velocity after the collision. Let $m_B$ be the mass of the cart that is originally at rest and $\vec{v}_B(t_2)$ be its velocity after the collision. Suppose cart A is initially moving in the positive $x$ direction. Then, according to Eq. 10–35,

$$v_{Ax}(t_2) = \frac{m_A - m_B}{m_A + m_B}v_{Ax}(t_1).$$

Solve for $m_B$:

$$m_B = \frac{v_{Ax}(t_1) - v_{Ax}(t_2)}{v_{Ax}(t_1) + v_{Ax}(t_2)}m_A = \left(\frac{1.2\,\text{m/s} - 0.66\,\text{m/s}}{1.2\,\text{m/s} + 0.66\,\text{m/s}}\right)(0.340\,\text{kg}) = 0.099\,\text{kg}.$$

(b) The velocity of the second cart is given by Eq. 10–36:

$$v_{Bx}(t_2) = \frac{2m_A}{m_A + m_B}v_{Ax}(t_1) = \left[\frac{2(0.340\,\text{kg})}{0.340\,\text{kg} + 0.099\,\text{kg}}\right](1.2\,\text{m/s}) = 1.9\,\text{m/s}.$$

(c) The speed of the center of mass is

$$v_{\text{com}} = \frac{m_Av_{Ax}(t_1) + m_Bv_{Bx}(t_1)}{m_A + m_B} = \frac{(0.340\,\text{kg})(1.2\,\text{m/s})}{0.340\,\text{kg} + 0.099\,\text{kg}} = 0.93\,\text{m/s}.$$

Values for the initial velocities were used but the same result is obtained if instead values for the final velocities are used.

**87**

(a) Time $t_1$ is just before the collision and time $t_2$ is just after. Let $m_A$ be the mass of the body that is originally moving, $\vec{v}_A(t_1)$ be its velocity before the collision, and $v_A(t_2)$ be its velocity after the collision. Let $m_B$ be the mass of the body that is originally at rest and $\vec{v}_B(t_2)$ be its velocity after the collision. Suppose body A is initially moving in the positive $x$ direction. Then, according to Eq. 10–35 of the text,

$$v_{Ax}(t_2) = \frac{m_A - m_B}{m_A + m_B} v_{Ax}(t_1).$$

Solve for $m_B$:

$$m_B = \frac{v_{Ax}(t_1) - v_{Ax}(t_2)}{v_A(t_2) + v_{Ax}(t_1)} m_A.$$

Substitute $v_{Ax}(t_2) = v_{Ax}(t_1)/4$ to obtain $m_B = 3m_A/5 = 3(2.0\,\text{kg})/5 = 1.2\,\text{kg}$.

(b) The speed of the center of mass is

$$v_{\text{com}} = \frac{m_A v_{Ax}(t_1) + m_B v_{Bx}(t_1)}{m_A + m_B} = \frac{(2.0\,\text{kg})(4.0\,\text{m/s})}{2.0\,\text{kg} + 1.2\,\text{kg}} = 2.5\,\text{m/s}.$$

**91**

(a) Time $t_1$ is just before the collision and time $t_2$ is just after. Let $m_A$ be the mass of the block on the left, $\vec{v}_A(t_1)$ be its velocity before the collision, and $\vec{v}_A(t_2)$ be its velocity after the collision. Let $m_B$ be the mass of the block on the right, $\vec{v}_B(t_1)$ be its velocity before the collision, and $\vec{v}_B(t_2)$ be its velocity after the collision. The momentum of the two-block system is conserved, so $m_A \vec{v}_A(t_1) + m_B \vec{v}_B(t_1) = m_A \vec{v}_A(t_2) + m_B \vec{v}_B(t_2)$ and

$$\vec{v}_A(t_2) = \frac{m_A \vec{v}_A(t_1) + m_B \vec{v}_B(t_1) - m_B \vec{v}_B(t_2)}{m_A}$$

$$= \frac{(1.6\,\text{kg})(5.5\,\text{m/s})\hat{\imath} + (2.4\,\text{kg})(2.5\,\text{m/s})\hat{\imath} - (2.4\,\text{kg})(4.9\,\text{m/s})\hat{\imath}}{1.6\,\text{kg}} = (1.9\,\text{m/s})\hat{\imath}.$$

The block continues going to the right after the collision.

(b) To see if the collision is elastic compare the total kinetic energy before the collision with the total kinetic energy after the collision. The total kinetic energy before is

$$K_2 = \frac{1}{2}m_A v_A^2((t_1) + \frac{1}{2}m_B v_B^2(t_1)$$

$$= \frac{1}{2}(1.6\,\text{kg})(5.5\,\text{m/s})^2 + \frac{1}{2}(2.4\,\text{kg})(2.5\,\text{m/s})^2 = 31.7\,\text{J}.$$

The total kinetic energy after is

$$K_2 = \frac{1}{2}m_A v_A^2(t_2) + \frac{1}{2}m_B v_B^2(t_2)$$

$$= \frac{1}{2}(1.6\,\text{kg})(1.9\,\text{m/s})^2 + \frac{1}{2}(2.4\,\text{kg})(4.9\,\text{m/s})^2 = 31.7\,\text{J}.$$

Since $K_i = K_f$ the collision is elastic.

(c) Now $\vec{v}_B(t_1) = -(2.5\,\text{m/s})\hat{\imath}$ and

$$\vec{v}_A(t_2) = \frac{(1.6\,\text{kg})(5.5\,\text{m/s})\hat{\imath} + (2.4\,\text{kg})(-2.5\,\text{m/s})\hat{\imath} - (2.4\,\text{kg})(4.9\,\text{m/s})\hat{\imath}}{1.6\,\text{kg}} = -(5.6\,\text{m/s})\hat{\imath}.$$

Block A must be traveling in the direction opposite to that shown in the figure.

# Chapter 11

## 3

(a) The time $T$ for one revolution is the circumference of the orbit divided by the speed $v$ of the Sun: $T = 2\pi R/v$, where $R$ is the radius of the orbit. Since

$$R = 2.3 \times 10^4 \text{ ly} = (2.3 \times 10^4 \text{ ly})(9.460 \times 10^{12} \text{ km/ly}) = 2.18 \times 10^{17} \text{ km},$$

$$T = \frac{2\pi(2.18 \times 10^{17} \text{ km})}{250 \text{ km/s}} = 5.5 \times 10^{15} \text{ s}.$$

(b) The number of revolutions is the total time $t$ divided by the time $T$ for one revolution: $N = t/T$. Convert the total time from years to seconds. The result for the number of revolutions is

$$N = \frac{(4.5 \times 10^9 \text{ y})(3.16 \times 10^7 \text{ s/y})}{5.5 \times 10^{15} \text{ s}} = 26.$$

## 5

(a) Evaluate

$$\theta(t) = 2 \text{ rad} + (4 \text{ rad/s}^2)t^2 + (2 \text{ rad/s}^3)t^3$$

for $t = 0$ to obtain $\theta(0) = 2$ rad.

(b) The component of the rotational velocity along the rotation axis is given by

$$\omega(t) = \frac{d\theta}{dt} = (8 \text{ rad/s}^2)t + (6 \text{ rad/s}^3)t^2.$$

Evaluate this expression for $t = 0$ to obtain $\omega(0) = 0$.

(c) For $t = 4.0$ s,

$$\omega = (8 \text{ rad/s}^2)(4.0 \text{ s}) + (6 \text{ rad/s}^3)(4.0 \text{ s})^2 = 130 \text{ rad/s}.$$

(d) The component of the rotational acceleration about the rotation axis is given by

$$\alpha = \frac{d\omega}{dt} = 8 \text{ rad/s}^2 + (12 \text{ rad/s}^3)t.$$

For $t = 2.0$ s,

$$\alpha = 8 \text{ rad/s}^2 + (12 \text{ rad/s}^3)(2.0 \text{ s}) = 32 \text{ rad/s}^2.$$

(e) The component of the rotational acceleration along the rotation axis, given by

$$\alpha = 8 \text{ rad/s}^2 + (12 \text{ rad/s}^3)t,$$

depends on the time and so is not constant.

**9**

(a) For constant rotational acceleration $\omega_2 = \omega_1 + \alpha \Delta t$, so

$$\alpha = \frac{\omega_2 - \omega_1}{\Delta t}.$$

Take $\omega_2 = 0$ and, to obtain the units requested, use

$$\Delta t = \frac{30\,\text{s}}{60\,\text{s/min}} = 0.50\,\text{min}.$$

Then

$$\alpha = -\frac{33.33\,\text{rev/min}}{0.50\,\text{min}} = -67\,\text{rev/min}^2.$$

The negative sign indicates that the direction of the rotational acceleration is opposite that of the rotational velocity.

(b) The angle through which the turntable turns is

$$\Delta\theta = \omega_1 \Delta t + \frac{1}{2}\alpha(\Delta t)^2 = (33.33\,\text{rev/min})(0.50\,\text{min}) + \frac{1}{2}(-66.7\,\text{rev/min}^2)(0.50\,\text{min})^2$$

$$= 8.3\,\text{rev}.$$

**13**

Take the time at the start of the interval to be $t_1 = 0$. Then the time at the end of the interval is $t_2 = 4.0\,\text{s}$, and the angle of rotation is

$$\Delta\theta = \omega_1 t_2 + \tfrac{1}{2}\alpha t_2^2.$$

Solve for $\omega_1$:

$$\omega_1 = \frac{\Delta\theta - \tfrac{1}{2}\alpha t_2^2}{t_2} = \frac{120\,\text{rad} - \tfrac{1}{2}(3.0\,\text{rad/s}^2)(4.0\,\text{s})^2}{4.0\,\text{s}} = 24\,\text{rad/s}.$$

Now use

$$\omega_2 = \omega_1 + \alpha t_2$$

to find the time when the wheel is at rest ($\omega_2 = 0$):

$$t_2 = -\frac{\omega_1}{\alpha} = -\frac{24\,\text{rad/s}}{3.0\,\text{rad/s}^2} = -8.0\,\text{s}.$$

That is, the wheel started from rest 8.0 s before the start of the 4.0 s interval.

**19**

The magnitude of the translational acceleration is given by $a = \omega^2 r$, where $r$ is the distance from the center of rotation and $\omega$ is the rotational speed. You must convert the given rotational velocity to rad/s:

$$\omega = \frac{(33.33\,\text{rev/min})(2\pi\,\text{rad/rev})}{60\,\text{s/min}} = 3.49\,\text{rad/s}.$$

Thus
$$a = \left(3.49\,\text{rad/s}^2\right)^2 (0.15\,\text{m}) = 1.8\,\text{m/s}^2 .$$

The acceleration vector is from the point toward the center of the record.

## 21

Use $v = \omega r$. First convert $50\,\text{km/h}$ to m/s: $(50\,\text{km/h})(1000\,\text{m/km})/(3600\,\text{s/h}) = 13.9\,\text{m/s}$. Then

$$\omega = \frac{v}{r} = \frac{13.9\,\text{m/s}}{110\,\text{m}} = 0.13\,\text{rad/s} .$$

## 29

Since the belt does not slip, a point on the rim of wheel C has the same tangential acceleration as a point on the rim of wheel A. This means that $|\alpha_A| r_A = |\alpha_C| r_C$, where $\alpha_A$ is the component of the rotational acceleration of wheel A along its rotation axis and $\alpha_C$ is the component of the rotational acceleration of wheel C along its rotation axis. Thus

$$|\alpha_C| = \left(\frac{r_A}{r_C}\right) |\alpha_A| = \left(\frac{10\,\text{cm}}{25\,\text{cm}}\right) (1.6\,\text{rad/s}^2) = 0.64\,\text{rad/s}^2 .$$

Since the rotational speed of wheel C is given by $|\omega_C| = |\alpha_C| t$, the time for it to reach an rotational speed of $100\,\text{rev/min}$ ($= 10.5\,\text{rad/s}$) from rest is

$$t = \frac{|\omega_C|}{|\alpha_C|} = \frac{10.5\,\text{rad/s}}{0.64\,\text{rad/s}^2} = 16\,\text{s} .$$

## 35

According to Table 11–2 the rotational inertia of a cylinder of mass $M$ and radius $R$ is $I = \frac{1}{2} M R^2$. Thus the kinetic energy of a cylinder when it rotates with rotational speed $|\omega|$ is

$$K = \frac{1}{2} I |\omega|^2 = \frac{1}{4} M R^2 |\omega|^2 .$$

For the first cylinder

$$K = \frac{1}{4}(1.25\,\text{kg})(0.25\,\text{m})^2 (235\,\text{rad/s})^2 = 1.1 \times 10^3\,\text{J} .$$

For the second cylinder

$$K = \frac{1}{4}(1.25\,\text{kg})(0.75\,\text{m})^2 (235\,\text{rad/s})^2 = 9.7 \times 10^3\,\text{J} .$$

## 39

Use the parallel axis theorem: $I = I_{\text{com}} + M h^2$, where $I_{\text{com}}$ is the rotational inertia about a parallel axis through the center of mass, $M$ is the mass, and $h$ is the distance between the two axes. In this case the axis through the center of mass is at the 0.50 m mark, so $h = 0.50\,\text{m} - 0.20\,\text{m} = 0.30\,\text{m}$. Now according to Table 11–2

$$I_{\text{com}} = \frac{1}{12}M\ell^2 = \frac{1}{12}(0.56\,\text{kg})(1.0\,\text{m})^2 = 4.67 \times 10^{-2}\,\text{kg} \cdot \text{m}^2 \,,$$

so

$$I = 4.67 \times 10^{-2}\,\text{kg} \cdot \text{m}^2 + (0.56\,\text{kg})(0.30\,\text{m})^2 = 9.7 \times 10^{-2}\,\text{kg} \cdot \text{m}^2 \,.$$

## 43

(a) According to Table 11–2, the rotational inertia of a uniform solid cylinder about its central axis is given by

$$I_C = \frac{1}{2}MR^2 \,,$$

where $M$ is its mass and $R$ is its radius. For a hoop with mass $M$ and radius $R_H$ Table 11–2 gives

$$I_H = MR_H^2 \,.$$

If the two bodies have the same mass, then they will have the same rotational inertia if $R^2/2 = R_H^2$, or $R_H = R/\sqrt{2}$.

(b) You want the rotational inertia to be given by $I = Mk^2$, where $M$ is the mass of the arbitrary body and $k$ is the radius of the equivalent hoop. Thus

$$k = \sqrt{\frac{I}{M}} \,.$$

## 51

(a) Use $\tau^{\text{net}} = I\alpha$, where $\tau^{\text{net}}$ is the net torque acting on the shell, $I$ is the rotational inertia of the shell, and $\alpha$ is its rotational acceleration. This gives

$$I = \frac{\tau}{\alpha} = \frac{960\,\text{N} \cdot \text{m}}{6.20\,\text{rad/s}^2} = 155\,\text{kg} \cdot \text{m}^2 \,.$$

(b) The rotational inertia of the shell is given by $I = (2/3)MR^2$ (see Table 11–2 of the text). This means

$$M = \frac{3I}{2R^2} = \frac{3(155\,\text{kg} \cdot \text{m}^2)}{2(1.90\,\text{m})^2} = 64.4\,\text{kg} \,.$$

## 65

Use conservation of mechanical energy. The center of mass is at the midpoint of the cross bar of the H and it drops by $\ell/2$, where $\ell$ is the length of any one of the rods. The gravitational potential energy decreases by $Mg\ell/2$, where $M$ is the mass of the body. The initial kinetic energy is zero and the final kinetic energy may be written $\frac{1}{2}I|\omega|^2$, where $I$ is the rotational inertia of the body and $|\omega|$ is its rotational speed when it is vertical. Thus $0 = -Mg\ell/2 + \frac{1}{2}I|\omega|^2$ and $|\omega| = \sqrt{Mg\ell/I}$.

Since the rods are thin, the one along the axis of rotation does not contribute to the rotational inertia. All points on the other leg are the same distance from the axis of rotation, so that leg contributes $(1/3)M\ell^2$ since $M/3$ is its mass.

The cross bar is a rod that rotates around one end. Use the parallel axis theorem to find its rotational inertia. According to Table 11–2 of the text the rotational inertia about an axis through the center of mass and perpendicular to the rod is $I_{\text{com}} = (1/12)m\ell^2$, where $m$ is its mass. The axis through an end is separated from the center of mass by $\ell/2$, so

$$I = I_{\text{com}} + m(\ell/2)^2 = (1/12)m\ell^2 + (1/4)m\ell^2 = (1/3)m\ell^2 .$$

The contribution to the rotational inertia of the H is $(1/3)(M/3)\ell^2 = (1/9)M\ell^2$.

The total rotational inertia of the H is

$$I = \frac{1}{3}M\ell^2 + \frac{2}{9}M\ell^2 = \frac{4}{9}M\ell^2 .$$

The rotational speed is

$$|\omega| = \sqrt{\frac{Mg\ell}{I}} = \sqrt{\frac{Mg\ell}{4M\ell^2/9}} = \sqrt{\frac{9g}{4\ell}} .$$

## 67

(a) Use conservation of energy to find the rotational speed $|\omega|$ as a function of the angle $\theta$ that the chimney makes with the vertical. If we take the gravitational potential energy to be zero when the chimney hits the ground, it is $U^{\text{grav}} = Mgh$ when the center of mass is a distance $h$ above the ground. Here $M$ is the mass of the chimney. When the chimney makes the angle $\theta$ with the vertical, $h = (H/2)\cos\theta$, so $U^{\text{grav}} = (mgH/2)\cos\theta$. Initially the kinetic energy is zero. Write $\frac{1}{2}I|\omega|^2$ for the kinetic energy when the chimney makes the angle $\theta$ with the vertical. Here $I$ is its rotational inertia. Conservation of energy then leads to

$$\frac{MgH}{2} = \frac{MgH}{2}\cos\theta + \frac{I|\omega|^2}{2} ,$$

so

$$|\omega| = \sqrt{\frac{MgH}{I}(1 - \cos\theta)} .$$

The chimney is rotating about its base. Use the parallel axis theorem to find its rotational inertia. According to Table 11–2 of the text the rotational inertia about an axis through the center of mass and perpendicular to the rod is $I_{\text{com}} = (1/12)MH^2$, where $m$ is its mass. The axis through an end is separated from the center of mass by $H/2$, so

$$I = I_{\text{com}} + M(H/2)^2 = \frac{1}{12}MH^2 + \frac{1}{4}MH^2 = \frac{1}{3}MH^2 .$$

Thus

$$|\omega| = \sqrt{\frac{3g}{H}(1 - \cos\theta)} .$$

(b) The magnitude of the radial component of the acceleration of the top is given by $a_r = H|\omega|^2$, and since $|\omega|^2 = (3g/H)(1 - \cos\theta)$, this is $a_r = 3g(1 - \cos\theta)$.

(c) The magnitude of the tangential component of the acceleration of the chimney top is given by $a_t = H|\alpha|$, where $|\alpha|$ is the magnitude of the rotational acceleration. Differentiate $|\omega|^2 = (MgH/I)(1 - \cos\theta)$ with respect to time, replace $d|\omega|/dt$ with $|\alpha|$, and replace $|\theta/dt|$ with $|\omega|$ to obtain

$$2|\omega||\alpha| = \frac{MgH}{I}|\omega|\sin\theta$$

or

$$|\alpha| = \frac{MgH}{2I}\sin\theta\,.$$

Thus

$$a_t = H|\alpha| = \frac{MgH^2}{2I}\sin\theta = \frac{3g}{2}\sin\theta\,,$$

where $I = (1/3)MH^2$ was used to obtain the last result.

(d) The angle $\theta$ for which $a_t = g$ is the solution to $(3g/2)\sin\theta = g$ or $\sin\theta = 2/3$. It is $\theta = 41.8°$.

# Chapter 12

## 3

The work required to stop the hoop is the negative of the initial kinetic energy of the hoop. The initial kinetic energy is given by

$$K = \tfrac{1}{2}I|\omega|^2 + \tfrac{1}{2}mv^2,$$

where $I$ is its rotational inertia, $m$ is its mass, $|\omega|$ is its rotational speed about its center of mass, and $v$ is the speed of its center of mass. The rotational inertia of the hoop is given by $I = mR^2$, where $R$ is its radius. Since the hoop rolls without sliding the rotational speed and the speed of the center of mass are related by $|\omega| = v/R$. Thus

$$K = \frac{1}{2}mR^2\left(\frac{v^2}{R^2}\right) + \frac{1}{2}mv^2 = mv^2 = (140\,\text{kg})(0.150\,\text{m/s})^2 = 3.15\,\text{J}.$$

The work required is $W = -3.15\,\text{J}$.

## 5

Let $M$ be the mass of the car and $v$ be its speed. Let $I$ be the rotational inertia of one wheel and $|\omega|$ be the rotational speed of each wheel. The total kinetic energy is given by

$$K = \frac{1}{2}Mv^2 + 4\frac{1}{2}I|\omega|^2 = \frac{1}{2}Mv^2 + 2I|\omega|^2,$$

where the factor 4 appears because there are four wheels. The kinetic energy of rotation is $K^{\text{rot}} = 2I|\omega|^2$ and the fraction of the total energy that is due to rotation is

$$f = \frac{K^{\text{rot}}}{K} = \frac{2I|\omega|^2}{\frac{1}{2}Mv^2 + 2I|\omega|^2} = \frac{4I|\omega|^2}{Mv^2 + 4I|\omega|^2}.$$

For a uniform wheel $I = \tfrac{1}{2}mR^2$, where $R$ is the radius of a wheel and $m$ is its mass. Since the wheels roll without sliding $|\omega| = v/R$. Thus

$$I|\omega|^2 = \tfrac{1}{2}mR^2v^2/R^2 = \tfrac{1}{2}mv^2$$

and

$$f = \frac{2mv^2}{Mv^2 + 2mv^2} = \frac{2m}{M + 2m} = \frac{2(10\,\text{kg})}{1000\,\text{kg} + 2(10\,\text{kg})} = 0.020.$$

Notice that the radius of the wheel cancels from the equations. The rotational inertia is proportional to $R^2$ and when it is multiplied by $|\omega|^2 = v^2/R^2$ the result is independent of $R$.

## 11

The area of a triangle is half the product of its base and altitude. Take the base to be the side formed by vector $\vec{a}$. Then the altitude is $b \sin \phi$ and the area is $A = \frac{1}{2} ab \sin \phi$. Since the magnitude of the vector product is $|\vec{a} \times \vec{b}| = ab \sin \phi$, the area is $A = \frac{1}{2} |\vec{a} \times \vec{b}|$.

## 15

The vector product of two vectors is perpendicular to both vectors. If two nonparallel vectors are drawn with their tails at the same point this common point and the two heads define a plane. The vector product is perpendicular to that plane.

## 19

(a) Let $\vec{F} = F_x \hat{\imath} + F_y \hat{\jmath}$ and $\vec{r} = x \hat{\imath} + y \hat{\jmath}$. Then

$$\vec{\tau} = \vec{r} \times \vec{F} = (x \hat{\imath} + y \hat{\jmath}) \times (F_x \hat{\imath} + F_y \hat{\jmath}) = (x F_y - y F_x) \hat{k}.$$

The last result can be obtained by multiplying out the quantities in parentheses and using $\hat{\imath} \times \hat{\jmath} = \hat{k}$, $\hat{\jmath} \times \hat{\imath} = -\hat{k}$, $\hat{\imath} \times \hat{\imath} = 0$, and $\hat{\jmath} \times \hat{\jmath} = 0$. Numerically,

$$\vec{\tau} = [(3.0\,\text{m})(6.0\,\text{N}) - (4.0\,\text{m})(-8.0\,\text{N})]\,\hat{k} = (50\,\text{N} \cdot \text{m})\,\hat{k}.$$

(b) Use the definition of the vector product. It gives $|\vec{r} \times \vec{F}| = rF \sin \phi$, where $\phi$ is the angle between $\vec{r}$ and $\vec{F}$ when they are drawn with their tails at the same point. Now

$$r = \sqrt{x^2 + y^2} = \sqrt{(3.0\,\text{m})^2 + (4.0\,\text{m})^2} = 5.0\,\text{m}$$

and

$$F = \sqrt{F_x^2 + F_y^2} = \sqrt{(-8.0\,\text{N})^2 + (6.0\,\text{N})^2} = 10\,\text{N}.$$

Thus

$$rF = (5.0\,\text{m})(10\,\text{N}) = 50\,\text{N} \cdot \text{m},$$

the same as the magnitude of the vector product. This means $\sin \phi = 1$ and $\phi = 90°$.

## 23

The net torque $\vec{\tau}^{\text{net}}$ is related to the rate of change of the rotational momentum $\vec{\ell}$ by

$$\vec{\tau}^{\text{net}} = \frac{d\vec{\ell}}{dt}.$$

For each of the given cases the rotational momentum is in the negative $z$ direction.

(a) $\vec{\tau}^{\text{net}} = \dfrac{d(-4.0\,\text{kg} \cdot \text{m}^2/\text{s})\,\hat{k}}{dt} = 0.$

(b) $\vec{\tau}^{\text{net}} = \dfrac{d(-4.0\,\text{kg} \cdot \text{m}^2/\text{s}^3)t^2\,\hat{k}}{dt} = (-8.0\,\text{kg} \cdot \text{m}^2/\text{s}^3)t\,\hat{k}.$

(c) $\vec{\tau}^{\mathrm{net}} = \dfrac{d(-4.0\,\mathrm{kg\cdot m^2\cdot s^{-3/2}})t^{1/2}\,\hat{k}}{dt} = (-4.0\,\mathrm{kg\cdot m^2\cdot s^{-3/2}})t^{-1/2}\,\hat{k}.$

(d) $\vec{\tau}^{\mathrm{net}} = \dfrac{d(-4.0\,\mathrm{kg\cdot m^2\cdot s})t^{-2}\,\hat{k}}{dt} = (8.0\,\mathrm{kg\cdot m^2\cdot s})t^{-3}\,\hat{k}.$

## 29

(a) and (b) The diagram on the right shows the particles and their lines of motion. The origin is marked $O$ and may be anywhere. The rotational momentum of particle A has magnitude $\ell_A = mvr_A \sin\theta_A = mv(d + h)$ and it is into the page. The rotational momentum of particle B has magnitude $\ell_B = mvr_B \sin\theta_B = mvh$ and it is out of the page. The net rotational momentum has magnitude $L = mv(d + h) - mvh = mvd$ and is into the page. This result is independent of the location of the origin.

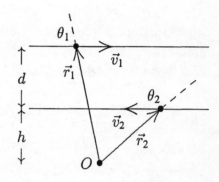

(c) Suppose particle B is traveling to the right. Then $L = mv(d + h) + mvh = mv(d + 2h)$. This result depends on $h$, the distance from the origin to one of the lines of motion. If the origin is midway between the lines of motion, then $h = -d/2$ and $L = 0$.

## 35

Suppose cylinder A exerts a uniform force of magnitude $F$ on cylinder B, tangent to the cylinder's surface at the point of contact. The magnitude of the torque applied to cylinder B is $|\tau_B| = R_B F$ and the rotational acceleration of that cylinder has magnitude $|\alpha_B| = |\tau_B|/I_B = R_B F/I_B$. Assume the cylinder starts from rest at time zero. Then its rotational speed as a function of time $t$ is

$$|\omega_B| = |\alpha_B|t = \frac{R_B F t}{I_B}.$$

The forces of the cylinders on each other obey Newton's third law, so the magnitude of the force of cylinder B on cylinder A is also $F$. The torque exerted by cylinder B on cylinder A has magnitude $|\tau_A| = R_A F$ and the rotational acceleration of cylinder A has magnitude $|\alpha_A| = |\tau_A|/I_A = R_A F/I_A$. This torque slows the cylinder. As a function of time its rotational speed is $|\omega_A| = |\omega_1| - R_A F t/I_A$. The force ceases and the cylinders continue rotating with constant rotational speeds when the translational speeds of points on their rims are the same. This means when $R_A|\omega_A| = R_B|\omega_B|$. Thus

$$R_A|\omega_1| - \frac{R_A^2 F t}{I_A} = \frac{R_B^2 F t}{I_B}.$$

When this equation is solved for $Ft$, the result is

$$Ft = \frac{R_A I_A I_B}{I_A R_B^2 + I_B R_A^2}|\omega_1|.$$

Substitute this expression for $Ft$ into $|\omega_B| = R_B Ft/I_B$ to obtain

$$|\omega_B| = \frac{R_A R_B I_A}{I_A R_B^2 + I_B R_A^2} |\omega_1| \, .$$

The angular velocity $\vec{\omega}_B$ is out of the page.

## 39

(a) No external torques act on the system consisting of the two wheels, so its total rotational momentum is conserved. Let $I_A$ be the rotational inertia of the wheel that is originally spinning and $I_B$ be the rotational inertia of the wheel that is initially at rest. If $|\omega_1|$ is the initial rotational speed of the first wheel and $|\omega_2|$ is the common final rotational velocity of each wheel, then $I_A |\omega_1| = (I_A + I_B)|\omega_2|$ and

$$|\omega_2| = \frac{I_A}{I_A + I_B} |\omega_1| \, .$$

Substitute $I_B = 2I_A$ and $|\omega_1| = 800 \, \text{rev/min}$ to obtain $|\omega_2| = 267 \, \text{rev/min}$.

(b) The initial kinetic energy is $K_1 = \frac{1}{2} I_A |\omega_1|^2$ and the final kinetic energy is $K_2 = \frac{1}{2}(I_A + I_B)|\omega_2|^2$. The fraction lost is

$$\frac{\Delta K}{K_1} = \frac{K_1 - K_2}{K_1} = \frac{I_A |\omega_1|^2 - (I_A + I_B)|\omega_2|^2}{I_A |\omega_1|^2} = \frac{|\omega_1|^2 - 3|\omega_2|^2}{|\omega_1|^2}$$

$$= \frac{(800 \, \text{rev/min})^2 - 3(267 \, \text{rev/min})^2}{(800 \, \text{rev/min})^2} = 0.67 \, .$$

## 41

(a) In terms of the radius of gyration $k$ the rotational inertia of the merry-go-round is $I = Mk^2$ and its value is $(180 \, \text{kg})(0.910 \, \text{m})^2 = 149 \, \text{kg} \cdot \text{m}^2$.

(b) Recall that an object moving along a straight line has rotational momentum about any point that is not on the line. Its magnitude is $mvd$, where $m$ is the mass of the object, $v$ is the speed of the object, and $d$ is the distance from the origin to the line of motion. In particular, the rotational momentum of the child about the center of the merry-go-round has magnitude $\ell^{\text{child}} = mvR$, where $R$ is the radius of the merry-go-round. Its value is $(44.0 \, \text{kg})(3.00 \, \text{m/s})(1.20 \, \text{m}) = 158 \, \text{kg} \cdot \text{m}^2/\text{s}$.

(c) No external torques act on the system consisting of the child and the merry-go-round, so the total rotational momentum of the system is conserved. The initial rotational momentum is given by $mvR$; the final rotational momentum is given by $(I + mR^2)|\omega|$, where $|\omega|$ is the final common rotational speed of the merry-go-round and child. Thus $mvR = (I + mR^2)|\omega|$ and

$$|\omega| = \frac{mvR}{I + mR^2} = \frac{158 \, \text{kg} \cdot \text{m}^2/\text{s}}{149 \, \text{kg} \cdot \text{m}^2 + (44.0 \, \text{kg})(1.20 \, \text{m})^2} = 0.744 \, \text{rad/s} \, .$$

## 49

(a) If we consider a short time interval from just before the wad hits to just after it hits and sticks, we may use the principle of conservation of rotational momentum. The initial rotational momentum is the rotational momentum of the falling putty wad. The wad initially moves along a line that is $d/2$ distant from the axis of rotation, where $d$ is the length of the rod. The rotational momentum of the wad has magnitude $mvd/2$. After the wad sticks, the rod has rotational speed $|\omega|$ and rotational momentum of magnitude $I|\omega|$, where $I$ is the rotational inertia of the system consisting of the rod with the two balls and the wad at its end. Conservation of rotational momentum yields $mvd/2 = I|\omega|$. If $M$ is the mass of one of the balls, $I = (2M + m)(d/2)^2$. When $mvd/2 = (2M + m)(d/2)^2|\omega|$ is solved for $|\omega|$, the result is

$$|\omega| = \frac{2mv}{(2M+m)d} = \frac{2(0.0500\,\text{kg})(3.00\,\text{m/s})}{[2(2.00\,\text{kg}) + 0.0500\,\text{kg}](0.500\,\text{m})} = 0.148\,\text{rad/s}.$$

(b) The initial kinetic energy is $K_1 = \frac{1}{2}mv^2$, the final kinetic energy is $K_2 = \frac{1}{2}I|\omega|^2$, and their ratio is $K_2/K_1 = I|\omega|^2/mv^2$. When $I = (2M + m)d^2/4$ and $|\omega| = 2mv/(2M + m)d$ are substituted, this becomes

$$\frac{K_2}{K_1} = \frac{m}{2M + m} = \frac{0.0500\,\text{kg}}{2(2.00\,\text{kg}) + 0.0500\,\text{kg}} = 0.0123.$$

(c) As the rod and wad rotate together the sum of the kinetic and potential energies is conserved. If one of the balls is lowered a distance $h$, the other is raised the same distance and the sum of the potential energies of the balls does not change. We need consider only the potential energy of the putty wad. It moves through a $90°$ arc to reach the lowest point on its path, gaining kinetic energy and losing gravitational potential energy as it goes. It then swings up through an angle $\theta$, losing kinetic energy and gaining potential energy, until it momentarily comes to rest. Take the lowest point on the path to be the zero of potential energy. It starts a distance $d/2$ above this point, so its initial potential energy is $U_1^{\text{grav}} = mgd/2$.

If it swings through the angle $\theta$, measured from its lowest point, then its final position is $(d/2)(1 - \cos\theta)$ above the lowest point and its final potential energy is $U_2^{\text{grav}} = mg(d/2)(1 - \cos\theta)$. The initial kinetic energy is the sum of the kinetic energies of the balls and wad: $K_1 = \frac{1}{2}I|\omega|^2 = \frac{1}{2}(2M + m)(d/2)^2|\omega|^2$. At its final position the rod is instantaneously stopped, so the final kinetic energy is $K_2 = 0$. Conservation of energy yields $mgd/2 + \frac{1}{2}(2M + m)(d/2)^2|\omega|^2 = mg(d/2)(1 - \cos\theta)$. When this equation is solved for $\cos\theta$, the result is

$$\cos\theta = -\frac{1}{2}\left(\frac{2M+m}{mg}\right)\left(\frac{d}{2}\right)|\omega|^2$$

$$= -\frac{1}{2}\left[\frac{2(2.00\,\text{kg}) + 0.0500\,\text{kg}}{(0.0500\,\text{kg})(9.8\,\text{m/s}^2)}\right]\left(\frac{0.500\,\text{m}}{2}\right)(0.148\,\text{rad/s})^2 = -0.0226.$$

The result for $\theta$ is $91.3°$. The total angle of the swing is $90° + 91.3° = 181.3°$.


## 55

(a) The rotational inertia of the system is the sum of the contributions from the bullet, the rod, and the block. Let $d$ be the length of the rod and $m$ be the mass of the bullet. Then the contribution of the bullet to the rotational inertia is $md^2 = (1.0 \times 10^{-3}\,\text{kg})(0.60))^2 = 3.6 \times 10^{-4}\,\text{kg}\cdot\text{m}^2$. If $M$ is the mass of the block then its contribution is $Md^2 = (0.5\,\text{kg})(0.60\,\text{m})^2 = 1.8 \times 10^{-1}\,\text{kg}\cdot\text{m}^2$. The contribution of the rod is given as $0.060\,\text{kg}\cdot\text{m}^2$. The total rotational inertia of the system is $I = 3.6 \times 10^{-4}\,\text{kg}\cdot\text{m}^2 + 1.8 \times 10^{-1}\,\text{kg}\cdot\text{m}^2 + 0.060\,\text{kg}\cdot\text{m}^2 = 2.4 \times 10^{-1}\,\text{kg}\cdot\text{m}^2$.

(b) Use conservation of rotational momentum. Before impact only the bullet has rotational momentum about A and its magnitude is $mvd$. After impact the rotational momentum of the system has magnitude $I|\omega|$, where $|\omega|$ is the rotational speed of the rod and bullet. Since rotational momentum is conserved $mvd = I|\omega|$ and

$$v = \frac{I|\omega|}{md} = \frac{(2.4 \times 10^{-1}\,\text{kg}\cdot\text{m}^2)(4.5\,\text{rad/s})}{(1.0 \times 10^{-3}\,\text{kg})(0.60\,\text{m})} = 1.8 \times 10^3\,\text{m/s}.$$

## 59

(a) The falling object has angular momentum of magnitude $L = 2mRv$, about the origin and kinetic energy of $K = \frac{1}{2}mv^2$. The angular momentum vector is into the page.

(b) The rotational inertia of a hoop is $I = mR^2$ (see Table 11–2), so its angular momentum has magnitude $L = I|\omega| = mR^2(v/R) = mRv$. The angular momentum vector is out of the page. The kinetic energy is $K = \frac{1}{2}I|\omega|^2 = \frac{1}{2}mR^2(v/R)^2 = \frac{1}{2}mv^2$.

(c) The rotational inertia of the disk is $I = \frac{1}{2}mR^2$ (see table 11–2), so its angular momentum has magnitude $L = I|\omega| = \frac{1}{2}mR^2(v/R) = \frac{1}{2}mRv$. The angular momentum vector is out of the page. The kinetic energy is $K = \frac{1}{2}I|\omega|^2 = \frac{1}{2}\frac{1}{2}mR^2(v/R)^2 = \frac{1}{4}mv^2$.

(d) The rotational inertia of the mass on the string is $I = mR^2$, so its angular momentum has magnitude $L = I|\omega| = mR^2(v/R) = mRv$. The angular momentum vector is out of the page. The kinetic energy is $K = \frac{1}{2}I|\omega|^2 = \frac{1}{2}mR^2(v/R)^2 = \frac{1}{2}mv^2$.

The falling object has the greatest angular momentum and the disk has the least. The falling object is a distance $2R$ from the line through the origin while the mass in the hoop and mass on a string are a distance $R$ from the rotation axes. The mass of the disk is all less then $R$ from the rotation axis. Furthermore the translational speeds of the falling object, the hoop, and the mass on a string are all the same. Only the mass at the rim of the disk has the same translational speed. The rest of the mass is moving more slowly. Since the angular momentum is the product of a translation speed and a distance from the mass to the rotation axis (or origin), we expect the falling object to have the greatest angular momentum and the disk to have the least.

The falling object, hoop, and mass on a string are all tied with the greatest kinetic energy. The disk has the least. All of the mass in the first three is traveling at the same speed $v$. Only the mass at the rim of the disk as this speed. The rest of its mass is slower, so we expect its kinetic energy to be less.

# Chapter 13

## 7

Three forces act on the sphere: the tension force $\vec{T}$ of the rope (along the rope), the force of the wall $\vec{N}$ (horizontally away from the wall), and the force of gravity $\vec{F}^{\text{grav}}$ (downward). Since the sphere is in equilibrium they sum to zero. If $\theta$ is the angle between the rope and the vertical, then the vertical component of Newton's second is $T\cos\theta - mg = 0$ and the horizontal component is $N - T\sin\theta = 0$, where $F^{\text{grav}} = mg$ was used.

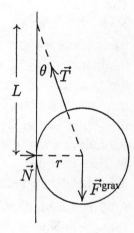

(a) Solve the first equation for $T$: $T = mg/\cos\theta$. Substitute $\cos\theta = L/\sqrt{L^2 + r^2}$, from the right triangle on the diagram on the right, to obtain $T = mg\sqrt{L^2 + r^2}/L$.

(b) Solve the second equation for $N$: $N = T\sin\theta$. Use $\sin\theta = r/\sqrt{L^2 + r^2}$, from the triangle, to obtain

$$N = \frac{Tr}{\sqrt{L^2 + r^2}} = \frac{mg\sqrt{L^2 + r^2}}{L}\frac{r}{\sqrt{L^2 + r^2}} = \frac{mgr}{L}.$$

The last form was obtained by substituting $mg\sqrt{L^2 + r^2}/L$ for $T$.

## 9

The board is in equilibrium, so the sum of the forces and the sum of the torques acting on it are each zero. Take the upward direction to be positive. Take the force of the left pedestal to be $\vec{F}_1$ and suppose the coordinate of its point of application is $x = x_1$, where the $x$ axis is along the diving board. Take the force of the right pedestal to be $\vec{F}_2$ and suppose the coordinate of its point of application is $x = x_2$. Let $F^{\text{grav}}$ be the weight of the diver, applied at $x = x_3$. Set the expression for the sum of the forces equal to zero. In terms of components $F_{1y} + F_{2y} - F^{\text{grav}} = 0$. Set the expression for the torque about $x_2$ equal to zero: $F_{1y}(x_2 - x_1) + F^{\text{grav}}(x_3 - x_2) = 0$.

(a) The second equation gives

$$F_{1y} = -\frac{x_3 - x_2}{x_2 - x_1}F^{\text{grav}} = -\left(\frac{3.0\,\text{m}}{1.5\,\text{m}}\right)(580\,\text{N}) = -1160\,\text{N}.$$

The result is negative, indicating that this force is downward.

(b) The first equation gives

$$F_{2y} = F^{\text{grav}} - F_{1y} = 580\,\text{N} + 1160\,\text{N} = 1740\,\text{N}.$$

The result is positive, indicating that this force is upward.

(c) and (d) The force of the diving board on the left pedestal is upward (opposite to the force of the pedestal on the diving board), so this pedestal is being stretched. The force of the diving board on the right pedestal is downward, so this pedestal is being compressed.

## 13

The forces on the ladder are shown in the diagram to the right. $\vec{F}_1$ is the force of the window, horizontal because the window is frictionless. $\vec{F}_2$ and $\vec{F}_3$ are components of the force of the ground on the ladder. $F_{man}^{grav}$ is the gravitational force on the man and $\vec{F}_{ladder}^{grav}$ is the gravitational force on the ladder. The gravitational force on the man acts at a point 3.0 m up the ladder and the gravitational force on the ladder acts at the center of the ladder. Let $\theta$ be the angle between the ladder and the ground. Use $\cos\theta = d/L$ or $\sin\theta = \sqrt{L^2 - d^2}/L$ to find $\theta = 60°$. Here $L$ is the length of the ladder (5.0 m) and $d$ is the distance from the wall to the foot of the ladder (2.5 m).

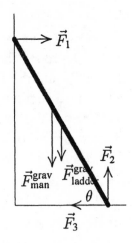

(a) Since the ladder is in equilibrium the sum of the torques about its foot (or any other point) vanishes. Let $\ell$ be the distance from the foot of the ladder to the position of the window cleaner. Use $F_{man}^{grav} = Mg$ and $F_{ladder}^{grav} = mg$, where $M$ is the mass of the man and $m$ is the mass of the ladder. Then $Mg\ell\cos\theta + mg(L/2)\cos\theta - F_1 L\sin\theta = 0$ and

$$F_1 = \frac{(M\ell + mL/2)g\cos\theta}{L\sin\theta}$$
$$= \frac{[(75\,\text{kg})(3.0\,\text{m}) + (10\,\text{kg})(2.5\,\text{m})]\,(9.8\,\text{N/kg})\cos 60°}{(5.0\,\text{m})\sin 60°} = 280\,\text{N}.$$

This force is outward, away from the wall. The force of the ladder on the window has the same magnitude but is in the opposite direction: it is 280 N, inward.

(b) The sum of the horizontal forces and the sum of the vertical forces also vanish: $F_1 - F_3 = 0$ and $F_2 - Mg - mg = 0$. The first of these equations gives $F_3 = F_1 = 280\,\text{N}$ and the second gives $F_2 = (M + m)g = (75\,\text{kg} + 10\,\text{kg})(9.8\,\text{N/kg}) = 830\,\text{N}$. The magnitude of the force of the ground on the ladder is given by the square root of the sum of the squares of its components: $F = \sqrt{F_2^2 + F_3^2} = \sqrt{(280\,\text{N})^2 + (830\,\text{N})^2} = 880\,\text{N}$. The angle $\phi$ between the force and the horizontal is given by $\tan\phi = F_3/F_2 = (830\,\text{N})/(280\,\text{N}) = 2.94$, so $\phi = 71°$. The force points to the left and upward, 71° above the horizontal. Note that it is not along the ladder.

## 17

Label the pulleys 1, 2, 3 from left to right and let $T_1$ be the magnitude of the tension in cable that goes around pulley 1, $T_2$ be the magnitude of the tension in the cable that goes around pulley 2, and $T_3$ be the magnitude of the tension in the cable that goes around pulley 3. The equilibrium conditions yield $T = 2T_1$, $T_1 = 2T_2$, and $T_2 = 2T_3$. Thus $T_1 = T/2$, $T_2 = T_1/2 = T/4$, and $T_3 = T_2/2 = T/8$. Since the block is in equilibrium $T_1 + T_2 + T_3 = mg$, where $m$ is the mass of the block. Thus

$$\frac{T}{2} + \frac{T}{4} + \frac{T}{8} = mg.$$

The solution for $T$ is

$$T = \frac{8mg}{7} = \frac{8(6.40\,\text{kg})(9.8\,\text{N/kg})}{7} = 71.7\,\text{N}.$$

## 21

Consider the wheel as it leaves the lower floor. The floor no longer exerts a force on the wheel, and the only forces acting are the force $\vec{F}$ applied horizontally at the axle, the force of gravity $\vec{F}^{\text{grav}}$ acting vertically at the center of the wheel, and the force of the step corner, shown as the two components $\vec{f}_h$ and $\vec{f}_v$. If the minimum force is applied the wheel does not accelerate, so both the total force and the total torque acting on it are zero.

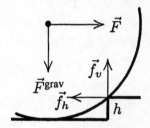

Calculate the torque around the step corner. Look at the second diagram to see that the distance from the line of $\vec{F}$ to the corner is $r - h$, where $r$ is the radius of the wheel and $h$ is the height of the step. The distance from the line of $\vec{F}^{\text{grav}}$ to the corner is $\sqrt{r^2 + (r-h)^2} = \sqrt{2rh - h^2}$. Thus $F(r-h) - F^{\text{grav}}\sqrt{2rh - h^2} = 0$. The solution for $F$ is

$$F = \frac{\sqrt{2rh - h^2}}{r - h}\,F^{\text{grav}}.$$

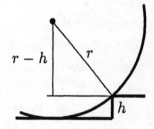

## 23

The beam is in equilibrium: the sum of the forces and the sum of the torques acting on it each vanish. As you can see from Fig. 13–35, the beam makes an angle of 60° with the vertical and the wire makes an angle of 30° with the vertical.

(a) Calculate the torques around the hinge. Their sum is $TL\sin 30° - F^{\text{grav}}(L/2)\sin 60° = 0$. Here $F^{\text{grav}}$ is the force of gravity, acting at the center of the beam, and $T$ is the magnitude of the tension force of the wire. Solve for $T$:

$$T = \frac{F^{\text{grav}}\sin 60°}{2\sin 30°} = \frac{(222\,\text{N})\sin 60°}{2\sin 30°} = 192.3\,\text{N}.$$

(b) Let $F_h$ be the horizontal component of the force exerted by the hinge and take it to be positive if the force is outward from the wall. Then the vanishing of the horizontal component of the net force on the beam yields $F_h - T\sin 30° = 0$ or $F_h = T\sin 30° = (192.3\,\text{N})\sin 30° = 96.1\,\text{N}$.

(c) Let $F_v$ be the vertical component of the force exerted by the hinge and take it to be positive if it is upward. Then the vanishing of the vertical component of the net force on the beam yields $F_v + T\cos 30° - F^{\text{grav}} = 0$ or $F_v = F^{\text{grav}} - T\cos 30° = 222\,\text{N} - (192.3\,\text{N})\cos 30° = 65.5\,\text{N}$.

**29**

The diagram on the right shows the forces acting on the plank. Since the roller is frictionless the force it exerts is normal to the plank and makes the angle $\theta$ with the vertical. It is designated $\vec{F}$. $\vec{F}^{\text{grav}}$ is the force of gravity; this force acts at the center of the plank, a distance $L/2$ from the point where the plank touches the floor. $\vec{N}$ is the normal force of the floor and $\vec{f}$ is the force of friction. The distance from the foot of the plank to the wall is denoted by $d$. This quantity is not given directly but it can be computed using $d = h/\tan\theta$.

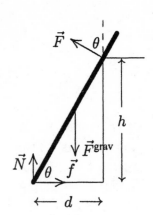

The equations of equilibrium are:

horizontal component of force:     $F\sin\theta - f = 0$

vertical component of force:     $F\cos\theta - F^{\text{grav}} + N = 0$

torque:     $Nd - fh - F^{\text{grav}}\left(d - \dfrac{L}{2}\cos\theta\right) = 0$.

The point of contact between the plank and the roller was used as the origin for writing the torque equation.

When $\theta = 70°$ the plank just begins to slip and $f = \mu^{\text{static}}N$, where $\mu^{\text{static}}$ is the coefficient of static friction. You want to use the equations of equilibrium to compute $N$ and $f$ for $\theta = 70°$, then use $\mu^{\text{static}} = f/N$ to compute the coefficient of friction.

The second equilibrium equation gives $F = (F^{\text{grav}} - N)/\cos\theta$ and this is substituted into the first to obtain $f = (F^{\text{grav}} - N)\sin\theta/\cos\theta = (F^{\text{grav}} - N)\tan\theta$. This is substituted into the third equilibrium equation and the result is solved for $N$:

$$N = \frac{d - (L/2)\cos\theta + h\tan\theta}{d + h\tan\theta}F^{\text{grav}}.$$

Now replace $d$ with $h/\tan\theta$ and multiply both numerator and denominator by $\tan\theta$. The result is

$$N = \frac{h(1 + \tan^2\theta) - (L/2)\sin\theta}{h(1 + \tan^2\theta)}F^{\text{grav}}.$$

Use the trigonometric identity $1 + \tan^2\theta = 1/\cos^2\theta$ and multiply both numerator and denominator by $\cos^2\theta$ to obtain

$$N = F^{\text{grav}}\left(1 - \frac{L}{2h}\cos^2\theta\sin\theta\right).$$

Now substitute the expression for $N$ into $f = (F^{\text{grav}} - N)\tan\theta$ to obtain

$$f = \frac{F^{\text{grav}}L}{2h}\sin^2\theta\cos\theta.$$

Substitute the expressions for $f$ and $N$ into $\mu^{\text{static}} = f/N$ to obtain

$$\mu^{\text{static}} = \frac{L\sin^2\theta\cos\theta}{2h - L\sin\theta\cos^2\theta}.$$

Evaluate this expression for $\theta = 70°$:

$$\mu^{\text{static}} = \frac{(6.1\,\text{m})\sin^2 70° \cos 70°}{2(3.05\,\text{m}) - (6.1\,\text{m})\sin 70° \cos^2 70°} = 0.34 \,.$$

### 31

The diagrams on the right show the forces on the two sides of the ladder, separated. $\vec{F}_A$ and $\vec{F}_E$ are the forces of the floor on the two feet, $T$ is the magnitude of the tension force of the tie rod, $W$ is the magnitude of the force of the man (equal to his weight), $F_h$ is the horizontal component of the force exerted by one side of the ladder on the other, and $F_v$ is the vertical component of that force. Note that the forces exerted by the floor are normal to the floor

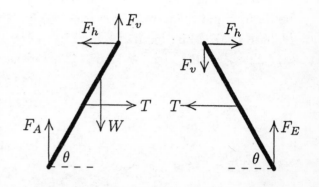

since the floor is frictionless. Also note that the force of the left side on the right and the force of the right side on the left are equal in magnitude and opposite in direction. The forces are labeled with symbols that represent their magnitudes to emphasize the third-law pairs. The arrows show their directions.

Since the ladder is in equilibrium, the vertical components of the forces on the left side of the ladder must sum to zero: $F_v + F_A - W = 0$. The horizontal components must sum to zero: $T - F_h = 0$. The torques must also sum to zero. Take the origin to be at the hinge and let $L$ be the length of a ladder side. Then $F_A L \cos\theta - W(L/4)\cos\theta - T(L/2)\sin\theta = 0$. Here we recognize that the man is one-fourth the length of the ladder side from the top and the tie rod is at the midpoint of the side.

The analogous equations for the right side are $F_E - F_v = 0$, $F_h - T = 0$, and $F_E L \cos\theta - T(L/2)\sin\theta = 0$.

There are 5 different equations:

$$F_v + F_A - W = 0 \,,$$
$$T - F_h = 0 \,,$$
$$F_A L \cos\theta - W(L/4)\cos\theta - T(L/2)\sin\theta = 0 \,,$$
$$F_E - F_v = 0 \,,$$
$$F_E L \cos\theta - T(L/2)\sin\theta = 0 \,.$$

The unknown quantities are $F_A$, $F_E$, $F_v$, $F_h$, and $T$.

(a) First solve for $T$ by systematically eliminating the other unknowns. The first equation gives $F_A = W - F_v$ and the fourth gives $F_v = F_E$. Use these to substitute into the remaining three equations to obtain

$$T - F_h = 0 \,,$$
$$W L \cos\theta - F_E L \cos\theta - W(L/4)\cos\theta - T(L/2)\sin\theta = 0 \,,$$
$$F_E L \cos\theta - T(L/2)\sin\theta = 0 \,.$$

The last of these gives $F_E = T \sin\theta / 2\cos\theta = (T/2)\tan\theta$. Substitute this expression into the second equation and solve for $T$. The result is

$$T = \frac{3W}{4\tan\theta} .$$

To find $\tan\theta$, consider the right triangle formed by the upper half of one side of the ladder, half the tie rod, and the vertical line from the hinge to the tie rod. The lower side of the triangle has a length of 0.381 m, the hypotenuse has a length of 1.22 m, and the vertical side has a length of $\sqrt{(1.22\,\text{m})^2 - (0.381\,\text{m})^2} = 1.16\,\text{m}$. This means $\tan\theta = (1.16\,\text{m})/(0.381\,\text{m}) = 3.04$. Thus,

$$T = \frac{3(854\,\text{N})}{4(3.04)} = 211\,\text{N} .$$

(b) Now solve for $F_A$. Since $F_v = F_E$ and $F_E = T\sin\theta/2\cos\theta$, $F_v = 3W/8$. Substitute this into $F_v + F_A - W = 0$ and solve for $F_A$. You should get $F_A = W - F_v = W - 3W/8 = 5W/8 = 5(884\,\text{N})/8 = 534\,\text{N}$.

(c) You have already obtained an expression for $F_E$: $F_E = 3W/8$. Evaluate it to obtain $F_E = 320\,\text{N}$.

## 35

(a) The force diagram shown on the right depicts the situation just before the crate tips, when the normal force acts at the front edge. However, it may also be used to calculate the angle for which the crate begins to slide. $\vec{F}^{\text{grav}}$ is the force of gravity on the crate, $\vec{N}$ is the normal force of the plane on the crate, and $\vec{f}$ is the force of friction. Take the $x$ axis to be down the plane and the $y$ axis to be in the direction of the normal force. Assume the acceleration is zero but the crate is on the verge of sliding. The $x$ component

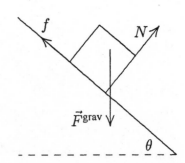

of Newton's second law is then $F^{\text{grav}}\sin\theta - f = 0$ and the $y$ component is $N - F^{\text{grav}}\cos\theta = 0$. The $y$ component equation gives $N = F^{\text{grav}}\cos\theta$. Since the crate is about to slide $f = \mu^{\text{static}} N = \mu^{\text{static}} F^{\text{grav}}\cos\theta$, where $\mu^{\text{static}}$ is the coefficient of static friction. Substitute into the $x$ component equation to obtain $F^{\text{grav}}\sin\theta - \mu^{\text{static}} F^{\text{grav}}\cos\theta = 0$, or $\tan\theta = \mu^{\text{static}}$. This means $\theta = \tan^{-1}\mu^{\text{static}} = \tan^{-1} 0.60 = 31.0°$.

Now develop an expression for the total torque about the center of mass when the crate is about to tip. Then the normal force and the force of friction act at the front edge. The torque associated with the force of friction tends to turn the crate clockwise and has magnitude $fh$, where $h$ is the perpendicular distance from the bottom of the crate to the center of gravity. The torque associated with the normal force tends to turn the crate counterclockwise and has magnitude $N\ell/2$, where $\ell$ is the length of a edge. Since the total torque vanishes, $fh = N\ell/2$. When the crate is about to tip, the acceleration of the center of gravity vanishes, so $f = F^{\text{grav}}\sin\theta$ and $N = F^{\text{grav}}\cos\theta$. Substitute these expressions into the torque equation and solve for $\theta$. You should get

$$\theta = \tan^{-1}\frac{\ell}{2h} = \tan^{-1}\frac{1.2\,\text{m}}{2(0.90\,\text{m})} = 33.7° .$$

As $\theta$ is increased from zero the crate slides before it tips. It starts to slide when $\theta = 31.0°$.

(b) The analysis is the same. The crate begins to slide when $\theta = \tan^{-1} \mu^{\text{static}} = \tan^{-1} 0.70 = 35.0°$ and begins to tip when $\theta = 33.7°$. Thus it tips first as the angle is increased. Tipping begins at $\theta = 33.7°$.

## 39

(a) Let $\vec{F}_A$ and $\vec{F}_B$ be the forces of the wires on the log and let $\vec{F}^{\text{grav}}$ be the gravitational force on the log. Since the log is in equilibrium $F_A + F_B - F^{\text{grav}} = 0$. Information given about the stretching of the wires allows us to find a relationship between $F_A$ and $F_B$. If wire $A$ originally had a length $L_A$ and stretches by $\Delta L_A$, then $\Delta L_A = F_A L_A / AE$, where $A$ is the cross-sectional area of the wire and $E$ is Young's modulus for steel ($200 \times 10^9 \, \text{N/m}^2$). Similarly, $\Delta L_B = F_B L_B / AE$. If $\ell$ is the amount by which $B$ was originally longer than $A$ then, since they have the same length after the log is attached, $\Delta L_A = \Delta L_B + \ell$. This means

$$\frac{F_A L_A}{AE} = \frac{F_B L_B}{AE} + \ell.$$

Solve for $F_B$:

$$F_B = \frac{F_A L_A}{L_B} - \frac{AE\ell}{L_B}.$$

Substitute into $F_A + F_B - F^{\text{grav}} = 0$ and solve for $F_A$:

$$F_A = \frac{F^{\text{grav}} L_B + AE\ell}{L_A + L_B}.$$

The cross-sectional area of a wire is $A = \pi r^2 = \pi (1.20 \times 10^{-3} \, \text{m})^2 = 4.52 \times 10^{-6} \, \text{m}^2$. Both $L_A$ and $L_B$ may be taken to be 2.50 m without loss of significance. Thus

$$F_A = \frac{(103 \, \text{kg})(9.8 \, \text{N/kg})(2.50 \, \text{m}) + (4.52 \times 10^{-6} \, \text{m}^2)(200 \times 10^9 \, \text{N/m}^2)(2.0 \times 10^{-3} \, \text{m})}{2.50 \, \text{m} + 2.50 \, \text{m}}$$

$$= 866 \, \text{N}.$$

(b) Solve $F_A + F_B - F^{\text{grav}} = 0$ for $F_B$: $F_B = F^{\text{grav}} - F_A = (103 \, \text{kg})(9.8 \, \text{N/kg}) - 866 \, \text{N} = 143 \, \text{N}$.

(c) The net torque must also vanish. Put the origin on the surface of the log at a point directly above the center of mass. The force of gravity does not exert a torque about this point. Then the torque equation is $F_A d_A - F_B d_B = 0$ and $d_A / d_B = F_B / F_A = (143 \, \text{N})/(866 \, \text{N}) = 0.165$.

## 51

There are two unknown quantities here: the weight of the diver and the position of the diver when the force of the left-hand support is zero. To find the weight of the diver write the expression for the net torque about the middle support when the diver is at the right end of the board. Since the system is in equilibrium the net torque must be zero. Let $F^{\text{grav}}_{\text{diver}}$ be the weight of the diver, $F^{\text{grav}}_{\text{board}}$ be the weight of the board, $L$ be the length of the board, $\ell$ be the distance between the supports, and $F$ be the magnitude of the force of the left-hand support. Take the force of gravity to act at the center of the board, a distance $(L/2) - \ell$ from the middle support. The torque equation becomes

$$F\ell - F^{\text{grav}}_{\text{diver}}(L - \ell) - F^{\text{grav}}_{\text{board}}[(L/2) - \ell] = 0.$$

The solution is

$$F_{\text{diver}}^{\text{grav}} = \frac{F\ell - F_{\text{board}}^{\text{grav}}[(L/2) - \ell]}{L - \ell} = \frac{(1200\,\text{N})(1.0\,\text{m}) - (40\,\text{kg})(9.8\,\text{N/kg})(0.75\,\text{m})}{2.5\,\text{m}} = 362\,\text{N}\,.$$

Now suppose the diver is a distance $x$ from the left-hand support. Again consider the net torque about that support. The torque equation is

$$F\ell - F_{\text{diver}}^{\text{grav}}(x - \ell) - F_{\text{board}}^{\text{grav}}[(L/2) - \ell] = 0\,.$$

The solution with $F = 0$ is

$$x = \frac{F_{\text{diver}}^{\text{grav}}\ell - F_{\text{board}}^{\text{grav}}[(L/2) - \ell]}{W} = \frac{362\,\text{N})(1.0\,\text{m}) - (40\,\text{kg})(9.8\,\text{N/kg})(0.75\,\text{m})}{362\,\text{N}} = 0.19\,\text{m}\,.$$

# Chapter 14

## 3

Use $F^{\text{grav}} = G m_s m_m / r^2$, where $m_s$ is the mass of the satellite, $m_m$ is the mass of the meteor, and $r$ is the distance between their centers. The distance between centers is $r = R + d = 15\,\text{m} + 3\,\text{m} = 18\,\text{m}$. Here $R$ is the radius of the satellite and $d$ is the distance from its surface to the center of the meteor. Thus

$$F^{\text{grav}} = \frac{(6.67 \times 10^{-11}\,\text{N} \cdot \text{m}^2 / \text{kg}^2)(20\,\text{kg})(7.0\,\text{kg})}{(18\,\text{m})^2} = 2.9 \times 10^{-11}\,\text{N}.$$

## 13

If the lead sphere were not hollowed the magnitude of its force on the small sphere would be $F_1^{\text{grav}} = GMm / d^2$. Part of this force is due to material that is removed. Calculate the force on the small sphere by a sphere that just fills the cavity, at the position of the cavity, and subtract it from the force of the solid sphere. The cavity has a radius $r = R/2$. The material that fills it has the same density (mass to volume ratio) as the solid sphere. That is $M_c / r^3 = M / R^3$, where $M_c$ is the mass that fills the cavity. The common factor $4\pi/3$ has been canceled. Thus

$$M_c = \left( \frac{r^3}{R^3} \right) M = \left( \frac{R^3}{8R^3} \right) M = \frac{M}{8}.$$

The center of the cavity is $d - r = d - R/2$ from $m$, so the force it exerts on $m$ is

$$F_2^{\text{grav}} = \frac{G(M/8)m}{(d - R/2)^2}.$$

The force of the hollowed sphere on $m$ is

$$F^{\text{grav}} = F_1^{\text{grav}} - F_2^{\text{grav}} = GMm \left[ \frac{1}{d^2} - \frac{1}{8(d - R/2)^2} \right] = \frac{GMm}{d^2} \left[ 1 - \frac{1}{8(1 - R/2d)^2} \right].$$

## 17

If the rotational speed were any greater, loose objects on the surface would not go around with the planet but would travel out into space. (a) The magnitude of the gravitational force exerted by the planet on an object of mass $m$ at its surface is given by $F^{\text{grav}} = GmM / R^2$, where $M$ is the mass of the planet and $R$ is its radius. According to Newton's second law this must equal $mv^2 / R$, where $v$ is the speed of the object. Thus

$$\frac{GM}{R^2} = \frac{v^2}{R}.$$

Replace $M$ with $(4\pi/3)\rho R^3$, where $\rho$ is the density of the planet, and $v$ with $2\pi R/T$, where $T$ is the period of revolution. The result is

$$\frac{4\pi}{3}G\rho R = \frac{4\pi^2 R}{T^2}.$$

Solve for $T$ to obtain

$$T = \sqrt{\frac{3\pi}{G\rho}}.$$

(b) The density is $3.0 \times 10^3 \, \text{kg/m}^3$. Evaluate the equation for $T$:

$$T = \sqrt{\frac{3\pi}{(6.67 \times 10^{-11}\,\text{m}^3/\text{s}^2 \cdot \text{kg})(3.0 \times 10^3\,\text{kg/m}^3)}} = 6.86 \times 10^3\,\text{s} = 1.9\,\text{h}.$$

## 25

(a) The magnitude of the force on a particle with mass $m$ at the surface of Earth is given by $F^{\text{grav}} = GMm/R^2$, where $M$ is the total mass of Earth and $R$ is Earth's radius. The local gravitational strength is

$$G_{\text{local}} = \frac{F^{\text{grav}}}{m} = \frac{GM}{R^2} = \frac{(6.67 \times 10^{-11}\,\text{m}^3/\text{s}^2 \cdot \text{kg})(5.98 \times 10^{24}\,\text{kg})}{(6.37 \times 10^6\,\text{m})^2} = 9.83\,\text{N/kg}.$$

(b) Now $g_{\text{local}} = GM/R^2$, where $M$ is the total mass contained in the core and mantle together and $R$ is the outer radius of the mantle ($6.345 \times 10^6$ m, according to Fig. 14–28). The total mass is $M = 1.93 \times 10^{24}$ kg $+ 4.01 \times 10^{24}$ kg $= 5.94 \times 10^{24}$ kg. The first term is the mass of the core and the second is the mass of the mantle. Thus

$$g_{\text{local}} = \frac{(6.67 \times 10^{-11}\,\text{m}^3/\text{s}^2 \cdot \text{kg})(5.94 \times 10^{24}\,\text{kg})}{(6.345 \times 10^6\,\text{m})^2} = 9.84\,\text{m/s}^2.$$

(c) A point 25 km below the surface is at the mantle-crust interface and is on the surface of a sphere with a radius of $R = 6.345 \times 10^6$ m. Since the mass is now assumed to be uniformly distributed the mass within this sphere can be found by multiplying the mass per unit volume by the volume of the sphere: $M = (R^3/R_e^3)M_e$, where $M_e$ is the total mass of Earth and $R_e$ is the radius of Earth. Thus

$$M = \left[\frac{6.345 \times 10^6\,\text{m}}{6.37 \times 10^6\,\text{m}}\right]^3 (5.98 \times 10^{24}\,\text{kg}) = 5.91 \times 10^{24}\,\text{kg}.$$

The local gravitational strength is

$$g_{\text{local}} = \frac{GM}{R^2} = \frac{(6.67 \times 10^{-11}\,\text{m}^3/\text{s}^2 \cdot \text{kg})(5.91 \times 10^{24}\,\text{kg})}{(6.345 \times 10^6\,\text{m})^2} = 9.79\,\text{m/s}^2.$$

**29**

(a) The density of a uniform sphere is given by $\rho = 3M/4\pi R^3$, where $M$ is its mass and $R$ is its radius. The ratio of the density of Mars to the density of Earth is

$$\frac{\rho_M}{\rho_E} = \frac{M_M}{M_E}\frac{R_E^3}{R_M^3} = 0.11\left(\frac{0.65 \times 10^4\,\text{km}}{3.45 \times 10^3\,\text{km}}\right)^3 = 0.74\,.$$

(b) The value of the gravitational acceleration $a$ at the surface of a planet is given by $a = GM/R^2$, so the value for Mars is

$$a_M = \frac{M_M}{M_E}\frac{R_E^2}{R_M^2}a_{gE} = 0.11\left(\frac{0.65 \times 10^4\,\text{km}}{3.45 \times 10^3\,\text{km}}\right)^2(9.8\,\text{m/s}^2) = 3.8\,\text{m/s}^2\,.$$

(c) If $v$ is the escape speed, then for a particle of mass $m$

$$\tfrac{1}{2}mv^2 = G\frac{mM}{R}$$

and

$$v = \sqrt{\frac{2GM}{R}}\,.$$

For Mars

$$v = \sqrt{\frac{2(6.67 \times 10^{-11}\,\text{m}^3/\text{s}^2\cdot\text{kg})(0.11)(5.98 \times 10^{24}\,\text{kg})}{3.45 \times 10^6\,\text{m}}} = 5.0 \times 10^3\,\text{m/s}\,.$$

**33**

(a) Use the principle of conservation of energy. Initially the rocket is at Earth's surface and the potential energy is $U_i = -GMm/R_e = -mgR_e$, where $M$ is the mass of Earth, $m$ is the mass of the rocket, and $R_e$ is the radius of Earth. The relationship $g = GM/R_e^2$ was used and the potential energy was taken to be zero when the rocket is far away from Earth. The initial kinetic energy is $\tfrac{1}{2}mv^2 = 2mgR_e$, where the substitution $v = 2\sqrt{gR_e}$ was made. If the rocket can escape then conservation of energy must lead to a positive kinetic energy no matter how far from Earth it gets. Take the final potential energy to be zero and let $K_f$ be the final kinetic energy. Then $U_i + K_i = U_f + K_f$ leads to $K_f = U_i + K_i = -mgR_e + 2mgR_e = mgR_e$. The result is positive and the rocket has enough kinetic energy to escape the gravitational pull of Earth.

(b) Write $\tfrac{1}{2}mv_f^2$ for the final kinetic energy. Then $\tfrac{1}{2}mv_f^2 = mgR_e$ and $v_f = \sqrt{2gR_e}$.

**35**

(a) Use the principle of conservation of energy. Initially the particle is at the surface of the asteroid and the gravitational potential energy is $U_i = -GMm/R$, where $M$ is the mass of the asteroid, $R$ is its radius, and $m$ is the mass of the particle being fired upward. The potential energy was taken to be zero when the particle is far away from the asteroid. The initial kinetic energy is $\tfrac{1}{2}mv^2$. The particle just escapes if its kinetic energy is zero when it is infinitely far from the asteroid. Thus the final potential and kinetic energies are both zero. Conservation of energy yields $-GMm/R + \frac{1}{2}mv^2 = 0$. Replace $GM/R$ with $aR$, where $a$ is the gravitational acceleration at the surface. Then the energy equation becomes $-aR + \frac{1}{2}v^2 = 0$. Solve for $v$:

$$v = \sqrt{2aR} = \sqrt{2(3.0\,\text{m/s}^2)(500 \times 10^3\,\text{m})} = 1.7 \times 10^3\,\text{m/s}.$$

(b) Initially the particle is at the surface; the potential energy is $U_i = -GMm/R$ and the kinetic energy is $K_i = \frac{1}{2}mv^2$. Suppose the particle is a distance $h$ above the surface when it momentarily comes to rest. The final potential energy is $U_f = -GMm/(R + h)$ and the final kinetic energy is $K_f = 0$. Conservation of energy yields

$$-\frac{GMm}{R} + \frac{1}{2}mv^2 = -\frac{GMm}{R+h}.$$

Replace $GM$ with $aR^2$ and cancel $m$ in the energy equation to obtain

$$-aR + \frac{1}{2}v^2 = -\frac{aR^2}{(R+h)}.$$

The solution for $h$ is

$$h = \frac{2aR^2}{2aR - v^2} - R$$

$$= \frac{2(3.0\,\text{m/s}^2)(500 \times 10^3\,\text{m})^2}{2(3.0\,\text{m/s}^2)(500 \times 10^3\,\text{m}) - (1000\,\text{m/s})^2} - (500 \times 10^3\,\text{m})$$

$$= 2.5 \times 10^5\,\text{m}.$$

(c) Initially the particle is a distance $h$ above the surface and is at rest. The potential energy is $U_i = -GMm/(R + h)$ and the initial kinetic energy is $K_i = 0$. Just before the particle hits the asteroid the potential energy is $U_f = -GMm/R$. Write $\frac{1}{2}mv_f^2$ for the final kinetic energy. Conservation of energy yields

$$-\frac{GMm}{R+h} = -\frac{GMm}{R} + \frac{1}{2}mv^2.$$

Replace $GM$ with $aR^2$ and cancel $m$ to obtain

$$-\frac{aR^2}{R+h} = -aR + \frac{1}{2}v^2.$$

The solution for $v$ is

$$v = \sqrt{2aR - \frac{2aR^2}{R+h}}$$

$$= \sqrt{2(3.0\,\text{m/s}^2)(500 \times 10^3\,\text{m}) - \frac{2(3.0\,\text{m/s}^2)(500 \times 10^3\,\text{m})^2}{500 \times 10^3\,\text{m} + 1000 \times 10^3\,\text{m}}}$$

$$= 1.4 \times 10^3\,\text{m/s}.$$

**37**

(a) The momentum of the two-star system is conserved, and since the stars have the same mass, their speeds and kinetic energies are the same. Use the principle of conservation of energy. The initial potential energy is $U_i = -GM^2/r_i$, where $M$ is the mass of either star and $r_i$ is their initial center-to-center separation. The potential energy was taken to be zero when the stars are far apart. The initial kinetic energy is zero since the stars are at rest. The final potential energy is $U_f = -2GM^2/r_i$ since the final separation is $r_i/2$. Write $Mv^2$ for the final kinetic energy of the system. This is the sum of two terms, each of which is $\frac{1}{2}Mv^2$. Conservation of energy yields

$$-\frac{GM^2}{r_i} = -\frac{2GM^2}{r_i} + Mv^2.$$

The solution for $v$ is

$$v = \sqrt{\frac{GM}{r_i}} = \sqrt{\frac{(6.67 \times 10^{-11}\,\text{m}^3/\text{s}^2 \cdot \text{kg})(10^{30}\,\text{kg})}{10^{10}\,\text{m}}} = 8.2 \times 10^4\,\text{m/s}.$$

(b) Now the final separation of the star centers is $r_f = 2R = 2 \times 10^5\,\text{m}$, where $R$ is the radius of either of the stars. The final potential energy is given by $U_f = -GM^2/r_f$ and the energy equation becomes $-GM^2/r_i = -GM^2/r_f + Mv^2$. The solution for $v$ is

$$v = \sqrt{GM\left(\frac{1}{r_f} - \frac{1}{r_i}\right)}$$

$$= \sqrt{(6.67 \times 10^{-11}\,\text{m}^3/\text{s}^2 \cdot \text{kg})(10^{30}\,\text{kg})\left(\frac{1}{2 \times 10^5\,\text{m}} - \frac{1}{10^{10}\,\text{m}}\right)}$$

$$= 1.8 \times 10^7\,\text{m/s}.$$

**41**

Center-of-mass frame. Conservation of translational momentum requires that the speed of the two spheres be the same at each instant of time. Just after the explosion each moves with a speed $v_1 = v^{\text{rel}}/2 = (1.05 \times 10^{-4}\,\text{m/s})/2 = 5.25 \times 10^{-5}\,\text{m/s}$. The kinetic energy is $K_1 = 2\frac{1}{2}mv_1^2 = (2.00\,\text{kg})(5.25 \times 10^{-5}\,\text{m/s})^2 = 5.51 \times 10^{-9}\,\text{J}$. The initial separation of the sphere centers is $2R$, so the initial potential energy is $U_1 = -Gm^2/2R$, where the potential energy is taken to be zero for infinite separation. When the centers are separated by $10R$ the potential energy is $U_2 = -Gm^2/10R$. Let $K_2$ be the kinetic energy when the separation is $10R$. Conservation of mechanical energy then gives

$$K_1 - G\frac{m^2}{2R} = K_2 - G\frac{m^2}{10R},$$

so

$$K_2 = K_1 - 0.400G\frac{m^2}{R} = 5.51 \times 10^{-9}\,\text{J} - (0.400)(6.67 \times 10^{-11}\,\text{m}^3/\text{s}^2 \cdot \text{kg})\frac{(2.00\,\text{kg})^2}{0.0200\,\text{m}}$$

$$= 1.72 \times 10^{-10}\,\text{J}.$$

Each sphere has kinetic energy $K = (1.72 \times 10^{-10}\,\text{J})/2 = 8.60 \times 10^{-11}\,\text{J}$, so each has a speed of

$$v_2 = \sqrt{\frac{2K}{m}} = \sqrt{\frac{2(8.60 \times 10^{-11}\,\text{J})}{2.00\,\text{kg}}} = 9.27 \times 10^{-6}\,\text{m/s}.$$

The speed of sphere B relative to sphere A is twice this or $1.85 \times 10^{-5}\,\text{m/s}$.

Sphere frame. The kinetic energy of sphere B just after the explosion is $K_1 = \frac{1}{2}m(v^{\text{rel}})^2 = \frac{1}{2}(2.00\,\text{kg})(1.05 \times 10^{-4}\,\text{m/s})^2 = 1.08 \times 10^{-8}\,\text{J}$. The kinetic energy of sphere A is zero. The potential energy is $U_1 = -Gm^2/2R$. Let $K_2$ be the kinetic energy of sphere B when the separation is $10R$. The kinetic energy of sphere $A$ is again zero and the potential energy is $-Gm^2/10R$. Conservation of mechanical energy gives

$$K_1 - G\frac{m^2}{2R} = K_2 - G\frac{m^2}{10R},$$

so

$$K_2 = K_1 - 0.400G\frac{m^2}{R} = 1.08 \times 10^{-8}\,\text{J} - (0.400)(6.67 \times 10^{-11}\,\text{m}^3/\text{s}^2 \cdot \text{kg})\frac{(2.00\,\text{kg})^2}{0.0200\,\text{m}}$$
$$= 5.46 \times 10^{-9}\,\text{J}.$$

The speed of sphere B is

$$v_2 = \sqrt{\frac{2K_2}{m}} = \sqrt{\frac{2(5.46 \times 10^{-9}\,\text{J})}{2.00\,\text{kg}}} = 1.39 \times 10^{-5}\,\text{m/s}.$$

The change in potential energy is the same for the two calculations, so the change in the total kinetic energy of the two-sphere system is the same. In the center-of-mass frame the kinetic energy is shared by the two spheres while in the sphere frame it all belongs to sphere B. Both sets of answers are correct. The speed and kinetic energy of an object depend on the frame used to measure them

# Chapter 15

**3**

The air inside pushes outward with a force of magnitude $P_iA$, where $P_i$ is the pressure inside the room and $A$ is the area of the window. Similarly, the air on the outside pushes inward with a force of magnitude $P_oA$, where $P_o$ is the pressure outside. The magnitude of the net force is $F^{\text{net}} = (P_i - P_o)A$. Since $1\,\text{atm} = 1.013 \times 10^5\,\text{Pa}$,

$$F^{\text{net}} = (1.0\,\text{atm} - 0.96\,\text{atm})(1.013 \times 10^5\,\text{Pa/atm})(3.4\,\text{m})(2.1\,\text{m}) = 2.9 \times 10^4\,\text{N}.$$

**7**

(a) At every point on the surface there is a net inward force, normal to the surface, due to the difference in pressure between the air inside and outside the sphere. The diagram to the right shows half the sphere and some of the force vectors. We suppose a team of horses is pulling to the right. To pull the sphere apart it must exert a force at least as great as the horizontal component of the net force of the air.

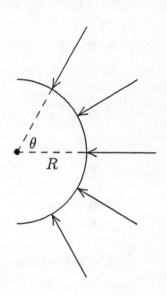

Consider the force acting at the angle $\theta$ shown. Its horizontal component is $\Delta P \cos\theta\, dA$, where $dA$ is an infinitesimal area element at the point where the force is applied. We take the area to be that of a ring of constant $\theta$ on the surface. The radius of the ring is $R\sin\theta$, where $R$ is the radius of the sphere. If the angular width of the ring is $d\theta$, in radians, then its width is $R\,d\theta$ and its area is $dA = 2\pi R^2 \sin\theta\, d\theta$. Thus the net horizontal component of the force of the air is given by

$$F_h = 2\pi R^2\, \Delta P \int_0^{\pi/2} \sin\theta \cos\theta\, d\theta$$

$$= \pi R^2\, \Delta p \sin^2\theta \,\Big|_0^{\pi/2} = \pi R^2\, \Delta P.$$

This is the force that must be exerted by each team of horses to pull the sphere apart.

(b) Use $1\,\text{atm} = 1.00 \times 10^5\,\text{Pa}$ to show that $\Delta P = 0.90\,\text{atm} = 9.00 \times 10^4\,\text{Pa}$. The sphere radius is $0.30\,\text{m}$, so $F_h = \pi(0.30\,\text{m})^2(9.00 \times 10^4\,\text{Pa}) = 2.5 \times 10^4\,\text{N}$.

(c) One team of horses could be used if half of the sphere is attached to a sturdy wall. The force of the wall on the sphere would balance the force of the horses. Two teams were probably used to heighten the dramatic effect.

**13**

The pressure $P$ at the depth $d$ of the hatch cover is $P^{\text{atm}} + \rho g d$, where $\rho$ is the density of ocean water and $P^{\text{atm}}$ is atmospheric pressure. The downward force of the water on the hatch cover has magnitude $(P^{\text{atm}} + \rho g d)A$, where $A$ is the area of the cover. If the air in the submarine is at atmospheric pressure then it exerts an upward force of magnitude $P^{\text{atm}}A$. The minimum force that must be applied by the crew to open the cover has magnitude

$$F = (P^{\text{atm}} + \rho g d)A - P^{\text{atm}}A = \rho g d A$$
$$= (1025 \text{ kg/m}^3)(9.8 \text{ N/kg})(100 \text{ m})(1.2 \text{ m})(0.60 \text{ m}) = 7.2 \times 10^5 \text{ N}.$$

**17**

Assume that the pressure is the same at all points that are the distance $d = 20$ km below the surface. For points on the left side of Fig. 15–37 this pressure is given by $P = P^{\text{atm}} + \rho_o g d_o + \rho_c g d_c + \rho_m g d_m$, where $P^{\text{atm}}$ is atmospheric pressure, $\rho_o$ is the density of ocean water, $d_o$ is the depth of the ocean, $\rho_c$ is the density of the crust, $d_c$ is the thickness of the crust, $\rho_m$ is the density of the mantle, and $d_m$ is the thickness of the mantle (to a depth of 20 km). For points on the right side of the figure $P$ is given by $P = P^{\text{atm}} + \rho_c g d$. Equate the two expressions for $P$ and note that $g$ cancels to obtain $\rho_c d = \rho_o d_o + \rho_c d_c + \rho_m d_m$. Substitute $d_m = d - d_o - d_c$ to obtain

$$\rho_c d = \rho_o d_o + \rho_c d_c + \rho_m d - \rho_m d_o - \rho_m d_c.$$

Solve for $d_o$:

$$d_o = \frac{\rho_c d_c - \rho_c d + \rho_m d - \rho_m d_c}{\rho_m - \rho_o} = \frac{(\rho_m - \rho_c)(d - d_c)}{\rho_m - \rho_o}$$
$$= \frac{(3.3 \text{ g/cm}^3 - 2.8 \text{ g/cm}^3)(20 \text{ km} - 12 \text{ km})}{3.3 \text{ g/cm}^3 - 1.0 \text{ g/cm}^3}$$
$$= 1.7 \text{ km}.$$

**21**

(a) Use the expression for the variation of pressure with height in an incompressible fluid: $P_2 = P_1 - \rho g (y_2 - y_1)$. Take $y_1$ to be at the surface of Earth, where the pressure is $P_1 = 1.01 \times 10^5$ Pa, and $y_2$ to be at the top of the atmosphere, where the pressure is $P_2 = 0$. Take the density to be $1.3 \text{ kg/m}^3$. Then

$$y_2 - y_1 = \frac{P_1}{\rho g} = \frac{1.01 \times 10^5 \text{ Pa}}{(1.3 \text{ kg/m}^3)(9.8 \text{ M/kg})} = 7.9 \times 10^3 \text{ m} = 7.9 \text{ km}.$$

(b) Let $h$ be the height of the atmosphere. Since the density varies with altitude, you must use the integral

$$P_2 = P_1 - \int_0^h \rho g \, dy.$$

Take $\rho = \rho_0(1 - y/h)$, where $\rho_0$ is the density at Earth's surface. This expression predicts that $\rho = \rho_0$ at $y = 0$ and $\rho = 0$ at $y = h$. Assume $g$ is uniform from $y = 0$ to $y = h$. Now the integral can be evaluated:

$$P_2 = P_1 - \int_0^h \rho_0 g \left(1 - \frac{y}{h}\right) dy = P_1 - \tfrac{1}{2}\rho_0 g h.$$

Since $P_2 = 0$, this means

$$h = \frac{2P_1}{\rho_0 g} = \frac{2(1.01 \times 10^5 \, \text{Pa})}{(1.3 \, \text{kg/m}^3)(9.8 \, \text{N/kg})} = 16 \times 10^3 \, \text{m} = 16 \, \text{km} \, .$$

## 25

(a) The anchor is completely submerged. It appears lighter than its actual weight because the water is pushing up on it with a buoyant force of $\rho_w g V$, where $\rho_w$ is the density of water and $V$ is the volume of the anchor. Its apparent weight (in water) is $W^{\text{app}} = F^{\text{grav}} - \rho_w g V$, where $F^{\text{grav}}$ is its actual weight (the force of gravity). Thus

$$V = \frac{F^{\text{grav}} - W^{\text{app}}}{\rho_w g} = \frac{200 \, \text{N}}{(998 \, \text{kg/m}^3)(9.8 \, \text{N/kg})} = 2.045 \times 10^{-2} \, \text{m}^3 \, .$$

The density of water was obtained from Table 15–2 of the text.

(b) The mass of the anchor is $m = \rho V$, where $\rho$ is the density of iron. Its weight in air is $F^{\text{grav}} = mg = \rho g V = (7870 \, \text{kg/m}^3)(9.8 \, \text{m/s}^2)(2.045 \times 10^{-2} \, \text{m}^3) = 1.58 \times 10^3 \, \text{N}$.

## 29

(a) The force of gravity $\vec{F}^{\text{grav}}$ is balanced by the buoyant force of the liquid $\vec{F}^{\text{buoy}}$. Since $F^{\text{grav}} = mg$ and $F^{\text{buoy}} = \rho g V_s$, $mg = \rho g V_s$. Here $m$ is the mass of the sphere, $\rho$ is the density of the liquid, and $V_s$ is the submerged volume. Thus $m = \rho V_s$. The submerged volume is half the volume enclosed by the outer surface of the sphere, or $V_s = \frac{1}{2}(4\pi/3)r_o^3$, where $r_o$ is the outer radius. This means

$$m = \frac{4\pi}{6}\rho r_o^3 = \left(\frac{4\pi}{6}\right)(800 \, \text{kg/m}^3)(0.090 \, \text{m})^3 = 1.22 \, \text{kg} \, .$$

Air in the hollow sphere, if any, has been neglected.

(b) The density $\rho_m$ of the material, assumed to be uniform, is given by $\rho_m = m/V$, where $m$ is the mass of the sphere and $V$ is its volume. If $r_i$ is the inner radius, the volume is

$$V = \frac{4\pi}{3}\left(r_o^3 - r_i^3\right) = \frac{4\pi}{3}\left[(0.090 \, \text{m})^3 - (0.080 \, \text{m})^3\right] = 9.09 \times 10^{-4} \, \text{m}^3 \, .$$

The density is

$$\rho = \frac{1.22 \, \text{kg}}{9.09 \times 10^{-4} \, \text{m}^3} = 1.3 \times 10^3 \, \text{kg/m}^3 \, .$$

## 33

The volume $V_{\text{cav}}$ of the cavities is the difference between the volume $V_{\text{cast}}$ of the casting as a whole and the volume $V_{\text{iron}}$ of iron in the casting: $V_{\text{cav}} = V_{\text{cast}} - V_{\text{iron}}$. The volume of the iron is given by $V_{\text{iron}} = F^{\text{grav}}/g\rho_{\text{iron}}$, where $F^{\text{grav}}$ is the weight of the casting and $\rho_{\text{iron}}$ is the density of iron. The effective weight in water can be used to find the volume of the casting. It is less than the actual weight $F^{\text{grav}}$ because the water pushes up on it with a force of magnitude $g\rho_w V_{\text{cast}}$, where $\rho_w$ is the density of water. That is, $F^{\text{app}} = F^{\text{grav}} - g\rho_w V_{\text{cast}}$. Thus $V_{\text{cast}} = (F^{\text{grav}} - F^{\text{app}})/g\rho_w$ and

$$
\begin{aligned}
V_{\text{cav}} &= \frac{F^{\text{grav}} - F^{\text{app}}}{g\rho_w} - \frac{F^{\text{grav}}}{g\rho_{\text{iron}}} \\
&= \frac{6000\,\text{N} - 4000\,\text{N}}{(9.8\,\text{N/kg})(998\,\text{kg/m}^3)} - \frac{6000\,\text{N}}{(9.8\,\text{m/s}^2)(7.87 \times 10^3\,\text{kg/m}^3)} \\
&= 0.127\,\text{m}^3\,.
\end{aligned}
$$

The density of water was obtained from Table 15–2 of the text.

## 35

(a) Assume that the top surface of the slab is at the surface of the water and that the force of gravity is balanced by the buoyant force of the water on the slab. Then the area is a minimum. Also assume that the automobile is at the center of the ice surface. Then the ice slab does not tilt. Let $M$ be the mass of the automobile, $\rho_i$ be the density of ice, and $\rho_w$ be the density of water. Suppose the ice slab has area $A$ and thickness $h$. Since the volume of ice is $Ah$, the downward force of gravity on the automobile and ice is $(M + \rho_i Ah)g$. The buoyant force of the water is $\rho_w Ahg$, so the condition of equilibrium is $(M + \rho_i Ah)g - \rho_w Ahg = 0$ and

$$
A = \frac{M}{(\rho_w - \rho_i)h} = \frac{1100\,\text{kg}}{(998\,\text{kg/m}^3 - 0.917 \times 10^3\,\text{kg/m}^3)(0.30\,\text{m})} = 45\,\text{m}^2\,.
$$

The densities are found in Table 15–2 of the text.

(b) It matters where the car is placed since the ice tilts if the automobile is not at the center of its surface.

## 39

Use the equation of continuity. Let $v_1$ be the speed of the water in the hose and $v_2$ be its speed as it leaves one of the holes. Let $A_1$ be the cross-sectional area of the hose. If there are $N$ holes you may think of the water in the hose as $N$ tubes of flow, each of which goes through a single hole. The area of each tube of flow is $A_1/N$. If $A_2$ is the area of a hole the equation of continuity becomes $v_1 A_1/N = v_2 A_2$. Thus $v_2 = (A_1/N A_2)v_1 = (R^2/Nr^2)v_1$, where $R$ is the radius of the hose and $r$ is the radius of a hole. Thus

$$
v_2 = \frac{R^2}{Nr^2}v_1 = \frac{(0.95\,\text{cm})^2}{24(0.065\,\text{cm})^2}(0.91\,\text{m/s}) = 8.1\,\text{m/s}\,.
$$

## 43

(a) Use the equation of continuity: $A_1 v_1 = A_2 v_2$. Here $A_1$ is the area of the pipe at the top and $v_1$ is the speed of the water there; $A_2$ is the area of the pipe at the bottom and $v_2$ is the speed of the water there. Thus $v_2 = (A_1/A_2)v_1 = [(4.0\,\text{cm}^2)/(8.0\,\text{cm}^2)](5.0\,\text{m/s}) = 2.5\,\text{m/s}$.

(b) Use the Bernoulli equation: $P_1 + \frac{1}{2}\rho v_1^2 + \rho g h_1 = P_2 + \frac{1}{2}\rho v_2^2 + \rho g h_2$, where $\rho$ is the density of water, $h_1$ is its initial altitude, and $h_2$ is its final altitude. Thus

$$P_2 = P_1 + \frac{1}{2}\rho(v_1^2 - v_2^2) + \rho g(h_1 - h_2)$$

$$= 1.5 \times 10^5 \,\text{Pa} + \frac{1}{2}(998 \,\text{kg/m}^3)\left[(5.0 \,\text{m/s})^2 - (2.5 \,\text{m/s})^2\right]$$

$$+ (998 \,\text{kg/m}^3)(9.8 \,\text{m/s}^2)(10 \,\text{m})$$

$$= 2.6 \times 10^5 \,\text{Pa}.$$

The density of water was obtained from Table 15–2 of the text.

## 47

(a) Use the Bernoulli equation: $P_1 + \frac{1}{2}\rho v_1^2 + \rho g h_1 = P_2 + \frac{1}{2}\rho v_2^2 + \rho g h_2$, where $h_1$ is the height of the water in the tank, $P_1$ is the pressure there, and $v_1$ is the speed of the water there; $h_2$ is the altitude of the hole, $P_2$ is the pressure there, and $v_2$ is the speed of the water there. $\rho$ is the density of water. The pressure at the top of the tank and at the hole is atmospheric, so $P_1 = P_2$. Since the tank is large we may neglect the water speed at the top; it is much smaller than the speed at the hole. The Bernoulli equation then becomes $\rho g h_1 = \frac{1}{2}\rho v_2^2 + \rho g h_2$ and

$$v_2 = \sqrt{2g(h_1 - h_2)} = \sqrt{2(9.8 \,\text{m/s}^2)(0.30 \,\text{m})} = 2.42 \,\text{m/s}.$$

The flow rate is $A_2 v_2 = (6.5 \times 10^{-4} \,\text{m}^2)(2.42 \,\text{m/s}) = 1.6 \times 10^{-3} \,\text{m}^3/\text{s}$.

(b) Use the equation of continuity: $A_2 v_2 = A_3 v_3$, where $A_3 = A_2/2$ and $v_3$ is the water speed where the area of the stream is half its area at the hole. Thus $v_3 = (A_2/A_3)v_2 = 2v_2 = 4.84 \,\text{m/s}$. The water is in free fall and we wish to know how far it has fallen when its speed is doubled to $4.84 \,\text{m/s}$. Since the pressure is the same throughout the fall, $\frac{1}{2}\rho v_2^2 + \rho g h_2 = \frac{1}{2}\rho v_3^2 + \rho g h_3$. Thus

$$h_2 - h_3 = \frac{v_3^2 - v_2^2}{2g} = \frac{(4.84 \,\text{m/s})^2 - (2.42 \,\text{m/s})^2}{2(9.8 \,\text{m/s}^2)} = 0.90 \,\text{m}.$$

## 49

Use the Bernoulli equation: $P_\ell + \frac{1}{2}\rho v_\ell^2 = P_u + \frac{1}{2}\rho v_u^2$, where $P_\ell$ is the pressure at the lower surface, $P_u$ is the pressure at the upper surface, $v_\ell$ is the air speed at the lower surface, $v_u$ is the air speed at the upper surface, and $\rho$ is the density of air. The two tubes of flow are essentially at the same altitude. We want to solve for $v_u$ such that $P_\ell - P_u = 900 \,\text{Pa}$. That is,

$$v_u = \sqrt{\frac{2(P_\ell - P_u)}{\rho} + v_\ell^2} = \sqrt{\frac{2(900 \,\text{Pa})}{1.30 \,\text{kg/m}^3} + (110 \,\text{m/s})^2} = 116 \,\text{m/s}.$$

## 55

(a) The continuity equation yields $A V_A = B V_B$ and Bernoulli's equation yields $\frac{1}{2}\rho V_A^2 = \Delta P + \frac{1}{2}\rho V_B^2$, where $\Delta P = P_B - P_A$. The first equation gives $V_B = (A/B)V_B$. Use this to substitute for $V_B$ in the second equation. You should obtain $\frac{1}{2}\rho V_A^2 = \Delta P + \frac{1}{2}\rho(A/B)^2 V_B^2$. Solve for $V_A$. The result is

$$V_A = \sqrt{\frac{2\,\Delta P}{\rho\left(1 - \frac{A^2}{B^2}\right)}} = \sqrt{\frac{2B^2\,\Delta P}{\rho(B^2 - A^2)}}.$$

(b) Substitute values to obtain

$$V_A = \sqrt{\frac{2(32 \times 10^{-4}\,\text{m}^2)^2(41 \times 10^3\,\text{Pa} - 55 \times 10^3\,\text{Pa})}{(998\,\text{kg/m}^3)\left[(32 \times 10^{-4}\,\text{m}^2)^2 - (64 \times 10^{-4}\,\text{m}^2)^2\right]}} = 3.06\,\text{m/s}.$$

The density of water was obtained from Table 15–2 of the text. The flow rate is $Av = (64 \times 10^{-4}\,\text{m}^2)(3.06\,\text{m/s}) = 1.96 \times 10^{-2}\,\text{m}^3/\text{s}$.

# Chapter 16

## 3

(a) The motion repeats every 0.500 s so the period must be $T = 0.500$ s.

(b) The frequency is the reciprocal of the period: $f = 1/T = 1/(0.500 \text{ s}) = 2.00$ Hz.

(c) The angular frequency $\omega$ is $\omega = 2\pi f = 2\pi(2.00 \text{ Hz}) = 12.57$ rad/s.

(d) The angular frequency is related to the spring constant $k$ and the mass $m$ by $\omega = \sqrt{k/m}$. Solve for $k$: $k = m\omega^2 = (0.500 \text{ kg})(12.57 \text{ rad/s})^2 = 79.0$ N/m.

(e) Let $X$ be the amplitude. The maximum speed is $V = \omega X = (12.57 \text{ rad/s})(0.350 \text{ m}) = 4.40$ m/s.

(f) The maximum force is exerted when the displacement is a maximum and its magnitude is given by $F_{\max} = kX = (79.0 \text{ N/m})(0.350 \text{ m}) = 27.6$ N.

## 13

Use $V = \omega X = 2\pi f X$, where $V$ is the maximum speed, $\omega$ is the angular frequency, $f$ is the frequency, and $X$ is amplitude. The frequency is $(180)/(60 \text{ s}) = 3.0$ Hz and the amplitude is half the stroke, or 0.38 m. Thus $V = 2\pi(3.0 \text{ Hz})(0.38 \text{ m}) = 7.2$ m/s.

## 17

The magnitude of the maximum force that can be exerted by the surface must be less than $\mu^{\text{static}} N$ or else the block will not follow the surface in its motion. Here $\mu^{\text{static}}$ is the coefficient of static friction and $N$ is the magnitude of the normal force exerted by the surface on the block. Since the block does not accelerate vertically, you know that $N = mg$, where $m$ is the mass of the block. If the block follows the table and moves in simple harmonic motion, the magnitude of the maximum force exerted on it is given by $F_{\max} = mA = m\omega^2 X = m(2\pi f)^2 X$, where $A$ is the magnitude of the maximum acceleration, $\omega$ is the angular frequency, and $f$ is the frequency. The relationship $\omega = 2\pi f$ was used to obtain the last form.

Substitute $F_{\max} = m(2\pi f)^2 X$ and $N = mg$ into $F_{\max} < \mu_s N$ to obtain $m(2\pi f)^2 X < \mu_s mg$. The largest amplitude for which the block does not slip is

$$X = \frac{\mu^{\text{static}} g}{(2\pi f)^2} = \frac{(0.50)(9.8 \text{ m/s}^2)}{(2\pi \times 2.0 \text{ Hz})^2} = 0.031 \text{ m} .$$

A larger amplitude requires a larger force at the end points of the motion. The surface cannot supply the larger force and the block slips.

## 21

(a) The object oscillates about its equilibrium point, where the downward force of gravity is balanced by the upward force of the spring. If $\ell$ is the elongation of the spring at equilibrium, then $k\ell = mg$, where $k$ is the spring constant and $m$ is the mass of the object. Thus $k/m = g/\ell$ and

$$f = \frac{\omega}{2\pi} = \frac{1}{2\pi}\sqrt{\frac{k}{m}} = \frac{1}{2\pi}\sqrt{\frac{g}{\ell}}.$$

Now the equilibrium point is halfway between the points where the object is momentarily at rest. One of these points is where the spring is unstretched and the other is the lowest point, 10 cm below. Thus $\ell = 5.0$ cm $= 0.050$ m and

$$f = \frac{1}{2\pi}\sqrt{\frac{9.8\,\text{m/s}^2}{0.050\,\text{m}}} = 2.23\,\text{Hz}.$$

(b) Use conservation of energy. Take the zero of gravitational potential energy to be at the initial position of the object, where the spring is unstretched. Then both the initial potential and kinetic energies are zero. Take the $y$ axis to be positive in the downward direction and let $y = 0.080$ m. The potential energy when the object is at this point is $U = \frac{1}{2}ky^2 - mgy$. The energy equation becomes $0 = \frac{1}{2}ky^2 - mgy + \frac{1}{2}mv^2$. Solve for $v$:

$$v = \sqrt{2gy - \frac{k}{m}y^2} = \sqrt{2gy - \frac{g}{\ell}y^2}$$

$$= \sqrt{2(9.8\,\text{m/s}^2)(0.080\,\text{m}) - \left(\frac{9.8\,\text{m/s}^2}{0.050\,\text{m}}\right)(0.080\,\text{m})^2} = 0.56\,\text{m/s}.$$

(c) Let $m$ be the original mass and $\Delta m$ be the additional mass. The new angular frequency is $\omega' = \sqrt{k/(m + \Delta m)}$. This should be half the original angular frequency, or $\frac{1}{2}\sqrt{k/m}$. Solve $\sqrt{k/(m + \Delta m)} = \frac{1}{2}\sqrt{k/m}$ for $m$. Square both sides of the equation, then take the reciprocal to obtain $m + \Delta m = 4m$. This gives $m = \Delta m/3 = (300\,\text{g})/3 = 100\,\text{g}$.

(d) The equilibrium position is determined by the balancing of the gravitational and spring forces, so $ky = (m + \Delta m)g$. Thus $y = (m + \Delta m)g/k$. You will need to find the value of the spring constant $k$. Use $k = m\omega^2 = m(2\pi f)^2$. Then

$$y = \frac{(m + \Delta m)g}{m(2\pi f)^2} = \frac{(0.10\,\text{kg} + 0.30\,\text{kg})(9.8\,\text{m/s}^2)}{(0.10\,\text{kg})(2\pi \times 2.24\,\text{Hz})^2} = 0.20\,\text{m}.$$

This is measured from the initial position.

## 23

(a) Let

$$x_A = \frac{A}{2}\cos\left(\frac{2\pi t}{T}\right)$$

be the coordinate as a function of time for particle A and

$$x_B = \frac{A}{2}\cos\left(\frac{2\pi t}{T} + \frac{\pi}{6}\right)$$

be the coordinate as a function of time for particle B. Here $T$ is the period. Note that since the range of the motion is $A$, the amplitudes are both $A/2$. The arguments of the cosine functions are in radians.

Particle A is at one end of its path ($x_A = A/2$) when $t = 0$. Particle B is at $A/2$ when $2\pi t/T + \pi/6 = 0$ or $t = -T/12$. That is, particle A lags particle B by one-twelfth a period. We want the coordinates of the particles 0.50 s later; that is, at $t = 0.50$ s. They are

$$x_A = \frac{A}{2} \cos\left(\frac{2\pi \times 0.50\,\text{s}}{1.5\,\text{s}}\right) = -0.250A$$

and

$$x_B = \frac{A}{2} \cos\left(\frac{2\pi \times 0.50\,\text{s}}{1.5\,\text{s}} + \frac{\pi}{6}\right) = -0.433A.$$

Their separation at that time is $x_A - x_B = -0.250A + 0.433A = 0.183A$.

(b) The velocities of the particles are given by

$$v_A = \frac{dx_A}{dt} = \frac{\pi A}{T} \sin\left(\frac{2\pi t}{T}\right)$$

and

$$v_B = \frac{dx_B}{dt} = \frac{\pi A}{T} \sin\left(\frac{2\pi t}{T} + \frac{\pi}{6}\right).$$

Evaluate these expressions for $t = 0.50$ s. You will find they are both negative, indicating that the particles are moving in the same direction.

## 27

We wish to find the effective spring constant for the combination of springs shown in Fig. 16–34. We do this by finding the magnitude $F$ of the force exerted on the mass when the total elongation of the springs is $\Delta x$. Then $k_{\text{eff}} = F/\Delta x$.

Suppose the left-hand spring is elongated by $\Delta x_L$ and the right-hand spring is elongated by $\Delta x_R$. The left-hand spring exerts a force of magnitude $k\,\Delta x_L$ on the right-hand spring and the right-hand spring exerts a force of magnitude $k\,\Delta x_R$ on the left-hand spring. According to Newton's third law these must be equal, so $\Delta x_L = \Delta x_R$. The two elongations must be the same and the total elongation is twice the elongation of either spring: $\Delta x = 2\Delta x_L$. The left-hand spring exerts a force on the block and its magnitude is $F = k\,\Delta x_L$. Thus

$$k_{\text{eff}} = k\,\frac{\Delta x_L}{2\Delta x_R} = \frac{k}{2}.$$

The block behaves as if it were subject to the force of a single spring with spring constant $k/2$. To find the frequency of its motion replace $k_{\text{eff}}$ in $f = (1/2\pi)\sqrt{k_{\text{eff}}/m}$ with $k/2$ to obtain

$$f = \frac{1}{2\pi}\sqrt{\frac{k}{2m}}.$$

## 31

(a) Take the angular displacement of the wheel to be $\theta = \Theta \cos(2\pi t/T)$, where $\Theta$ is the amplitude and $T$ is the period. Differentiate with respect to time to find the angular velocity:

$\Omega = -(2\pi/T)\Theta \sin(2\pi t/T)$. The symbol $\Omega$ is used for the angular velocity of the wheel so it is not confused with the angular frequency $\omega$. The maximum angular speed is

$$\Omega_{max} = \frac{2\pi\Theta}{T} = \frac{(2\pi)(\pi\,\text{rad})}{0.500\,\text{s}} = 39.5\,\text{rad/s}.$$

(b) When $\theta = \pi/2$, then $\theta/\Theta = 1/2$, $\cos(2\pi t/T) = 1/2$, and

$$\sin(2\pi t/T) = \sqrt{1 - \cos^2(2\pi t/T)} = \sqrt{1 - (1/2)^2} = \sqrt{3}/2,$$

where the trigonometric identity $\cos^2 A + \sin^2 A = 1$ was used. Thus

$$\Omega = -\frac{2\pi}{T}\Theta \sin\left(\frac{2\pi t}{T}\right) = -\left(\frac{2\pi}{0.500\,\text{s}}\right)(\pi\,\text{rad})\left(\frac{\sqrt{3}}{2}\right) = -34.2\,\text{rad/s}.$$

The rotational speed is 34.2 rad/s.

(c) The angular acceleration is

$$\alpha = \frac{d^2\theta}{dt^2} = -\left(\frac{2\pi}{T}\right)^2 \Theta \cos(2\pi t/T) = -\left(\frac{2\pi}{T}\right)^2 \theta.$$

When $\theta = \pi/4$,

$$\alpha = -\left(\frac{2\pi}{0.500\,\text{s}}\right)^2 \left(\frac{\pi}{4}\right) = -124\,\text{rad/s}^2.$$

The magnitude is $124\,\text{rad/s}^2$.

## 37

(a) The period of the pendulum is given by $T = 2\pi\sqrt{I/mgd}$, where $I$ is its rotational inertia, $m$ is its mass, and $d$ is the distance from the center of mass to the pivot point. The rotational inertia of a rod pivoted at its center is $mL^2/12$ and, according to the parallel-axis theorem, its rotational inertia when it is pivoted a distance $d$ from the center is $I = mL^2/12 + md^2$. Thus

$$T = 2\pi\sqrt{\frac{m(L^2/12 + d^2)}{mgd}} = 2\pi\sqrt{\frac{L^2 + 12d^2}{12gd}}.$$

(b) $(L^2 + 12d^2)/12gd$, considered as a function of $d$, has a minimum at $d = L/\sqrt{12}$, so the period increases as $d$ decreases if $d < L/\sqrt{12}$ and decreases as $d$ decreases if $d > L/\sqrt{12}$. You can prove this by setting the derivative of $T$ (or $T^2$) with respect to $d$ equal to zero and solving for $d$.

(c) $L$ occurs only in the numerator of the expression for the period, so $T$ increases as $L$ increases.

(d) The period does not depend on the mass of the pendulum, so $T$ does not change when $m$ increases.

**41**

If the torque exerted by the spring on the rod is proportional to the angle of rotation of the rod and if the torque tends to pull the rod toward its equilibrium orientation, then the rod will oscillate in simple harmonic motion. If $\tau = -C\theta$, where $\tau$ is the torque, $\theta$ is the angle of rotation, and $C$ is a constant of proportionality, then the angular frequency of oscillation is $\omega = \sqrt{C/I}$ and the period is $T = 2\pi/\omega = 2\pi\sqrt{I/C}$, where $I$ is the rotational inertia of the rod. The plan is to find the torque as a function of $\theta$ and identify the constant $C$ in terms of given quantities. This immediately gives the period in terms of given quantities.

Let $\ell_0$ be the distance from the pivot point to the wall. This is also the equilibrium length of the spring. Suppose the rod turns through the angle $\theta$, with the left end moving away from the wall. This end is now $(L/2)\sin\theta$ further from the wall and has moved $(L/2)(1 - \cos\theta)$ to the right. The length of the spring is now $\sqrt{(L/2)^2(1 - \cos\theta)^2 + [\ell_0 + (L/2)\sin\theta]^2}$. If the angle $\theta$ is small we may approximate $\cos\theta$ with 1 and $\sin\theta$ with $\theta$ in radians. Then the length of the spring is given by $\ell_0 + L\theta/2$ and its elongation is $\Delta x = L\theta/2$. The force it exerts on the rod has magnitude $F = k\,\Delta x = kL\theta/2$. Since $\theta$ is small we may approximate the torque exerted by the spring on the rod by $\tau = -FL/2$, where the pivot point was taken as the origin. Thus $\tau = -(kL^2/4)\theta$. The constant of proportionality $C$ that relates the torque and angle of rotation is $C = kL^2/4$.

The rotational inertia for a rod pivoted at its center is $I = mL^2/12$, where $m$ is its mass. See Table 11–2. Thus the period of oscillation is

$$T = 2\pi\sqrt{\frac{I}{C}} = 2\pi\sqrt{\frac{mL^2/12}{kL^2/4}} = 2\pi\sqrt{\frac{m}{3k}}\,.$$

**49**

When the block is at the end of its path and is momentarily stopped, its displacement is equal to the amplitude and all the mechanical energy is potential in nature. If the spring potential energy is taken to be zero when the block is at its equilibrium position, then

$$E^{\text{mec}} = \frac{1}{2}kX^2 = \frac{1}{2}(1.3 \times 10^2\,\text{N/m})(0.024\,\text{m})^2 = 3.7 \times 10^{-2}\,\text{J}\,.$$

**55**

(a) Assume the bullet becomes embedded and moves with the block before the block moves a significant distance. Then the momentum of the bullet-block system is conserved during the collision. Let $m$ be the mass of the bullet, $M$ be the mass of the block, $v_0$ be the initial speed of the bullet, and $v$ be the final speed of the block and bullet, just after the collision. Conservation of momentum yields $mv_0 = (m + M)v$, so

$$v = \frac{mv_0}{m + M} = \frac{(0.050\,\text{kg})(150\,\text{m/s})}{0.050\,\text{kg} + 4.0\,\text{kg}} = 1.85\,\text{m/s}\,.$$

When the block is in its initial position the spring and gravitational forces balance, so the spring is elongated by $Mg/k$. After the collision, however, the block oscillates with simple harmonic motion about the point where the spring and gravitational forces balance with the bullet embedded. At this point the spring is elongated a distance $\ell = (M + m)g/k$, which is not the same as the initial elongation.

Mechanical energy is conserved during the oscillation. At the initial position, just after the bullet is embedded, the kinetic energy is $\frac{1}{2}(M + m)v^2$ and the elastic potential energy is $\frac{1}{2}k(Mg/k)^2$. Take the gravitational potential energy to be zero at this point. When the block and bullet reach the highest point in their motion the kinetic energy is zero. The block is then a distance $Y$ above the position where the spring and gravitational forces balance. Note that $Y$ is the amplitude of the motion. The spring is compressed by $Y - \ell$, so the elastic potential energy is $\frac{1}{2}k(Y - \ell)^2$. The gravitational potential energy is $(M + m)gY$. Conservation of mechanical energy yields

$$\frac{1}{2}(M + m)v^2 + \frac{1}{2}k\left(\frac{Mg}{k}\right)^2 = \frac{1}{2}k(Y - \ell)^2 + (M + m)gY.$$

Substitute $\ell = (M + m)g/k$. A little algebra reveals that

$$Y = \sqrt{\frac{(m + M)v^2}{k} - \frac{mg^2}{k^2}(2M + m)}$$

$$= \sqrt{\frac{(0.050\,\text{kg} + 4.0\,\text{kg})(1.85\,\text{m/s})^2}{500\,\text{N/m}} - \frac{(0.050\,\text{kg})(9.8\,\text{m/s}^2)^2}{(500\,\text{N/m})^2}[2(4.0\,\text{kg}) + 0.050\,\text{kg}]}$$

$$= 0.166\,\text{m}$$

(b) The original energy of the bullet is $E_0 = \frac{1}{2}mv_0^2 = \frac{1}{2}(0.050\,\text{kg})(150\,\text{m/s})^2 = 563\,\text{J}$. The kinetic energy of the bullet-block system just after the collision is $K = \frac{1}{2}(m + M)v^2 = \frac{1}{2}(0.050\,\text{kg} + 4.0\,\text{kg})(1.85\,\text{m/s})^2 = 6.94\,\text{J}$. Since the block does not move significantly during the collision the elastic and gravitational potential energies do not change. Thus $K$ is the energy that is transferred. The ratio is $K/E_0 = (6.94\,\text{J})/(563\,\text{J}) = 0.0123$ or $1.23\%$.

## 59

(a) You want to solve $e^{-bt/2m} = 1/3$ for $t$. Take the natural logarithm of both sides to obtain $-bt/2m = \ln(1/3)$. Now solve for $t$: $t = -(2m/b)\ln(1/3) = (2m/b)\ln 3$, where the sign was reversed when the argument of the logarithm was replaced by its reciprocal. Thus

$$t = \frac{2(1.50\,\text{kg})}{0.230\,\text{kg/s}}\ln 3 = 14.3\,\text{s}.$$

(b) The angular frequency is

$$\omega' = \sqrt{\frac{k}{m} - \frac{b^2}{4m^2}} = \sqrt{\frac{8.00\,\text{N/m}}{1.50\,\text{kg}} - \frac{(0.230\,\text{kg/s})^2}{4(1.50\,\text{kg})^2}} = 2.31\,\text{rad/s}.$$

The period is $T = 2\pi/\omega' = (2\pi)/(2.31\,\text{rad/s}) = 2.72\,\text{s}$ and the number of oscillations is $t/T = (14.3\,\text{s})/(2.72\,\text{s}) = 5.27$.

# Chapter 17

## 3

(a) The motion from maximum displacement to zero is one-fourth of a cycle so 0.170 s is one-fourth of a period. The period is $T = 4(0.170 \text{ s}) = 0.680 \text{ s}$.

(b) The frequency is the reciprocal of the period: $f = 1/T = 1/(0.680 \text{ s}) = 1.47 \text{ Hz}$.

(c) A sinusoidal wave travels one wavelength in one period:

$$v^{\text{wave}} = \lambda/T = (1.40 \text{ m})/(0.680 \text{ s}) = 2.06 \text{ m/s}.$$

## 7

(a) Write the expression for the displacement in the form $y(x,t) = Y \sin(kx - \omega t)$. A negative sign is used before the $\omega t$ term in the argument of the sine function because the wave is traveling in the positive $x$ direction. The wave number $k$ is $k = 2\pi/\lambda = 2\pi/(0.10 \text{ m}) = 62.8 \text{ m}^{-1}$ and the angular frequency is $\omega = 2\pi f = 2\pi(400 \text{ Hz}) = 2510 \text{ rad/s}$. Here $\lambda$ is the wavelength and $f$ is the frequency. The amplitude is $Y = 2.0$ cm. Thus

$$y(x,t) = (2.0 \text{ cm}) \sin[(62.8 \text{ m}^{-1})x - (2510 \text{ s}^{-1})t].$$

(b) The speed of a point on the cord is given by $v_y^{\text{cord}}(x,t) = \partial y/\partial t = -\omega Y \cos(kx - \omega t)$ and the maximum speed is $V^{\text{cord}} = \omega Y = (2510 \text{ rad/s})(0.020 \text{ m}) = 50 \text{ m/s}$.

(c) The wave speed is $v^{\text{wave}} = \lambda/T = \omega/k = (2510 \text{ rad/s})/(62.8 \text{ m}^{-1}) = 40 \text{ m/s}$.

## 11

The wave speed $v^{\text{wave}}$ is given by $v^{\text{wave}} = \sqrt{F^{\text{tension}}/\mu}$, where $F^{\text{tension}}$ is the tension in the rope and $\mu$ is the linear mass density of the rope. The linear mass density is the mass per unit length of rope: $\mu = m/L = (0.0600 \text{ kg})/(2.00 \text{ m}) = 0.0300 \text{ kg/m}$. Thus

$$v^{\text{wave}} = \sqrt{\frac{500 \text{ N}}{0.0300 \text{ kg/m}}} = 129 \text{ m/s}.$$

## 15

Write the string displacement in the form $y = Y \sin(kx + \omega t)$. The positive sign is used since the wave is traveling in the negative $x$ direction. The frequency is $f = 100 \text{ Hz}$, so the angular frequency is $\omega = 2\pi f = 2\pi(100 \text{ Hz}) = 628 \text{ rad/s}$. The wave speed is given by $v^{\text{wave}} = \sqrt{F^{\text{tension}}/\mu}$, where $F^{\text{tension}}$ is the tension in the string and $\mu$ is the linear mass density of the string, so the wavelength is $\lambda = v^{\text{wave}}/f = \sqrt{F^{\text{tension}}/\mu}/f$ and the wave number is

$$k = \frac{2\pi}{\lambda} = 2\pi f \sqrt{\frac{\mu}{F^{\text{tension}}}} = 2\pi(100 \text{ Hz})\sqrt{\frac{0.50 \text{ kg/m}}{10 \text{ N}}} = 141 \text{ m}^{-1}.$$

The amplitude is $Y = 0.12\,\text{mm}$. Thus

$$y = (0.12\,\text{mm})\sin[(141\,\text{m}^{-1})x + (628\,\text{s}^{-1})t]\,.$$

## 19

(a) Read the amplitude from the graph. It is about 5.0 cm.

(b) Read the wavelength from the graph. The curve crosses $y = 0$ at about $x = 15$ cm and again with the same slope at about $x = 55$ cm, so $\lambda = 55\,\text{cm} - 15\,\text{cm} = 40\,\text{cm} = 0.40\,\text{m}$.

(c) The wave speed is $v^{\text{wave}} = \sqrt{F^{\text{tension}}/\mu}$, where $F^{\text{tension}}$ is the tension in the string and $\mu$ is the linear mass density of the string. Thus

$$v^{\text{wave}} = \sqrt{\frac{3.6\,\text{N}}{25 \times 10^{-3}\,\text{kg/m}}} = 12\,\text{m/s}\,.$$

(d) The frequency is $f = v^{\text{wave}}/\lambda = (12\,\text{m/s})/(0.40\,\text{m}) = 30\,\text{Hz}$ and the period is $T = 1/f = 1/(30\,\text{Hz}) = 0.033$ s.

(e) The maximum string speed is $V_y = \omega Y = 2\pi f Y = 2\pi(30\,\text{Hz})(5.0\,\text{cm}) = 940\,\text{cm/s} = 9.4\,\text{m/s}$.

(f) Assume the string displacement has the form $y(x,t) = Y\sin(kx + \omega t + \phi_0)$. A positive sign appears in the argument of the trigonometric function because the wave is moving in the negative $x$ direction. The amplitude is $Y = 5.0 \times 10^{-2}$ m, the angular frequency is $\omega = 2\pi f = 2\pi(30\,\text{Hz}) = 190\,\text{rad/s}$, and the wave number is $k = 2\pi/\lambda = 2\pi/(0.40\,\text{m}) = 16\,\text{m}^{-1}$. According to the graph, the displacement at $x = 0$ and $t = 0$ is $4.0 \times 10^{-2}$ m. The formula for the displacement gives $y(0,0) = Y\sin\phi_0$. We wish to select $\phi_0$ so that $5.0 \times 10^{-2}\sin\phi_0 = 4.0 \times 10^{-2}$. The solution is either 0.93 rad or 2.21 rad. In the first case the function has a positive slope at $x = 0$ and matches the graph. In the second case it has negative slope and does not match the graph. We select $\phi_0 = 0.93$ rad. The expression for the displacement is

$$y(x,t) = (5.0 \times 10^{-2}\,\text{m})\sin\left[(16\,\text{m}^{-1})x + (190\,\text{s}^{-1})t + 0.93\right]\,.$$

## 23

(a) The wave speed at any point on the rope is given by $v^{\text{wave}} = \sqrt{F^{\text{tension}}/\mu}$, where $F^{\text{tension}}$ is the tension at that point and $\mu$ is the linear mass density. Because the rope is hanging the tension and speed vary from point to point. Consider a point on the rope a distance $y$ from the bottom end. The forces acting on it are the weight of the rope below it, pulling down, and the tension, pulling up. Since the rope is in equilibrium these balance. The weight of the rope below is given by $\mu g y$, so the tension is $F^{\text{tension}} = \mu g y$. The wave speed is $v^{\text{wave}} = \sqrt{\mu g y/\mu} = \sqrt{g y}$.

(b) The time $dt$ for the wave to move past a length $dy$, a distance $y$ from the bottom end, is $dt = dy/v^{\text{wave}} = dy/\sqrt{g y}$ and the total time for the wave to move the entire length of the rope is

$$t = \int_0^L \frac{dy}{\sqrt{g y}} = 2\sqrt{\frac{y}{g}}\,\bigg|_0^L = 2\sqrt{\frac{L}{g}}\,.$$

## 25

The displacement of the string is given by $y' = Y\sin(kx - \omega t) + Y\sin(kx - \omega t + \phi_0) = 2Y\cos(\frac{1}{2}\phi_0)\sin(kx - \omega t + \frac{1}{2}\phi_0)$, where $\phi_0 = \pi/2$. Eq. 17–36 was used to sum the waves. The amplitude is $Y' = 2Y\cos(\frac{1}{2}\phi_0) = 2Y\cos(\pi/4) = 1.41Y$.

## 31

(a) The wave speed is given by $v^{\text{wave}} = \sqrt{F^{\text{tension}}/\mu}$, where $F^{\text{tension}}$ is the tension in the string and $\mu$ is the linear mass density of the string. Since the mass density is the mass per unit length, $\mu = M/L$, where $M$ is the mass of the string and $L$ is its length. Thus

$$v^{\text{wave}} = \sqrt{\frac{F^{\text{tension}}L}{M}} = \sqrt{\frac{(96.0\,\text{N})(8.40\,\text{m})}{0.120\,\text{kg}}} = 82.0\,\text{m/s}.$$

(b) The longest possible wavelength $\lambda$ for a standing wave is related to the length of the string by $L = \lambda/2$, so $\lambda = 2L = 2(8.40\,\text{m}) = 16.8\,\text{m}$.

(c) The frequency is $f = v^{\text{wave}}/\lambda = (82.0\,\text{m/s})/(16.8\,\text{m}) = 4.88\,\text{Hz}$.

## 35

(a) The resonant wavelengths are given by $\lambda = 2L/n$, where $L$ is the length of the string and $n$ is an integer, and the resonant frequencies are given by $f = v^{\text{wave}}/\lambda = nv^{\text{wave}}/2L$, where $v^{\text{wave}}$ is the wave speed. Suppose the lower frequency is associated with the integer $n$. Then since there are no resonant frequencies between, the higher frequency is associated with $n + 1$. That is, $f_1 = nv^{\text{wave}}/2L$ is the lower frequency and $f_2 = (n + 1)v^{\text{wave}}/2L$ is the higher. The ratio of the frequencies is

$$\frac{f_2}{f_1} = \frac{n + 1}{n}.$$

The solution for $n$ is

$$n = \frac{f_1}{f_2 - f_1} = \frac{315\,\text{Hz}}{420\,\text{Hz} - 315\,\text{Hz}} = 3.$$

Note that $v^{\text{wave}} = 2Lf_1/n$, so the lowest possible resonant frequency is $f = v^{\text{wave}}/2L = f_1/n = (315\,\text{Hz})/3 = 105\,\text{Hz}$.

(b) The longest possible wavelength is $\lambda = 2L$. If $f$ is the lowest possible frequency then $v^{\text{wave}} = \lambda f = 2Lf = 2(0.75\,\text{m})(105\,\text{Hz}) = 158\,\text{m/s}$.

## 39

(a) Since the standing wave has three loops the string is three half-wavelengths long. If $L$ is the length of the string and $\lambda$ is the wavelength, then $L = 3\lambda/2$, or $\lambda = 2L/3$. If $v^{\text{wave}}$ is the wave speed, then the frequency is $f = v^{\text{wave}}/\lambda = 3v^{\text{wave}}/2L = 3(100\,\text{m/s})/2(3.0\,\text{m}) = 50\,\text{Hz}$.

(b) The waves have the same amplitude, the same angular frequency, and the same wave number, but they travel in opposite directions. Take them to be $y_1 = Y\sin(kx - \omega t)$ and $y_2 = Y\sin(kx + \omega t)$. The amplitude $Y$ is half the maximum displacement of the standing wave, or $5.0 \times 10^{-3}\,\text{m}$. The angular frequency is the same as that of the standing wave, or $\omega = 2\pi f = 2\pi(50\,\text{Hz}) = 314\,\text{rad/s}$. The wave number is $k = 2\pi/\lambda = 2\pi/(2.0\,\text{m}) = 3.14\,\text{m}^{-1}$. Thus

$$y_1 = (5.0 \times 10^{-3}\,\text{m})\sin[(3.14\,\text{m}^{-1})x - (314\,\text{s}^{-1})t]$$

and

$$y_2 = (5.0 \times 10^{-3}\,\text{m})\sin[(3.14\,\text{m}^{-1})x + (314\,\text{s}^{-1})t].$$

## 43

(a) The angular frequency is $\omega = 8.0\pi/2 = 4.0\pi\,\text{rad/s}$, so the frequency is $f = \omega/2\pi = (4.0\pi\,\text{rad/s})/2\pi = 2.0\,\text{Hz}$.

(b) Since the wave number is $k = 2.0\pi/2 = 1.0\pi\,\text{m}^{-1}$, the wavelength is $\lambda = 2\pi/k = 2\pi/(1.0\pi\,\text{m}^{-1}) = 2.0\,\text{m}$.

(c) The wave speed is $v^{\text{wave}} = \lambda f = (2.0\,\text{m})(2.0\,\text{Hz}) = 4.0\,\text{m/s}$.

(d) You need to add two cosine functions. First convert them to sine functions using $\cos\alpha = \sin(\alpha + \pi/2)$, then apply Eq. 17–36. Here are the steps:

$$\cos\alpha + \cos\beta = \sin\left(\alpha + \frac{\pi}{2}\right) + \sin\left(\beta + \frac{\pi}{2}\right) = 2\sin\left(\frac{\alpha+\beta+\pi}{2}\right)\cos\left(\frac{\alpha-\beta}{2}\right)$$

$$= 2\cos\left(\frac{\alpha+\beta}{2}\right)\cos\left(\frac{\alpha-\beta}{2}\right).$$

Let $\alpha = kx$ and $\beta = \omega t$. Then

$$Y\cos(kx + \omega t) + Y\cos(kx - \omega t) = 2Y\cos(kx)\cos(\omega t).$$

Nodes occur where $\cos(kx) = 0$ or $kx = n\pi + \pi/2$, where $n$ is an integer (including zero). Since $k = 1.0\pi\,\text{m}^{-1}$, this means $x = (n + \frac{1}{2})(1.0\,\text{m})$. Nodes occur at $x = 0.50\,\text{m}$, $1.5\,\text{m}$, $2.5\,\text{m}$, etc.

(e) The displacement is a maximum where $\cos(kx) = \pm 1$. This means $kx = n\pi$, where $n$ is an integer. Thus $x = n(1.0\,\text{m})$. Antinodes occur at $x = 0$, $1.0\,\text{m}$, $2.0\,\text{m}$, $3.0\,\text{m}$, etc.

## 45

Consider an infinitesimal segment of a string oscillating in a standing wave pattern. Its length is $dx$ and its mass is $dm = \mu\,dx$, where $\mu$ is its linear mass density. If it is moving with speed $v$ its kinetic energy is $dK = \frac{1}{2}v^2\,dm = \frac{1}{2}\mu v^2\,dx$. If the segment is located at $x$ its displacement at time $t$ is $y = 2Y\sin(kx)\cos(\omega t)$ and its speed is $v = |\partial y/\partial t| = |-2\omega Y\sin(kx)\sin(\omega t)|$, so its kinetic energy is

$$dK = \frac{1}{2}\left(4\mu\omega^2 Y^2\right)\sin^2(kx)\sin^2(\omega t)\,dx = 2\mu\omega^2 Y^2\sin^2(kx)\sin^2(\omega t)\,dx.$$

Here $Y$ is the amplitude of either one of the traveling waves that combine to form the standing wave.

The infinitesimal segment has maximum kinetic energy when $\sin^2(\omega t) = 1$ and the maximum kinetic energy is given by

$$dK_{max} = 2\mu\omega^2 Y^2 \sin^2(kx)\,dx\,.$$

Note that every portion of the string has its maximum kinetic energy at the same time although the values of these maxima are different for different parts of the string.

If the string is oscillating with $n$ loops, the length of string in any one loop is $L/n$ and the kinetic energy the loop is given by the integral

$$K_{max} = 2\mu\omega^2 Y^2 \int_0^{L/n} \sin^2(kx)\,dx\,.$$

Use the trigonometric identity $\sin^2(kx) = \frac{1}{2}[1 + 2\cos(2kx)]$ to write this

$$K_{max} = \mu\omega^2 Y^2 \int_0^{L/n} [1 + 2\cos(2kx)]\,dx = \mu\omega^2 Y^2 \left[\frac{L}{n} + \frac{1}{k}\sin\frac{2kL}{n}\right]\,.$$

For a standing wave of $n$ loops the wavelength is $\lambda = 2L/n$ and the wave number is $k = 2\pi/\lambda = n\pi/L$, so $2kL/n = 2\pi$ and $\sin(2kL/n) = 0$, no matter what the value of $n$. Thus

$$K_{max} = \frac{\mu\omega^2 Y^2 L}{n}\,.$$

To obtain the expression given in the problem statement, first make the substitutions $\omega = 2\pi f$ and $L/n = \lambda/2$, where $f$ is the frequency and $\lambda$ is the wavelength. This produces $K_{max} = 2\pi^2 \mu Y^2 f^2 \lambda$. Now substitute the wave speed $v^{wave}$ for $f\lambda$ to obtain $K_m = 2\pi^2 \mu Y^2 f v^{wave}$.

## 47

(a) The frequency of the wave is the same for both sections of the wire. The wave speed and wavelength, however, are both different in different sections. Suppose there are $n_1$ loops in the aluminum section of the wire. Then $L_1 = n_1\lambda_1/2 = n_1 v_1^{wave}/2f$, where $\lambda_1$ is the wavelength and $v_1^{wave}$ is the wave speed in that section. The substitution $\lambda_1 = v_1^{wave}/f$, where $f$ is the frequency, was made. Thus $f = n_1 v_1^{wave}/2L_1$. A similar expression holds for the steel section: $f = n_2 v_2^{wave}/2L_2$. Since the frequency is the same for the two sections, $n_1 v_1^{wave}/L_1 = n_2 v_2^{wave}/L_2$. Now the wave speed in the aluminum section is given by $v_1^{wave} = \sqrt{F^{tension}/\mu_1}$, where $\mu_1$ is the linear mass density of the aluminum wire and $F^{tension}$ is the tension in the wire. The mass of aluminum in the wire is given by $m_1 = \rho_1 A L_1$, where $\rho_1$ is the mass density (mass per unit volume) for aluminum and $A$ is the cross-sectional area of the wire. Thus $\mu_1 = \rho_1 A L_1/L_1 = \rho_1 A$ and $v_1^{wave} = \sqrt{F^{tension}/\rho_1 A}$. A similar expression holds for the wave speed in the steel section: $v_2^{wave} = \sqrt{F^{tension}/\rho_2 A}$. Note that the cross-sectional area and the tension are the same for the two sections.

The equality of the frequencies for the two sections now leads to $n_1/L_1\sqrt{\rho_1} = n_2/L_2\sqrt{\rho_2}$, where $A$ has been canceled from both sides. The ratio of the integers is

$$\frac{n_2}{n_1} = \frac{L_2\sqrt{\rho_2}}{L_1\sqrt{\rho_1}} = \frac{(0.866\,\text{m})\sqrt{7.80 \times 10^3\,\text{kg/m}^3}}{(0.600\,\text{m})\sqrt{2.60 \times 10^3\,\text{kg/m}^3}} = 2.5\,.$$

The smallest integers that have this ratio are $n_1 = 2$ and $n_2 = 5$. The frequency is $f = n_1 v_1^{\text{wave}}/2L_1 = (n_1/2L_1)\sqrt{F^{\text{tension}}/\rho_1 A}$. The tension is provided by the hanging block and is $F^{\text{tension}} = mg$, where $m$ is the mass of the block. Thus

$$f = \frac{n_1}{2L_1}\sqrt{\frac{mg}{\rho_1 A}} = \frac{2}{2(0.600\,\text{m})}\sqrt{\frac{(10.0\,\text{kg})(9.8\,\text{m/s}^2)}{(2.60 \times 10^3\,\text{kg/m}^3)(1.00 \times 10^{-6}\,\text{m}^2)}} = 324\,\text{Hz}.$$

(b) The standing wave pattern has two loops in the aluminum section and five loops in the steel section, or seven loops in all. There are eight nodes, counting the end points.

# Chapter 18

## 3

(a) The time for the sound to travel from the kicker to a spectator is given by $d/v^{\text{wave}}$, where $d$ is the distance and $v^{\text{wave}}$ is the speed of sound. The time for light to travel the same distance is given by $d/c$, where $c$ is the speed of light. The delay between seeing and hearing the kick is $\Delta t = (d/v^{\text{wave}}) - (d/c)$. The speed of light is so much greater than the speed of sound that the delay can be approximated by $\Delta t = d/v^{\text{wave}}$. This means $d = v^{\text{wave}} \Delta t$. The distance from the kicker to the first spectator is $d_1 = v^{\text{wave}} \Delta t_1 = (343 \text{ m/s})(0.23 \text{ s}) = 79 \text{ m}$. The distance from the kicker to the second spectator is $d_2 = v^{\text{wave}} \Delta t_2 = (343 \text{ m/s})(0.12 \text{ s}) = 41 \text{ m}$.

(b) Lines from the kicker to each spectator and from one spectator to the other form a right triangle with the line joining the spectators as the hypotenuse, so the distance between the spectators is
$$D = \sqrt{d_1^2 + d_2^2} = \sqrt{(79 \text{ m})^2 + (41 \text{ m})^2} = 89 \text{ m}.$$

## 7

Let $t_f$ be the time for the stone to fall to the water and $t_s$ be the time for the sound of the splash to travel from the water to the top of the well. Then, the total time elapsed from dropping the stone to hearing the splash is $t = t_f + t_s$. If $d$ is the depth of the well, then the kinematics of free fall gives $d = \frac{1}{2}gt_f^2$, or $t_f = \sqrt{2d/g}$. The sound travels at a constant speed $v^{\text{wave}}$, so $d = v^{\text{wave}} t_s$, or $t_s = d/v^{\text{wave}}$. Thus the total time is $t = \sqrt{2d/g} + d/v^{\text{wave}}$. This equation is to be solved for $d$. Rewrite it as $\sqrt{2d/g} = t - d/v^{\text{wave}}$ and square both sides to obtain $2d/g = t^2 - 2(td/v^{\text{wave}}) + d^2/(v^{\text{wave}})^2$. Now multiply by $g(v^{\text{wave}})^2$ and rearrange to get $gd^2 - 2v^{\text{wave}}(gt + v^{\text{wave}})d + g(v^{\text{wave}})^2 t^2 = 0$. This is a quadratic equation for $d$. Its solutions are

$$d = \frac{2v^{\text{wave}}(gt + v^{\text{wave}}) \pm \sqrt{4(v^{\text{wave}})^2(gt + v^{\text{wave}})^2 - 4g^2(v^{\text{wave}})^2 t^2}}{2g}.$$

The physical solution must yield $d = 0$ for $t = 0$, so we use the negative sign in front of the square root. Once values are substituted the result $d = 40.7 \text{ m}$ is obtained.

## 13

Let $L_1$ be the distance from the closer speaker to the listener. The distance from the other speaker to the listener is $L_2 = \sqrt{L_1^2 + d^2}$, where $d$ is the distance between the speakers. The phase difference at the listener is $\Delta\phi = 2\pi(L_2 - L_1)/\lambda$, where $\lambda$ is the wavelength.

(a) For a minimum in intensity at the listener, $\Delta\phi = (2n + 1)\pi$, where $n$ is an integer. Thus $\lambda = 2(L_2 - L_1)/(2n + 1)$. The frequency is

$$f = \frac{v^{\text{wave}}}{\lambda} = \frac{(2n + 1)v^{\text{wave}}}{2\left[\sqrt{L_1^2 + d^2} - L_1\right]} = \frac{(2n + 1)(343 \text{ m/s})}{2\left[\sqrt{(3.75 \text{ m})^2 + (2.00 \text{ m})^2} - 3.75 \text{ m}\right]} = (2n + 1)(343 \text{ Hz}).$$

Now $20,000/343 = 58.3$, so $2n+1$ must range from 0 to 58 for the frequency to be in the audible range. This means $n$ ranges from 1 to 28 and $f = 1029, 1715, 2401, \ldots, 19550$ Hz.

(b) For a maximum in intensity at the listener, $\Delta\phi = 2n\pi$, where $n$ is any positive integer. Thus

$$\lambda = \frac{1}{n}\left[\sqrt{L_1^2 + d^2} - L_1\right]$$

and

$$f = \frac{v^{\mathrm{wave}}}{\lambda} = \frac{nv^{\mathrm{wave}}}{\sqrt{L_1^2 + d^2} - L_1} = \frac{n(343\ \mathrm{m/s})}{\sqrt{(3.75\ \mathrm{m})^2 + (2.00\ \mathrm{m})^2} - 3.75\ \mathrm{m}} = n(686\ \mathrm{Hz})\,.$$

Since $20,000/686 = 29.2$, $n$ must be in the range from 1 to 29 for the frequency to be audible and $f = 686, 1372, 2058, \ldots, 19890$ Hz

## 19

(a) Let $I_1$ be the original intensity and $I_2$ be the final intensity. The original sound level is $\beta_1 = (10\,\mathrm{dB})\log(I_1/I_0)$ and the final sound level is $\beta_2 = (10\,\mathrm{dB})\log(I_2/I_0)$, where $I_0$ is the reference intensity. Since $\beta_2 = \beta_1 + 30\,\mathrm{dB}$, $(10\,\mathrm{dB})\log(I_2/I_0) = (10\,\mathrm{dB})\log(I_1/I_0) + 30\,\mathrm{dB}$, or $(10\,\mathrm{dB})\log(I_2/I_0) - (10\,\mathrm{dB})\log(I_1/I_0) = 30\,\mathrm{dB}$. Divide by $10\,\mathrm{dB}$ and use $\log(I_2/I_0) - \log(I_1/I_0) = \log(I_2/I_1)$ to obtain $\log(I_2/I_1) = 3$. Now use each side as an exponent of 10 and recognize that $10^{\log(I_2/I_1)} = I_2/I_1$. The result is $I_2/I_1 = 10^3$. The intensity is multiplied by a factor of 1000.

(b) The pressure amplitude is proportional to the square root of the intensity so it is multiplied by a factor of $\sqrt{1000} = 32$.

## 25

(a) When the right side of the instrument is pulled out a distance $d$ the path length for sound waves increases by $2d$. Since the interference pattern changes from a minimum to the next maximum, this distance must be half a wavelength of the sound. So $2d = \lambda/2$, where $\lambda$ is the wavelength. Thus $\lambda = 4d$ and, if $v^{\mathrm{wave}}$ is the speed of sound, the frequency is $f = v^{\mathrm{wave}}/\lambda = v^{\mathrm{wave}}/4d = (343\ \mathrm{m/s})/4(0.0165\ \mathrm{m}) = 5.2 \times 10^3$ Hz.

(b) The pressure amplitude is proportional to the square root of the intensity (see Eq. 18–15). Write $\sqrt{I} = C\,\Delta P^{\mathrm{max}}$, where $I$ is the intensity, $\Delta P^{\mathrm{max}}$ is the pressure amplitude, and $C$ is a constant of proportionality. At the minimum, interference is destructive and the pressure amplitude is the difference in the amplitudes of the individual waves: $\Delta P^{\mathrm{max}} = \Delta P_{\mathrm{SAD}}^{\mathrm{max}} - \Delta P_{\mathrm{SBD}}^{\mathrm{max}}$, where the subscripts indicate the paths of the waves. At the maximum, the waves interfere constructively and the displacement amplitude is the sum of the amplitudes of the individual waves: $\Delta P^{\mathrm{max}} = \Delta P_{\mathrm{SAD}}^{\mathrm{max}} + \Delta P_{\mathrm{SBD}}^{\mathrm{max}}$. Solve $\sqrt{100} = C(\Delta P_{\mathrm{SAD}}^{\mathrm{max}} - \Delta P_{\mathrm{SBD}}^{\mathrm{max}})$ and $\sqrt{900} = C(\Delta P_{\mathrm{SAD}}^{\mathrm{max}} + \Delta P_{\mathrm{SBD}}^{\mathrm{max}})$ for $\Delta P_{\mathrm{SAD}}^{\mathrm{max}}$ and $\Delta p_{\mathrm{SBD}}^{\mathrm{max}}$. Add the equations to obtain $\Delta P_{\mathrm{SAD}}^{\mathrm{max}} = (\sqrt{100} + \sqrt{900})/2C = 20/C$, then subtract them to obtain $\Delta P_{\mathrm{SBD}}^{\mathrm{max}} = (\sqrt{900} - \sqrt{100})/2C = 10/C$. The ratio of the amplitudes is $\Delta P_{\mathrm{SAD}}^{\mathrm{max}}/\Delta P_{\mathrm{SBD}}^{\mathrm{max}} = 2$.

(c) Any energy losses, such as might be caused by frictional forces of the walls on the air in the tubes, result in a decrease in the displacement amplitude. Those losses are greater on path B since it is longer than path A.

**31**

(a) Since the pipe is open at both ends there is a pressure node at each end and an integer number of half-wavelengths fit into the length of the pipe. If $L$ is the pipe length and $\lambda$ is the wavelength then $\lambda = 2L/n$, where $n$ is an integer. If $v^{\text{wave}}$ is the speed of sound then the resonant frequencies are given by $f = v^{\text{wave}}/\lambda = nv/2L$. Now $L = 0.457$ m, so $f = n(344\,\text{m/s})/2(0.457\,\text{m}) = 376.4n$ Hz. To find the resonant frequencies that lie between 1000 Hz and 2000 Hz, first set $f = 1000$ Hz and solve for $n$, then set $f = 2000$ Hz and again solve for $n$. You should get 2.66 and 5.32. This means $n = 3$, 4, and 5 are the appropriate values of $n$. For $n = 3$, $f = 3(376.4\,\text{Hz}) = 1129$ Hz; for $n = 4$, $f = 4(376.4\,\text{Hz}) = 1526$ Hz; and for $n = 5$, $f = 5(376.4\,\text{Hz}) = 1882$ Hz.

(b) For any integer value of $n$ the displacement has $n+1$ nodes and $n$ antinodes, counting the ends. The nodes (N) and antinodes (A) are marked on the diagrams below for the three resonances found in part (a).

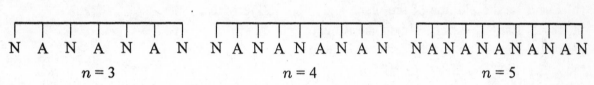

$$N \quad A \quad N \quad A \quad N \quad A \quad N \qquad N \quad A \quad N \quad A \quad N \quad A \quad N \quad A \quad N \qquad N \quad A \quad N \quad A \quad N \quad A \quad N \quad A \quad N \quad A \quad N$$
$$n = 3 \qquad\qquad\qquad n = 4 \qquad\qquad\qquad n = 5$$

**35**

(a) We expect the center of the star to be a displacement node. The star has spherical symmetry and the waves are spherical. If matter at the center moved it would move equally in all directions and this is not possible.

(b) Assume the oscillation is at the lowest resonance frequency. Then exactly one-fourth of a wavelength fits the star radius. If $\lambda$ is the wavelength and $R$ is the star radius then $\lambda = 4R$. The frequency is $f = \langle v \rangle/\lambda = \langle v \rangle/4R$. The period is $T = 1/f = 4R/\langle v \rangle$.

(c) Take the value of $\langle v \rangle$ to be $\sqrt{B/\rho}$, where $B$ is the bulk modulus and $\rho$ is the density of stellar material. The radius is $R = 9.0 \times 10^{-3} R_s$, where $R_s$ is the radius of the Sun ($6.96 \times 10^8$ m). Thus

$$T = 4R\sqrt{\frac{\rho}{B}} = 4(9.0 \times 10^{-3})(6.96 \times 10^8\,\text{m})\sqrt{\frac{1.0 \times 10^{10}\,\text{kg/m}^3}{1.33 \times 10^{22}\,\text{Pa}}} = 22\,\text{s}.$$

**39**

Since the beat frequency equals the difference between the frequencies of the two tuning forks, the frequency of the first fork is either 381 Hz or 387 Hz. When mass is added to this fork its frequency decreases (recall, for example, that the frequency of a mass-spring oscillator is proportional to $1/\sqrt{m}$). Since the beat frequency also decreases the frequency of the first fork must be greater than the frequency of the second. It must be 387 Hz.

**43**

The general expression for the Doppler shifted frequency is

$$f' = f\,\frac{v^{\text{wave}} \pm v_D}{v^{\text{wave}} \mp v_S},$$

where $f$ is the unshifted frequency, $v^{\text{wave}}$ is the speed of sound, $v_D$ is the speed of the detector, and $v_S$ is the speed of the source. All speeds are measured relative to the medium of propagation, the air in this case. The detector (the second plane) is moving toward the source (the first plane). This tends to increase the frequency, so we use the plus sign in the numerator. The source is moving away from the detector. This tends to decrease the frequency, so we use the plus sign in the denominator. Thus

$$f' = f\,\frac{v^{\text{wave}} + v_D}{v^{\text{wave}} + v_S} = (16,000\,\text{Hz})\left(\frac{343\,\text{m/s} + 250\,\text{m/s}}{343\,\text{m/s} + 200\,\text{m/s}}\right) = 17,500\,\text{Hz}.$$

## 55

(a) The half angle $\theta$ of the Mach cone is given by $\sin\theta = v^{\text{wave}}/v_S$, where $v^{\text{wave}}$ is the speed of sound and $v_S$ is the speed of the plane. Since $v_S = 1.5v^{\text{wave}}$, $\sin\theta = v/1.5v = 1/1.5$. This means $\theta = 42°$.

(b) Let $h$ be the altitude of the plane and suppose the Mach cone intersects Earth's surface a distance $d$ behind the plane. The situation is shown on the diagram to the right, with P indicating the plane and O indicating the observer. The cone angle is related to $h$ and $d$ by $\tan\theta = h/d$, so $d = h/\tan\theta$. The shock wave reaches O in the time the plane takes to fly the distance $d$: $t = d/v = h/v\tan\theta = (5000\,\text{m})/1.5(331\,\text{m/s})\tan 42° = 11\,\text{s}$.

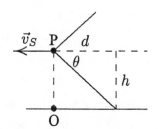

# Chapter 19

## 3

(a) Changes in temperature take place by means of radiation, conduction, and convection. The constant $A$ can be reduced by placing the object in isolation, by surrounding it with a vacuum jacket, for example. This reduces conduction and convection. Absorption of radiation can be reduced by polishing the surface to a mirror finish. Clearly $A$ depends on the condition of the surface and on the ability of the environment to conduct or convect energy to or from the object. $A$ has the dimensions of reciprocal time.

(b) Rearrange the equation to obtain

$$\frac{1}{\Delta T}\frac{d\Delta T}{dt} = -A.$$

Now integrate with respect to time and recognize that

$$\int \frac{1}{\Delta T}\frac{d\,\Delta T}{dt}\,dt = \int \frac{1}{\Delta T}\,d(\Delta T).$$

Thus

$$\int_{\Delta T_0}^{\Delta T}\frac{1}{\Delta T}\,d(\Delta T) = -\int_0^t A\,dt.$$

The integral on the right side yields $-At$ and the integral on the left yields $\ln \Delta T \big|_{\Delta T_0}^{\Delta T} = \ln(\Delta T) - \ln(\Delta T_0) = \ln(\Delta T/\Delta T_0)$, so

$$\ln \frac{\Delta T}{\Delta T_0} = -At.$$

Use each side as the exponent of $e$, the base of the natural logarithms, to obtain

$$\frac{\Delta T}{\Delta T_0} = e^{-At}$$

or

$$\Delta T = \Delta T_0\, e^{-At}.$$

## 5

(a) The specific heat is given by $c = Q/m(T_2 - T_1)$, where $Q$ is the thermal energy added, $m$ is the mass of the sample, $T_1$ is the initial temperature, and $T_2$ is the final temperature. Thus

$$c = \frac{314\,\text{J}}{(30.0 \times 10^{-3}\,\text{kg})(45.0^\circ\,\text{C} - 25.0^\circ\,\text{C})} = 523\,\text{J/kg}\cdot\text{K}.$$

(b) The molar specific heat is given by

$$c_m = \frac{Q}{N(T_2 - T_1)} = \frac{314\,\text{J}}{(0.600\,\text{mol})(45.0^\circ\,\text{C} - 25.0^\circ\,\text{C})} = 26.2\,\text{J/mol}\cdot\text{K}.$$

(c) If $N$ is the number of moles of the substance and $M$ is the mass per mole, then $m = NM$, so

$$N = \frac{m}{M} = \frac{30.0 \times 10^{-3}\,\text{kg}}{50 \times 10^{-3}\,\text{kg/mol}} = 0.600\,\text{mol}.$$

## 11

(a) The thermal energy generated is the power output of the drill multiplied by the time: $Q = Pt$. Use $1\,\text{hp} = 2545\,\text{Btu/h}$ to convert the given value of the power to Btu/h and $1\,\text{min} = (1/60)\,\text{h}$ to convert the given value of the time to hours. Then,

$$Q = \frac{(0.400\,\text{hp})(2545\,\text{Btu/h})(2.00\,\text{min})}{60\,\text{min/h}} = 33.9\,\text{Btu}.$$

(b) Use $0.750Q = cm\,\Delta T$ to compute the rise in temperature. Here $c$ is the specific heat of copper and $m$ is the mass of the copper block. Table 19–2 gives $c = 386\,\text{J/kg}\cdot\text{K}$. Use $1\,\text{J} = 9.481 \times 10^{-4}\,\text{Btu}$ and $1\,\text{kg} = 6.852 \times 10^{-2}\,\text{slug}$ to show that

$$c = \frac{(386\,\text{J/kg}\cdot\text{K})(9.481 \times 10^{-4}\,\text{Btu/J})}{6.852 \times 10^{-2}\,\text{slug/kg}} = 5.341\,\text{Btu/slug}\cdot\text{K}.$$

The mass of the block is its weight $W$ divided by $g$: $m = W/g = (1.60\,\text{lb})/(32\,\text{ft/s}^2) = 0.0500\,\text{slug}$. Thus

$$\Delta T = \frac{0.750Q}{cm} = \frac{(0.750)(33.9\,\text{Btu})}{(5.341\,\text{Btu/slug}\cdot\text{K})(0.0500\,\text{slug})} = 95.3\,\text{K}.$$

This is equivalent to $(9/5)(95.3\,\text{K}) = 172\,\text{F}°$.

## 15

Mass $m$ of water must be raised from an initial temperature $T_1$ ($= 59°\text{F} = 15°\text{C}$) to a final temperature $T_2$ ($= 100°\text{C}$). If $c$ is the specific heat of water then the thermal energy required is $Q = cm(T_2 - T_1)$. Each shake supplies energy $mgh$, where $h$ is the distance moved during the downward stroke of the shake. If $N$ is the total number of shakes then $Nmgh = Q$. If $t$ is the time taken to raise the water to its boiling point then $(N/t)mgh = Q/t$. Notice that $N/t$ is the rate $R$ of shaking (30 shakes/min). Thus $Rmgh = Q/t$. The distance $h$ is $1.0\,\text{ft} = 0.3048\,\text{m}$. Hence

$$t = \frac{Q}{Rmgh} = \frac{cm(T_f - T_i)}{Rmgh} = \frac{c(T_f - T_i)}{Rgh}$$
$$= \frac{(4190\,\text{J/kg}\cdot\text{K})(100°\text{C} - 15°\text{C})}{(30\,\text{shakes/min})(9.8\,\text{m/s}^2)(0.3048\,\text{m})} = 3.97 \times 10^3\,\text{min}.$$

This is $2.8\,d$.

## 17

Let $m$ be the mass of the ethyl alcohol, $c$ be its specific heat, $L_V$ be its heat of vaporization, and $L_F$ be its heat of fusion. The thermal energy required to liquify the gas is $Q_1 = mL_V = (0.510\,\text{kg})(879 \times 10^3\,\text{J/kg}) = 4.48 \times 10^5\,\text{J}$.

The thermal energy required to lower the temperature of the liquid from $T_1 = 78°C$ to the freezing point $T_2 = -114°C$ is $Q_2 = mc(T_2 - T_1) = (0.510\,\text{kg})(2.43 \times 10^3\,\text{J/kg} \cdot \text{C}°)(78°C + 114°C) = 2.38 \times 10^5\,\text{J}$.

The thermal energy required to freeze the liquid is $Q_3 = mL_F = (0.510\,\text{kg})(109 \times 10^3\,\text{J/kg}) = 5.56 \times 10^4\,\text{J}$.

The total thermal energy required is $Q_1 + Q_2 + Q_3$, which has the value $7.42 \times 10^5\,\text{J}$.

## 23

(a) There are three possibilities:

1.  None of the ice melts and the water-ice system reaches thermal equilibrium at a temperature that is at or below the melting point of ice.

2.  The system reaches thermal equilibrium at the melting point of ice, with some of the ice melted.

3.  All of the ice melts and the system reaches thermal equilibrium at a temperature at or above the melting point of ice.

First suppose that no ice melts. The temperature of the water decreases from $T_{Wi}$ (= 25°C) to some final temperature $T_f$ and the temperature of the ice increases from $T_{Ii}$ (= -15°C) to $T_f$. If $m_W$ is the mass of the water and $c_W$ is its specific heat then the water rejects thermal energy

$$Q = c_W m_W (T_{Wi} - T_f).$$

If $m_I$ is the mass of the ice and $c_I$ is its specific heat then the ice absorbs thermal energy

$$Q = c_I m_I (T_f - T_{Ii}).$$

Since no energy is lost these two thermal energies must be the same and

$$c_W m_W (T_{Wi} - T_f) = c_I m_I (T_f - T_{Ii}).$$

The solution for the final temperature is

$$
\begin{aligned}
T_f &= \frac{c_W m_W T_{Wi} + c_I m_I T_{Ii}}{c_W m_W + c_I m_I} \\
&= \frac{(4190\,\text{J/kg} \cdot \text{K})(0.200\,\text{kg})(25°C) + (2220\,\text{J/kg} \cdot \text{K})(0.100\,\text{kg})(-15°C)}{(4190\,\text{J/kg} \cdot \text{K})(0.200\,\text{kg}) + (2220\,\text{J/kg} \cdot \text{K})(0.100\,\text{kg})} \\
&= 16.6°C.
\end{aligned}
$$

This is above the melting point of ice, so at least some of the ice must have melted. The calculation just completed does not take into account the melting of the ice and is in error.

Now assume the water and ice reach thermal equilibrium at $T_f = 0°C$, with mass $m$ ($< m_I$) of the ice melted. The magnitude of the thermal energy rejected by the water is

$$Q = c_W m_W T_{Wi},$$

and the thermal energy absorbed by the ice is

$$Q = c_I m_I(0 - T_{Ii}) + mL_F,$$

where $L_F$ is the heat of fusion for water. The first term is the energy required to warm all the ice from its initial temperature to 0°C and the second term is the energy required to melt mass $m$ of the ice. The two thermal energies are equal, so

$$c_W m_W T_{Wi} = -c_I m_I T_{Ii} + mL_F.$$

This equation can be solved for the mass $m$ of ice melted:

$$\begin{aligned} m &= \frac{c_W m_W T_{Wi} + c_I m_I T_{Ii}}{L_F} \\ &= \frac{(4190\,\text{J/kg}\cdot\text{K})(0.200\,\text{kg})(25°\text{C}) + (2220\,\text{J/kg}\cdot\text{K})(0.100\,\text{kg})(-15°\text{C})}{333 \times 10^3\,\text{J/kg}} \\ &= 5.3 \times 10^{-2}\,\text{kg} = 53\,\text{g}. \end{aligned}$$

Since the total mass of ice present initially was 100 g, there is enough ice to bring the water temperature down to 0°C. This is the solution: the ice and water reach thermal equilibrium at a temperature of 0°C with 53 g of ice melted.

(b) Now there is less than 53 g of ice present initially. All the ice melts and the final temperature is above the melting point of ice. The thermal energy rejected by the water is

$$Q = c_W m_W(T_{Wi} - T_f)$$

and the thermal energy absorbed by the ice and the water it becomes when it melts is

$$Q = c_I m_I(0 - T_{Ii}) + c_W m_I(T_f - 0) + m_I L_F.$$

The first term is the energy required to raise the temperature of the ice to 0°C, the second term is the energy required to raise the temperature of the melted ice from 0°C to $T_f$, and the third term is the energy required to melt all the ice. Since the two thermal energies are equal,

$$c_W m_W(T_{Wi} - T_f) = c_I m_I(-T_{Ii}) + c_W m_I T_f + m_I L_F.$$

The solution for $T_f$ is

$$T_f = \frac{c_W m_W T_{Wi} + c_I m_I T_{Ii} - m_I L_F}{c_W(m_W + m_I)}.$$

Substitute given values to obtain $T_f = 2.5°\text{C}$.

## 29

(a) The change in internal energy $\Delta E^{\text{int}}$ is the same for path $iaf$ and path $ibf$. According to the first law of thermodynamics, $\Delta E^{\text{int}} = Q - W$, where $Q$ is the thermal energy absorbed and $W$ is the work done by the system. Along $iaf$ $\Delta E^{\text{int}} = Q - W = 50\,\text{cal} - 20\,\text{cal} = 30\,\text{cal}$. Along $ibf$ $W = Q - \Delta E^{\text{int}} = 36\,\text{cal} - 30\,\text{cal} = 6\,\text{cal}$.

(b) Since the curved path is traversed from $f$ to $i$ the change in internal energy is $-30\,\mathrm{cal}$ and $Q = \Delta E^{\mathrm{int}} + W = -30\,\mathrm{cal} - 13\,\mathrm{cal} = -43\,\mathrm{cal}$.

(c) Let $\Delta E^{\mathrm{int}} = E_f^{\mathrm{int}} - E_i^{\mathrm{int}}$. Then, $E_f^{\mathrm{int}} = \Delta E^{\mathrm{int}} + E_i^{\mathrm{int}} = 30\,\mathrm{cal} + 10\,\mathrm{cal} = 40\,\mathrm{cal}$.

(d) The work $W_{bf}$ for the path $bf$ is zero, so $Q_{bf} = E_f^{\mathrm{int}} - E_b^{\mathrm{int}} = 40\,\mathrm{cal} - 22\,\mathrm{cal} = 18\,\mathrm{cal}$. For the path $ibf$ $Q = 36\,\mathrm{cal}$ so $Q_{ib} = Q - Q_{bf} = 36\,\mathrm{cal} - 18\,\mathrm{cal} = 18\,\mathrm{cal}$.

## 35

When the temperature changes from $T$ to $T + \Delta T$ the diameter of the mirror changes from $D$ to $D + \Delta D$, where $\Delta D = \alpha D\,\Delta T$. Here $\alpha$ is the coefficient of linear expansion for Pyrex glass ($3.2 \times 10^{-6}/\mathrm{C}°$, according to Table 19–5). The range of values for the diameters can be found by setting $\Delta T$ equal to the temperature range. Thus $\Delta D = (3.2 \times 10^{-6}/\mathrm{C}°)(200\,\mathrm{in.})(60\,\mathrm{C}°) = 3.84 \times 10^{-2}\,\mathrm{in.}$ Since $1\,\mathrm{in.} = 2.50\,\mathrm{cm} = 2.50 \times 10^4\,\mu\mathrm{m}$, this is $960\,\mu\mathrm{m}$.

## 41

If $V_c$ is the original volume of the cup, $\alpha_a$ is the coefficient of linear expansion of aluminum, and $\Delta T$ is the temperature increase, then the change in the volume of the cup is $\Delta V_c = 3\alpha_a V_c\,\Delta T$. See Eq. 19–31. If $\beta$ is the coefficient of volume expansion for glycerin then the change in the volume of glycerin is $\Delta V_g = \beta V_c\,\Delta T$. Note that the original volume of glycerin is the same as the original volume of the cup. The volume of glycerin that spills is

$$\begin{aligned}
\Delta V_g - \Delta V_c &= (\beta - 3\alpha_a)V_c\,\Delta T \\
&= \left[(5.1 \times 10^{-4}/\mathrm{C}°) - 3(23 \times 10^{-6}/\mathrm{C}°)\right](100\,\mathrm{cm}^3)(6\,\mathrm{C}°) = 0.26\,\mathrm{cm}^3\,.
\end{aligned}$$

## 45

The change in volume of the liquid is given by $\Delta V = \beta V\,\Delta T$. If $A$ is the cross-sectional area of the tube and $h$ is the height of the liquid, then $V = Ah$ is the original volume and $\Delta V = A\,\Delta h$ is the change in volume. Since the tube does not change the cross-sectional area of the liquid remains the same. Therefore, $A\,\Delta h = \beta Ah\,\Delta T$ or $\Delta h = \beta h\,\Delta T$.

## 53

The rate of thermal energy flow is given by

$$P^{\mathrm{cond}} = kA\,\frac{T_H - T_C}{L}\,,$$

where $k$ is the thermal conductivity of copper ($401\,\mathrm{W/m \cdot K}$), $A$ is the cross-sectional area (in a plane perpendicular to the flow), $L$ is the distance along the direction of flow between the points where the temperature is $T_H$ and $T_C$. Thus

$$P^{\mathrm{cond}} = \frac{(401\,\mathrm{W/m \cdot K})(90.0 \times 10^{-4}\,\mathrm{m}^2)(125°\mathrm{C} - 10.0°\mathrm{C})}{0.250\,\mathrm{m}} = 1.66 \times 10^3\,\mathrm{J/s}\,.$$

The thermal conductivity was found in Table 19–6 of the text.

**61**

Let $h$ be the thickness of the slab and $A$ be its area. Then, the rate of thermal energy flow through the slab is

$$P^{\text{cond}} = \frac{kA(T_H - T_C)}{h},$$

where $k$ is the thermal conductivity of ice, $T_H$ is the temperature of the water (0°C), and $T_C$ is the temperature of the air above the ice (−10°C). The thermal energy leaving the water freezes it, the thermal energy required to freeze mass $m$ of water being $Q = L_F m$, where $L_F$ is the heat of fusion for water. Differentiate with respect to time and recognize that $dQ/dt = P^{\text{cond}}$. You should obtain

$$P^{\text{cond}} = L_F \frac{dm}{dt}.$$

Now the mass of the ice is given by $m = \rho A h$, where $\rho$ is the density of ice and $h$ is the thickness of the ice slab, so $dm/dt = \rho A (dh/dt)$ and

$$P^{\text{cond}} = L_F \rho A \frac{dh}{dt}.$$

Equate the two expressions for $P^{\text{cond}}$ and solve for $dh/dt$:

$$\frac{dh}{dt} = \frac{k(T_H - T_C)}{L_F \rho h}.$$

Since 1 cal = 4.186 J and 1 cm = $1 \times 10^{-2}$ m, the thermal conductivity of ice has the SI value $k = (0.0040 \, \text{cal/s} \cdot \text{cm} \cdot \text{K})(4.186 \, \text{J/cal})/(1 \times 10^{-2} \, \text{m/cm}) = 1.674 \, \text{W/m} \cdot \text{K}$. The SI value for the density of ice is $\rho = 0.92 \, \text{g/cm}^3 = 0.92 \times 10^3 \, \text{kg/m}^3$. Thus

$$\frac{dh}{dt} = \frac{(1.674 \, \text{W/m} \cdot \text{K})(0°\text{C} + 10°\text{C})}{(333 \times 10^3 \, \text{J/kg})(0.92 \times 10^3 \, \text{kg/m}^3)(0.050 \, \text{m})} = 1.1 \times 10^{-6} \, \text{m/s} = 0.40 \, \text{cm/h}.$$

# Chapter 20

## 7

(a) Solve $PV = nRT$ for $n$. First, convert the temperature to the Kelvin scale: $T = 40.0 + 273.15 = 313.15$ K. Also convert the volume to m$^3$: $1000 \text{ cm}^3 = 1000 \times 10^{-6} \text{ m}^3$. Then according to the ideal gas law,

$$n = \frac{PV}{RT} = \frac{(1.01 \times 10^5 \text{ Pa})(1000 \times 10^{-6} \text{ m}^3)}{(8.31 \text{ J/mol} \cdot \text{K})(313.15 \text{ K})} = 3.88 \times 10^{-2} \text{ mol}.$$

(b) Solve the ideal gas law $PV = nRT$ for $T$:

$$T = \frac{PV}{nR} = \frac{(1.06 \times 10^5 \text{ Pa})(1500 \times 10^{-6} \text{ m}^3)}{(3.88 \times 10^{-2} \text{ mol})(8.31 \text{ J/mol} \cdot \text{K})} = 493 \text{ K} = 220^\circ \text{ C}.$$

## 11

Since the pressure is constant the work is given by $W = P(V_2 - V_1)$. The initial volume is $V_1 = (AT_1 - BT_1^2)/P$, where $T_1$ is the initial temperature. The final volume is $V_2 = (AT_2 - BT_2^2)/P$. Thus $W = A(T_2 - T_1) - B(T_2^2 - T_1^2)$.

## 15

Assume that the pressure of the air in the bubble is essentially the same as the pressure in the surrounding water. If $d$ is the depth of the lake and $\rho$ is the density of water, then the pressure at the bottom of the lake is $P_1 = P_0 + \rho g d$, where $P_0$ is atmospheric pressure. Since $P_1 V_1 = nRT_1$, the number of moles of gas in the bubble is $n = P_1 V_1 / RT_1 = (P_0 + \rho g d) V_1 / RT_1$, where $V_1$ is the volume of the bubble at the bottom of the lake and $T_1$ is the temperature there. At the surface of the lake the pressure is $P_0$ and the volume of the bubble is $V_2 = nRT_2/P_0$. Substitute for $n$ to obtain

$$V_2 = \frac{T_2}{T_1} \frac{P_0 + \rho g d}{P_0} V_1$$

$$= \left(\frac{293 \text{ K}}{277 \text{ K}}\right) \left(\frac{1.013 \times 10^5 \text{ Pa} + (0.998 \times 10^3 \text{ kg/m}^3)(9.8 \text{ N/kg})(40 \text{ m})}{1.013 \times 10^5 \text{ Pa}}\right) (20 \text{ cm}^3)$$

$$= 100 \text{ cm}^3.$$

Each of the Celsius temperatures was converted to the corresponding Kelvin temperature.

## 23

On reflection only the normal component of the momentum changes, so for one molecule the change in momentum is $2mv \cos \theta$, where $m$ is the mass of the molecule, $v$ is its speed, and $\theta$

is the angle between its velocity and the normal to the wall. If $N$ molecules collide with the wall, then the change in their total momentum is $2Nmv\cos\theta$, and if the total time taken for the collisions is $\Delta t$, then the average rate of change of the total momentum is $2(N/\Delta t)mv\cos\theta$. This is the average force of the $N$ molecules on the wall. The pressure is the average force per unit area:

$$P = \frac{2}{A}\left(\frac{N}{\Delta t}\right)mv\cos\theta$$

$$= \left(\frac{2}{2.0\times10^{-4}\,\text{m}^2}\right)(1.0\times10^{23}\,\text{s}^{-1})(3.3\times10^{-27}\,\text{kg})(1.0\times10^3\,\text{m/s})\cos55°$$

$$= 1.9\times10^3\,\text{Pa}\,.$$

Notice that the value given for the mass was converted to kg and the value given for the area was converted to m².

## 35

(a) Use the ideal gas law in the form $PV = NkT$, where $P$ is the pressure, $V$ is the volume, $T$ is the temperature, and $N$ is the number of molecules. Since 1 cm of mercury = 1333 Pa, the pressure is $P = (10^{-7})(1333) = 1.333\times10^{-4}\,\text{Pa}$. Thus

$$\frac{N}{V} = \frac{P}{kT} = \frac{1.333\times10^{-4}\,\text{Pa}}{(1.38\times10^{-23}\,\text{J/K})(295\,\text{K})}$$

$$= 3.27\times10^{16}\,\text{molecules/m}^3 = 3.27\times10^{10}\,\text{molecules/cm}^3\,.$$

(b) The molecular diameter is $d = 2.00\times10^{-10}\,\text{m}$, so, according to Eq. 20–24, the mean free path is

$$\lambda = \frac{1}{\sqrt{2}\pi d^2 N/V} = \frac{1}{\sqrt{2}\pi(2.00\times10^{-10}\,\text{m})^2(3.27\times10^{16}\,\text{m}^{-3})} = 172\,\text{m}\,.$$

## 39

(a) The rms speed of molecules in a gas is given by $v^{\text{ms}} = \sqrt{3RT/M}$, where $T$ is the temperature and $M$ is the molar mass of the gas. See Eq. 20–33. The speed required for escape from Earth's gravitational pull is $v = \sqrt{2gr_e}$, where $g$ is the gravitational strength at Earth's surface and $r_e$ ($= 6.37\times10^6\,\text{m}$) is the radius of Earth. See Section 14–6.

Equate the expressions for the speeds to obtain $\sqrt{3RT/M} = \sqrt{2gr_e}$. The solution for $T$ is $T = 2gr_eM/3R$. According to Table 20–1 the molar mass of molecular hydrogen is $2.02\times10^{-3}\,\text{kg/mol}$, so for that gas

$$T = \frac{2(9.8\,\text{N/kg})(6.37\times10^6\,\text{m})(2.02\times10^{-3}\,\text{kg/mol})}{3(8.31\,\text{J/mol}\cdot\text{K})} = 1.0\times10^4\,\text{K}\,.$$

(b) According to Table 20–1 the molar mass of oxygen is $32.0\times10^{-3}\,\text{kg/mol}$, so for that gas

$$T = \frac{2(9.8\,\text{N/kg})(6.37\times10^6\,\text{m})(32.0\times10^{-3}\,\text{kg/mol})}{3(8.31\,\text{J/mol}\cdot\text{K})} = 1.6\times10^5\,\text{K}\,.$$

(c) Now $T = 2g_m r_m M / 3R$, where $r_m$ (= $1.74 \times 10^6$ m) is the radius of the Moon and $g_m$ (= $0.16g$) is the gravitational strength at the Moon's surface. For hydrogen

$$T = \frac{2(0.16)(9.8 \, \text{N/kg})(1.74 \times 10^6 \, \text{m})(2.02 \times 10^{-3} \, \text{kg/mol})}{3(8.31 \, \text{J/mol} \cdot \text{K})} = 4.4 \times 10^2 \, \text{K}.$$

For oxygen

$$T = \frac{2(0.16)(9.8 \, \text{N/kg})(1.74 \times 10^6 \, \text{m})(32.0 \times 10^{-3} \, \text{kg/mol})}{3(8.31 \, \text{J/mol} \cdot \text{K})} = 7.0 \times 10^3 \, \text{K}.$$

(d) The temperature high in Earth's atmosphere is great enough for a significant number of hydrogen atoms in the tail of the Maxwellian distribution to escape. As a result the atmosphere is depleted of hydrogen. On the other hand, very few oxygen atoms escape.

## 41

(a) The root-mean-square speed is given by $v^{\text{rms}} = \sqrt{3RT/M}$. See Eq. 20–33. According to Table 20–1 the molar mass of hydrogen is $2.02 \times 10^{-3}$ kg/mol, so

$$v_{\text{rms}} = \sqrt{\frac{3(8.31 \, \text{J/mol} \cdot \text{K})(4000 \, \text{K})}{2.02 \times 10^{-3} \, \text{kg/mol}}} = 7.0 \times 10^3 \, \text{m/s}.$$

(b) When the surfaces of the spheres that represent an $H_2$ molecule and an Ar atom are touching, the distance between their centers is the sum of their radii: $d = r_1 + r_2 = 0.5 \times 10^{-8}$ cm + $1.5 \times 10^{-8}$ cm = $2.0 \times 10^{-8}$ cm.

(c) The argon atoms are essentially at rest so in time $t$ the hydrogen atom collides with all the argon atoms in a cylinder of radius $d$ and length $vt$, where $v$ is its speed. That is, the number of collisions is $\pi d^2 vt N / V$, where $N/V$ is the concentration of argon atoms. The number of collisions per unit time is

$$\frac{\pi d^2 v N}{V} = \pi(2.0 \times 10^{-10} \, \text{m})^2(7.0 \times 10^3 \, \text{m/s})(4.0 \times 10^{25} \, \text{m}^{-3}) = 3.5 \times 10^{10} \, \text{collisions/s}.$$

## 45

According to the first law of thermodynamics, $\Delta E^{\text{int}} = Q - W$. Since the process is isothermal $\Delta E^{\text{int}} = 0$ (the internal energy of an ideal gas depends only on the temperature) and $Q = W$. The work done by the gas as its volume expands from $V_i$ to $V_f$ at temperature $T$ is

$$W = \int_{V_i}^{V_f} P \, dV = nRT \int_{V_i}^{V_f} \frac{dV}{V} = nRT \ln \frac{V_f}{V_i},$$

where the ideal gas law $PV = nRT$ was used to substitute for $P$. For 1 mole $Q = W = RT \ln(V_f/V_i)$.

## 53

(a) Since the process is at constant pressure the thermal energy transferred to the gas is given by $Q = nC_P \, \Delta T$, where $n$ is the number of moles in the gas, $C_P$ is the molar specific heat at constant pressure, and $\Delta T$ is the increase in temperature. For a diatomic ideal gas $C_P = \frac{7}{2}R$. Thus

$$Q = \frac{7}{2}nR\,\Delta T = \frac{7}{2}(4.00\,\text{mol})(8.314\,\text{J/mol}\cdot\text{K})(60.0\,\text{K}) = 6.98 \times 10^3 \,\text{J}.$$

(b) The change in the internal energy is given by $\Delta E^{\text{int}} = nC_V \, \Delta T$, where $C_V$ is the specific heat at constant volume. For a diatomic ideal gas $C_V = \frac{5}{2}R$, so

$$\Delta E^{\text{int}} = \frac{5}{2}nR\,\Delta T = \frac{5}{2}(4.00\,\text{mol})(8.314\,\text{J/mol}\cdot\text{K})(60.0\,\text{K}) = 4.99 \times 10^3 \,\text{J}.$$

(c) According to the first law of thermodynamics, $\Delta E^{\text{int}} = Q - W$, so

$$W = Q - \Delta E_{\text{int}} = 6.98 \times 10^3 \,\text{J} - 4.99 \times 10^3 \,\text{J} = 1.99 \times 10^3 \,\text{J}.$$

(d) The change in the total translational kinetic energy is

$$\Delta K = \frac{3}{2}nR\,\Delta T = \frac{3}{2}(4.00\,\text{mol})(8.314\,\text{J/mol}\cdot\text{K})(60.0\,\text{K}) = 2.99 \times 10^3 \,\text{J}.$$

## 55

(a) Let $P_i$, $V_i$, and $T_i$ represent the pressure, volume, and temperature of the initial state of the gas. Let $P_f$, $V_f$, and $T_f$ represent the pressure, volume, and temperature of the final state. Since the process is adiabatic $P_i V_i^{\gamma} = P_f V_f^{\gamma}$, so

$$P_f = \left(\frac{V_i}{V_f}\right)^{\gamma} P_i = \left(\frac{4.3\,\text{L}}{0.76\,\text{L}}\right)^{1.4}(1.2\,\text{atm}) = 13.6\,\text{atm}.$$

Notice that since $V_i$ and $V_f$ have the same units, their units cancel and $P_f$ has the same units as $P_i$.

(b) The gas obeys the ideal gas law $PV = nRT$, so $P_i V_i / P_f V_f = T_i / T_f$ and

$$T_f = \frac{P_f V_f}{P_i V_i}T_i = \left[\frac{(13.6\,\text{atm})(0.76\,\text{L})}{(1.2\,\text{atm})(4.3\,\text{L})}\right](310\,\text{K}) = 620\,\text{K}.$$

Note that the units of $p_i V_i$ and $p_f V_f$ cancel since they are the same.

## 61

In the following $C_V$ $(= \frac{3}{2}R)$ is the molar specific heat at constant volume, $C_P$ $(= \frac{5}{2}R)$ is the molar specific heat at constant pressure, $\Delta T$ is the temperature change, and $n$ is the number of moles.

(a) The process $1 \rightarrow 2$ takes place at constant volume. The thermal energy added is

$$Q = nC_V \, \Delta T = \frac{3}{2} nR \, \Delta T$$

$$= \frac{3}{2}(1.00 \, \text{mol})(8.314 \, \text{J/mol} \cdot \text{K})(600 \, \text{K} - 300 \, \text{K}) = 3.74 \times 10^3 \, \text{J} \,.$$

Since the process takes place at constant volume the work $W$ done by the gas is zero and the first law of thermodynamics tells us that the change in the internal energy is

$$\Delta E^{\text{int}} = Q = 3.74 \times 10^3 \, \text{J} \,.$$

The process $2 \rightarrow 3$ is adiabatic. The thermal energy absorbed is zero. The change in the internal energy is

$$\Delta E^{\text{int}} = nC_V \, \Delta T = \frac{3}{2} nR \, \Delta T$$

$$= \frac{3}{2}(1.00 \, \text{mol})(8.314 \, \text{J/mol} \cdot \text{K})(455 \, \text{K} - 600 \, \text{K}) = -1.81 \times 10^3 \, \text{J} \,.$$

According to the first law of thermodynamics the work done by the gas is

$$W = Q - \Delta E^{\text{int}} = +1.81 \times 10^3 \, \text{J} \,.$$

The process $3 \rightarrow 1$ takes place at constant pressure. The heat added is

$$Q = nC_P \, \Delta T = \frac{5}{2} nR \, \Delta T$$

$$= \frac{5}{2}(1.00 \, \text{mol})(8.314 \, \text{J/mol} \cdot \text{K})(300 \, \text{K} - 455 \, \text{K}) = -3.22 \times 10^3 \, \text{J} \,.$$

The change in the internal energy is

$$\Delta E^{\text{int}} = nC_V \, \Delta T = \frac{3}{2} nR \, \Delta T$$

$$= \frac{3}{2}(1.00 \, \text{mol})(8.314 \, \text{J/mol} \cdot \text{K})(300 \, \text{K} - 455 \, \text{K}) = -1.93 \times 10^3 \, \text{J} \,.$$

According to the first law of thermodynamics the work done by the gas is

$$W = Q - \Delta E^{\text{int}} = -3.22 \times 10^3 \, \text{J} + 1.93 \times 10^3 \, \text{J} = -1.29 \times 10^3 \, \text{J} \,.$$

For the entire process the thermal energy absorbed is

$$Q = 3.74 \times 10^3 \, \text{J} + 0 - 3.22 \times 10^3 \, \text{J} = 520 \, \text{J} \,,$$

the change in the internal energy is

$$\Delta E^{\text{int}} = 3.74 \times 10^3 \, \text{J} - 1.81 \times 10^3 \, \text{J} - 1.93 \times 10^3 \, \text{J} = 0 \,,$$

and the work done by the gas is

$$W = 0 + 1.81 \times 10^3 \, \text{J} - 1.29 \times 10^3 \, \text{J} = 520 \, \text{J} \,.$$

(b) First find the initial volume. Use the ideal gas law $P_1 V_1 = nRT_1$ to obtain

$$V_1 = \frac{nRT_1}{P_1} = \frac{(1.00 \, \text{mol})(8.314 \, \text{J/mol} \cdot \text{K})(300 \, \text{K})}{(1.013 \times 10^5 \, \text{Pa})} = 2.46 \times 10^{-2} \, \text{m}^3 \,.$$

Since $1 \rightarrow 2$ is a constant volume process $V_2 = V_1 = 2.46 \times 10^{-2} \, \text{m}^3$.

The pressure for state 2 is

$$P_2 = \frac{nRT_2}{V_2} = \frac{(1.00\,\text{mol})(8.314\,\text{J/mol} \cdot \text{K})(600\,\text{K})}{2.46 \times 10^{-2}\,\text{m}^3} = 2.02 \times 10^5\,\text{Pa}.$$

This is equivalent to 1.99 atm.

Since $3 \to 1$ is a constant pressure process, the pressure for state 3 is the same as the pressure for state 1: $P_3 = P_1 = 1.013 \times 10^5\,\text{Pa}$ (1.00 atm). The volume for state 3 is

$$V_3 = \frac{nRT_3}{P_3} = \frac{(1.00\,\text{mol})(8.314\,\text{J/mol} \cdot \text{K})(455\,\text{K})}{1.013 \times 10^5\,\text{Pa}} = 3.73 \times 10^{-2}\,\text{m}^3.$$

# Chapter 21

## 3

(a) Since the gas is ideal, its pressure $P$ is given in terms of the number of moles $n$, the volume $V$, and the temperature $T$ by $P = nRT/V$. The work done by the gas during the isothermal expansion is

$$W = \int_{V_1}^{V_2} P\, dV = nRT \int_{V_1}^{V_2} \frac{dV}{V} = nRT \ln \frac{V_2}{V_1}.$$

Substitute $V_2 = 2V_1$ to obtain

$$W = nRT \ln 2 = (4.00\,\text{mol})(8.314\,\text{J/mol} \cdot \text{K})(400\,\text{K}) \ln 2 = 9.22 \times 10^3\,\text{J}.$$

(b) Since the expansion is isothermal, the change in entropy is given by $\Delta S = \int (1/T)\, dQ = Q/T$, where $Q$ is the thermal energy absorbed. According to the first law of thermodynamics, $\Delta E^{\text{int}} = Q - W$. Now the internal energy of an ideal gas depends only on the temperature and not on the pressure and volume. Since the expansion is isothermal, $\Delta E^{\text{int}} = 0$ and $Q = W$. Thus

$$\Delta S = \frac{W}{T} = \frac{9.22 \times 10^3\,\text{J}}{400\,\text{K}} = 23.1\,\text{J/K}.$$

(c) $\Delta S = 0$ for all reversible adiabatic processes.

## 15

The ice warms to $0°\,\text{C}$, then melts, and the resulting water warms to the temperature of the lake water, which is $15°\,\text{C}$.

As the ice warms, the thermal energy it receives when the temperature changes by $dT$ is $dQ = mc_I\, dT$, where $m$ is the mass of the ice and $c_I$ is the specific heat of ice. If $T_i$ ($= 263\,\text{K}$) is the initial temperature and $T_f$ ($= 273\,\text{K}$) is the final temperature, then the change in its entropy is

$$\Delta S = \int \frac{dQ}{T} = mc_I \int_{T_i}^{T_f} \frac{dT}{T} = mc_I \ln \frac{T_f}{T_i}$$

$$= (0.010\,\text{kg})(2220\,\text{J/kg} \cdot \text{K}) \ln \frac{273\,\text{K}}{263\,\text{K}} = 0.828\,\text{J/K}.$$

Melting is an isothermal process. The thermal energy leaving the ice is $Q = mL_F$, where $L_F$ is the heat of fusion for ice. Thus $\Delta S = Q/T = mL_F/T = (0.010\,\text{kg})(333 \times 10^3\,\text{J/kg})/(273\,\text{K}) = 12.20\,\text{J/K}$.

For the warming of the water from the melted ice, the change in entropy is

$$\Delta S = mc_w \ln \frac{T_f}{T_i},$$

where $c_w$ is the specific heat of water ($4190\,\text{J/kg}\cdot\text{K}$). Thus

$$\Delta S = (0.010\,\text{kg})(4190\,\text{J/kg}\cdot\text{K})\ln\frac{(273+15)\,\text{K}}{273\,\text{K}} = 2.24\,\text{J/K}.$$

The total change in entropy for the ice and the water it becomes is $\Delta S = 0.828\,\text{J/K} + 12.20\,\text{J/K} + 2.24\,\text{J/K} = 15.27\,\text{J/K}$.

Since the temperature of the lake does not change significantly when the ice melts, the change in its entropy is $\Delta S = Q/T$, where $Q$ is the thermal energy it receives (the negative of the energy it supplies the ice) and $T$ is its temperature. When the ice warms to $0^\circ\,\text{C}$,

$$Q = -mc_I(T_f - T_i) = -(0.010\,\text{kg})(2220\,\text{J/kg}\cdot\text{K})(10\,\text{K}) = -222\,\text{J}.$$

When the ice melts,

$$Q = -mL_F = -(0.010\,\text{kg})(333\times10^3\,\text{J/kg}) = -3.33\times10^3\,\text{J}.$$

When the water from the ice warms,

$$Q = -mc_w(T_f - T_i) = -(0.010\,\text{kg})(4190\,\text{J/kg}\cdot\text{J})(15\,\text{K}) = -629\,\text{J}.$$

The total energy leaving the lake water is $Q = -222\,\text{J} - 3.33\times10^3\,\text{J} - 6.29\times10^2\,\text{J} = -4.18\times10^3\,\text{J}$. The change in entropy is

$$\Delta S = \frac{-4.18\times10^3\,\text{J}}{(273+15)\,\text{K}} = -14.51\,\text{J/K}.$$

The change in the entropy of the ice-lake system is $\Delta S = 15.27\,\text{J/K} - 14.51\,\text{J/K} = 0.76\,\text{J/K}$.

### 19

(a) Work is done only for the $ab$ portion of the process. This portion is at constant pressure, so the work done by the gas is

$$W = \int_{V_0}^{4V_0} P_0\,dV = P_0(4V_0 - V_0) = 3P_0V_0.$$

(b) Use the first law: $\Delta E^{\text{int}} = Q - W$. Since the process is at constant volume, the work done by the gas is zero and $E^{\text{int}} = Q$. The thermal energy $Q$ absorbed by the gas is $Q = nC_V\,\Delta T$, where $C_V$ is the molar specific heat at constant volume and $\Delta T$ is the change in temperature. Since the gas is a monatomic ideal gas, $C_V = \frac{3}{2}R$. Use the ideal gas law to find that the initial temperature is $T_b = P_bV_b/nR = 4P_0V_0/nR$ and that the final temperature is $T_c = P_cV_c/nR = (2P_0)(4V_0)/nR = 8p_0V_0/nR$. Thus

$$Q = \frac{3}{2}nR\left(\frac{8P_0V_0}{nR} - \frac{4P_0V_0}{nR}\right) = 6P_0V_0.$$

The change in the internal energy is $\Delta E^{\text{int}} = 6P_0V_0$. Since $n = 1\,\text{mol}$, this can also be written $Q = 6RT_0$.

Since the process is at constant volume, use $dQ = nC_V\, dT$ to obtain

$$\Delta S = \int \frac{dQ}{T} = nC_V \int_{T_b}^{T_c} \frac{dT}{T} = nC_V \ln \frac{T_c}{T_b}.$$

Substitute $C_V = \frac{3}{2}R$. Use the ideal gas law to write

$$\frac{T_c}{T_b} = \frac{P_c V_c}{P_b V_b} = \frac{(2P_0)(4V_0)}{P_0(4V_0)} = 2.$$

Thus $\Delta S = \frac{3}{2}nR\ln 2$. Since $n = 1$, this is $\Delta S = \frac{3}{2}R\ln 2$.

(c) For a complete cycle, $\Delta E^{\text{int}} = 0$ and $\Delta S = 0$.

## 25

For a Carnot engine, the efficiency is related to the reservoir temperatures by $\epsilon = (T_H - T_C)/T_H$. Thus $T_H = (T_H - T_C)/\epsilon = (75\,\text{K})/(0.22) = 341\,\text{K}\ (= 68°\,\text{C})$. The temperature of the cold reservoir is $T_C = T_H - 75 = 341\,\text{K} - 75\,\text{K} = 266\,\text{K}\ (= -7°\,\text{C})$.

## 31

(a) If $T_H$ is the temperature of the high-temperature reservoir and $T_C$ is the temperature of the low-temperature reservoir, then the maximum efficiency of the engine is

$$\epsilon = \frac{T_H - T_C}{T_H} = \frac{(800 + 40)\,\text{K}}{(800 + 273)\,\text{K}} = 0.78.$$

(b) The efficiency is defined by $\epsilon = |W|/|Q_H|$, where $W$ is the work done by the engine and $Q_H$ is the thermal energy input. $W$ is positive. Over a complete cycle, $Q_H = W + |Q_C|$, where $Q_C$ is the thermal energy output, so $\epsilon = W/(W + |Q_C|)$ and $|Q_C| = W[(1/\epsilon) - 1]$. Now $\epsilon = (T_H - T_C)/T_H$, where $T_H$ is the temperature of the high-temperature heat reservoir and $T_C$ is the temperature of the low-temperature reservoir. Thus $(1/\epsilon) - 1 = T_C/(T_H - T_C)$ and $|Q_C| = WT_C/(T_H - T_C)$.

The thermal energy output is used to melt ice at temperature $T_i\ (= -40°\,\text{C})$. The ice must be brought to $0°\,\text{C}$, then melted, so $|Q_C| = mc(T_f - T_i) + mL_F$, where $m$ is the mass of ice melted, $T_f$ is the melting temperature ($0°\,\text{C}$), $c$ is the specific heat of ice, and $L_F$ is the heat of fusion of ice. Thus $WT_C/(T_H - T_C) = mc(T_f - T_i) + mL_F$. Differentiate with respect to time and replace $dW/dt$ with $P^{\text{engine}}$, the power output of the engine. You should obtain $P^{\text{engine}}T_C/(T_H - T_C) = (dm/dt)[c(T_f - T_i) + L_F]$. Thus

$$\frac{dm}{dt} = \left(\frac{P^{\text{engine}}T_C}{T_H - T_C}\right)\left(\frac{1}{c(T_f - T_i) + L_F}\right).$$

Now $P^{\text{engine}} = 100 \times 10^6\,\text{W}$, $T_C = 0 + 273 = 273\,\text{K}$, $T_H = 800 + 273 = 1073\,\text{K}$, $T_i = -40 + 273 = 233\,\text{K}$, $T_f = 0 + 273 = 273\,\text{K}$, $c = 2220\,\text{J/kg} \cdot \text{K}$, and $L_F = 333 \times 10^3\,\text{J/kg}$, so

$$\frac{dm}{dt} = \left[\frac{(100 \times 10^6\,\text{J/s})(273\,\text{K})}{1073\,\text{K} - 273\,\text{K}}\right]\left[\frac{1}{(2220\,\text{J/kg} \cdot \text{K})(273\,\text{K} - 233\,\text{K}) + 333 \times 10^3\,\text{J/kg}}\right]$$

$$= 82\,\text{kg/s}.$$

Notice that the engine is now operated between $0°$ C and $800°$ C.

## 33

(a) The pressure at 2 is $P_2 = 3P_1$, as given in the problem statement. The volume is $V_2 = V_1 = nRT_1/P_1$. The temperature is $T_2 = P_2 V_2/nR = 3P_1 V_1/nR = 3T_1$.

The process $4 \rightarrow 1$ is adiabatic, so $P_4 V_4^\gamma = P_1 V_1^\gamma$ and

$$ P_4 = \left(\frac{V_1}{V_4}\right)^\gamma P_1 = \frac{P_1}{4^\gamma}, $$

since $V_4 = 4V_1$. The temperature at 4 is

$$ T_4 = \frac{P_4 V_4}{nR} = \left(\frac{P_1}{4^\gamma}\right)\left(\frac{4nRT_1}{P_1}\right)\left(\frac{1}{nR}\right) = \frac{T_1}{4^{\gamma-1}}. $$

The process $2 \rightarrow 3$ is adiabatic, so $P_2 V_2^\gamma = P_3 V_3^\gamma$ and $P_3 = (V_2/V_3)^\gamma P_2$. Substitute $V_3 = 4V_1$, $V_2 = V_1$, and $P_2 = 3P_1$ to obtain

$$ P_3 = \frac{3P_1}{4^\gamma}. $$

The temperature is

$$ T_3 = \frac{P_3 V_3}{nR} = \left(\frac{1}{nR}\right)\left(\frac{3P_1}{4^\gamma}\right)\left(\frac{4nRT_1}{P_1}\right) = \frac{3T_1}{4^{\gamma-1}}, $$

where $V_3 = V_4 = 4V_1 = 4nRT/P_1$ was used.

(b) The efficiency of the cycle is $\epsilon = W/Q_{12}$, where $W$ is the total work done by the gas during the cycle and $Q_{12}$ is the thermal energy absorbed during the $1 \rightarrow 2$ portion of the cycle, the only portion in which thermal energy is absorbed.

The work done during the portion of the cycle from 2 to 3 is $W_{23} = \int P\,dV$. Substitute $P = P_2 V_2^\gamma/V^\gamma$ to obtain

$$ W_{23} = P_2 V_2^\gamma \int_{V_2}^{V_3} V^{-\gamma}\,dV = \left(\frac{P_2 V_2^\gamma}{\gamma - 1}\right)\left(V_2^{1-\gamma} - V_3^{1-\gamma}\right). $$

Substitute $V_2 = V_1$, $V_3 = 4V_1$, and $P_3 = 3P_1$ to obtain

$$ W_{23} = \left(\frac{3P_1 V_1}{1-\gamma}\right)\left(1 - \frac{1}{4^{\gamma-1}}\right) = \left(\frac{3nRT_1}{\gamma-1}\right)\left(1 - \frac{1}{4^{\gamma-1}}\right). $$

Similarly, the work done during the portion of the cycle from 4 to 1 is

$$ W_{41} = \left(\frac{P_1 V_1^\gamma}{\gamma - 1}\right)\left(V_4^{1-\gamma} - V_1^{1-\gamma}\right) = -\left(\frac{P_1 V_1}{\gamma - 1}\right)\left(1 - \frac{1}{4^{\gamma-1}}\right) = -\left(\frac{nRT_1}{\gamma - 1}\right)\left(1 - \frac{1}{4^{\gamma-1}}\right). $$

No work is done during the $1 \rightarrow 2$ and $3 \rightarrow 4$ portions, so the total work done by the gas during the cycle is

$$ W = W_{23} + W_{41} = \left(\frac{2nRT_1}{\gamma - 1}\right)\left(1 - \frac{1}{4^{\gamma-1}}\right). $$

The thermal energy absorbed is $Q_{12} = nC_V(T_2 - T_1) = nC_V(3T_1 - T_1) = 2nC_V T_1$, where $C_V$ is the molar specific heat at constant volume. Now $\gamma = C_P/C_V = (C_V + R)/C_V = 1 + (R/C_V)$, so $C_V = R/(\gamma - 1)$. Here $C_P$ is the molar specific heat at constant pressure, which for an ideal gas is $C_P = C_V + R$. Thus $Q_{12} = 2nRT_1/(\gamma - 1)$. The efficiency is

$$\epsilon = \frac{2nRT_1}{\gamma - 1}\left(1 - \frac{1}{4^{\gamma-1}}\right)\frac{\gamma - 1}{2nRT_1} = 1 - \frac{1}{4^{\gamma-1}}.$$

## 37

The coefficient of performance for a refrigerator is given by $K = |Q_C|/|W|$, where $Q_C$ is the thermal energy absorbed from the cold reservoir and $W$ is the work done by the refrigerator, a negative value. The first law of thermodynamics yields $Q_H + Q_C - W = 0$ for an integer number of cycles. Here $Q_H$ is the thermal energy ejected to the hot reservoir. Thus $Q_C = W - Q_H$. $Q_H$ is negative and greater in magnitude than $W$, so $|Q_C| = |Q_H| - |W|$. Thus

$$K = \frac{|Q_H| - |W|}{|W|}.$$

The solution for $|W|$ is $|W| = |Q_H|/(K + 1)$. In one hour,

$$|W| = \frac{7.54\,\text{MJ}}{3.8 + 1} = 1.57\,\text{MJ}.$$

The rate at which work is done is $(1.57 \times 10^6\,\text{J})/(3600\,\text{s}) = 440\,\text{W}$.

## 41

The efficiency of the engine is defined by $\epsilon = W/Q_1$ and is shown in the text to be $\epsilon = (T_1 - T_2)/T_1$, so $W/Q_1 = (T_1 - T_2)/T_1$. The coefficient of performance of the refrigerator is defined by $K = Q_4/W$ and is shown in the text to be $K = T_4/(T_3 - T_4)$, so $Q_4/W = T_4/(T_3 - T_4)$. Now $Q_4 = Q_3 - W$, so $(Q_3 - W)/W = T_4/(T_3 - T_4)$. The work done by the engine is used to drive the refrigerator, so $W$ is the same for the two. Solve the engine equation for $W$ and substitute the resulting expression into the refrigerator equation. The engine equation yields $W = (T_1 - T_2)Q_1/T_1$ and the substitution yields

$$\frac{T_4}{T_3 - T_4} = \frac{Q_3}{W} - 1 = \frac{Q_3 T_1}{Q_1(T_1 - T_2)} - 1.$$

Solve for $Q_3/Q_1$:

$$\frac{Q_3}{Q_1} = \left(\frac{T_4}{T_3 - T_4} + 1\right)\left(\frac{T_1 - T_2}{T_1}\right) = \left(\frac{T_3}{T_3 - T_4}\right)\left(\frac{T_1 - T_2}{T_1}\right) = \frac{1 - (T_2/T_1)}{1 - (T_4/T_3)}.$$

## 49

Since the gas is ideal and the process is isothermal the change in the internal energy of the gas is zero and, according to the first law of thermodynamics, $Q = W$, where $Q$ is the thermal energy absorbed by the gas and $W$ is the work done by the gas. Thus

$$Q = \int_{V_i}^{V_f} P\,dV = nRT \int_{V_i}^{V_f} \frac{dV}{V} = nRT\,ln\frac{V_f}{V_i},$$

where $V_i$ is the initial volume, $T$ is the temperature, and $n$ is the number of moles of gas. The change in entropy is

$$\Delta S = \frac{Q}{T} = nR\ln\frac{V_f}{V_i}.$$

According to the graph $\Delta S = 0$ for $V_f = 0.70\,\mathrm{m}^3$. Now $\Delta S = 0$ for $V_f - V_i$, so $V_i = 0.40\,\mathrm{m}^3$. Now take $V_f$ to be equal to $3.6\,\mathrm{m}^3$, for which $\Delta S = 64\,\mathrm{J/K}$. The solution for $n$ is

$$n = \frac{\Delta S}{R\ln\dfrac{V_f}{V_i}} = \frac{64\,\mathrm{J/K}}{(8.31\,\mathrm{J/mol\cdot K})\ln\dfrac{3.6\,\mathrm{m}^3}{0.40\,\mathrm{m}^3}} = 3.5\,\mathrm{mol}.$$

# Chapter 22

## 9

The magnitude of the force of one particle on the other is given by Coulomb's law:

$$F^{elec} = k \frac{|q_A||q_B|}{r^2},$$

where $r$ is the distance between the particles. The solution for $r$ is

$$r = \sqrt{\frac{k|q_A||q_B|}{F^{elec}}} = \sqrt{\frac{(8.99 \times 10^9 \, \text{N} \cdot \text{m}^2/\text{C})(26.0 \times 10^{-6} \, \text{C})(47.0 \times 10^{-6} \, \text{C})}{5.70 \, \text{N}}} = 1.39 \, \text{m}.$$

## 15

Assume the spheres are far apart. Then the charge distribution on each of them is spherically symmetric and Coulomb's law can be used. Let $q_A$ and $q_B$ be the original charges and choose the coordinate system so the force on sphere A is positive if it is repelled by sphere B. Take the distance between the spheres to be $r$. Then the component of the force of sphere A on sphere B is

$$F_1^{elec} = -k \frac{q_A q_B}{r^2}.$$

The negative sign indicates that the spheres attract each other.

After the wire is connected, the spheres, being identical, have the same charge. Since charge is conserved, the total charge is the same as it was originally. This means the charge on each sphere is $(q_A + q_B)/2$. The force is now one of repulsion and is given by

$$F_2^{elec} = k \frac{(q_A + q_B)^2}{4r^2}.$$

Solve the two force equations simultaneously for $q_A$ and $q_B$. The first gives

$$q_A q_B = -\frac{r^2 F_1^{elec}}{k} = -\frac{(0.500 \, \text{m})^2 (0.108 \, \text{N})}{8.99 \times 10^9 \, \text{N} \cdot \text{m}^2/\text{C}^2} = -3.00 \times 10^{-12} \, \text{C}^2$$

and the second gives

$$q_A + q_B = 2r\sqrt{\frac{F_b}{k}} = 2(0.500 \, \text{m})\sqrt{\frac{0.0360 \, \text{N}}{8.99 \times 10^9 \, \text{N} \cdot \text{m}^2/\text{C}^2}} = 2.00 \times 10^{-6} \, \text{C}.$$

Thus

$$q_B = \frac{-(3.00 \times 10^{-12} \, \text{C}^2)}{q_A}$$

and

$$q_A - \frac{3.00 \times 10^{-12}\, \text{C}^2}{q_A} = 2.00 \times 10^{-6}\, \text{C}.$$

Multiply by $q_A$ to obtain the quadratic equation

$$q_A^2 - (2.00 \times 10^{-6}\, \text{C})q_1 - 3.00 \times 10^{-12}\, \text{C}^2 = 0.$$

The solutions are

$$q_A = \frac{2.00 \times 10^{-6}\, \text{C} \pm \sqrt{(-2.00 \times 10^{-6}\, \text{C})^2 + 4(3.00 \times 10^{-12}\, \text{C}^2)}}{2}.$$

If the positive sign is used, $q_A = 3.00 \times 10^{-6}\, \text{C}$ and if the negative sign is used, $q_A = -1.00 \times 10^{-6}\, \text{C}$. Use $q_B = (-3.00 \times 10^{-12})/q_A$ to calculate $q_B$. If $q_A = 3.00 \times 10^{-6}\, \text{C}$, then $q_B = -1.00 \times 10^{-6}\, \text{C}$ and if $q_A = -1.00 \times 10^{-6}\, \text{C}$, then $q_B = 3.00 \times 10^{-6}\, \text{C}$. Since the spheres are identical, the solutions are essentially the same: one sphere originally had charge $-1.00 \times 10^{-6}\, \text{C}$ and the other had charge $+3.00 \times 10^{-6}\, \text{C}$.

Another solution exists. If the signs of the charges are reversed, the forces remain the same, so a charge of $-1.00 \times 10^{-6}\, \text{C}$ on one sphere and a charge of $3.00 \times 10^{-6}\, \text{C}$ on the other also satisfies the conditions of the problem.

## 17

(a) If the system of three particles is to be in equilibrium, the force on each particle must be zero. Let the charge of the third particle be $q_0$. This particle must lie between the other two or else the forces acting on it due to the other charges would be in the same direction and the particle could not be in equilibrium. Suppose the third particle is a distance $x$ from the particle with charge $q$, as shown on the diagram to the right. The $x$ component of the force on the third particle is then given by

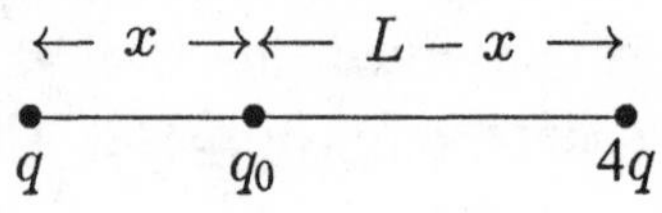

$$F_{0x}^{\text{elec}} = k \left[ \frac{qq_0}{x^2} - \frac{4qq_0}{(L-x)^2} \right] = 0,$$

where the positive direction was taken to be toward the right. Solve this equation for $x$. Canceling common factors yields $1/x^2 = 4/(L-x)^2$ and taking the square root yields $1/x = 2/(L-x)$. The solution is $x = L/3$.

The $x$ component of the force on the particle with charge $q$ is

$$F_{qx}^{\text{elec}} = k \left[ \frac{qq_0}{x^2} + \frac{4q^2}{L^2} \right] = 0.$$

The solution for $q_0$ is $q_0 = -4qx^2/L^2 = -(4/9)q$, where $x = L/3$ was used.

The $x$ component of the force on the particle with charge $4q$ is

$$F_{4qx}^{elec} = k \left[ \frac{4q^2}{L^2} + \frac{4qq_0}{(L-x)^2} \right] = \frac{1}{4\pi\epsilon_0} \left[ \frac{4q^2}{L^2} + \frac{4(-4/9)q^2}{(4/9)L^2} \right]$$

$$= \frac{1}{4\pi\epsilon_0} \left[ \frac{4q^2}{L^2} - \frac{4q^2}{L^2} \right] = 0 \,.$$

With $q_0 = -(4/9)q$ and $x = L/3$, all three charges are in equilibrium.

(b) If the third particle moves toward the particle with charge $q$ the force of attraction of that particle on the third particle is greater in magnitude than the force of attraction exerted by the particle with charge $4q$ and the third particle continues to move away from its initial position. The equilibrium is unstable.

## 21

(a) The magnitude of the force between the ions is given by $F^{elec} = kq^2/r^2$, where $q$ is the magnitude of the charge on either of them and $r$ is the distance between them. Solve for the magnitude of the charge:

$$q = r\sqrt{\frac{F^{elec}}{k}} = (5.0 \times 10^{-10}\,\mathrm{m})\sqrt{\frac{3.7 \times 10^{-9}\,\mathrm{N}}{8.99 \times 10^9\,\mathrm{N\cdot m^2/C^2}}} = 3.2 \times 10^{-19}\,\mathrm{C} \,.$$

(b) Let $N$ be the number of electrons missing from each ion. Then $Ne = q$, or

$$N = \frac{q}{e} = \frac{3.2 \times 10^{-19}\,\mathrm{C}}{1.60 \times 10^{-19}\,\mathrm{C}} = 2 \,.$$

## 27

When particle C is at $x = 0.40\,\mathrm{m}$ the net force on particle B is zero. This means that the magnitude of the force of C is the same as the magnitude of the force of A. The charge of A must be positive and A must be located at $x = -0.40\,\mathrm{m}$. Then the forces of A on B and C on B have the same magnitude and are in opposite directions.

When C is far away the only force on B is the force of A and its component is given by

$$F_{Bx}^{elec} = k\frac{q_A q_B}{x_A^2} \,,$$

where $x_A$ is the coordinate of particle A. The solution for $q_B$ is

$$q_B = \frac{x_A^2 F_{Bx}^{elec}}{kq_A} = \frac{(0.40\,\mathrm{m})^2(1.5 \times 10^{-25}\,\mathrm{N})}{(8.99 \times 10^9\,\mathrm{N\cdot m^2/C^2})(8.00)(1.60 \times 10^{-19}\,\mathrm{C})} = 2.086 \times 10^{-18}\,\mathrm{C} \,.$$

This is $(2.086 \times 10^{-18}\,\mathrm{C})e/(1.60 \times 10^{-19}\,\mathrm{C}) = 13.0e$.

## 29

(a) When particle C is close to particle A the force of particle A on it is much larger than the force of particle B. The graph tells us that this force is in the positive $x$ direction, so the force of A on C must be a repulsive force and the charges on A and C must have the same sign. The charge of C is positive, so the charge of A is positive.

(b) When C is at $x = 2.0$ cm the net force on C is zero, so the forces of A on C and B on C have the same magnitude. Thus

$$k\frac{|q_A|\,|q_C|}{x^2} = k\frac{|q_B|\,|q_C|}{(x_B - x)^2}.$$

This means

$$\frac{|q_B|}{|q_A|} = \frac{(x_B - x)^2}{x^2} = \frac{(8.00\,\text{cm} - 2.0\,\text{cm})^2}{(2.00\,\text{cm})^2} = 9.0.$$

The forces of A and B on C are in opposite directions. Thus the force of B on C is a repulsive force and the charge of B is positive. The ratio is $q_A/q_B = +9.0$.

## 31

Let $Q$ (= 6.0 $\mu$C) be the original charge and suppose that, after the split, one of particles has charge $q$ and the other has charge $Q - q$. Then the electrostatic force of one of the particles on the other has magnitude

$$F^{\text{elec}} = k\frac{q(Q - q)}{r^2},$$

where $r$ is their separation. The derivative of $F^{\text{elec}}$ with respect to $q$ is

$$\frac{dF^{\text{elec}}}{dq} = k\frac{Q - 2q}{r^2}.$$

If $F^{\text{elec}}$ is to have its maximum value the derivative must be zero, so $q = Q/2$. The second derivative of $F^{\text{elec}}$ is negative, indicating that the force is indeed maximum for $q = Q/2$. The magnitude of the maximum force is

$$F^{\text{elec}}_{\text{max}} = k\frac{(Q/2)^2}{r^2} = (8.99 \times 10^9\,\text{N}\cdot\text{m}^2/\text{C}^2)\frac{(3.0 \times 10^{-6}\,\text{C})^2}{(3.0 \times 10^{-3}\,\text{m})^2} = 9.0 \times 10^3\,\text{N}.$$

## 33

Assume the charge of the third particle is positive. If the particle is to the left of the origin the forces of the fixed particles on it are in opposite directions but it is closer to the particle with the larger magnitude charge so the forces cannot sum to zero. If the third particle is between the fixed charges both forces are to the left and cannot sum to zero. The third particle must be to the right of $x = a$. Place the origin of an $x$ axis at the fixed particle on the right and take $x$ to be the coordinate of the third particle. Let $q_A$ be the charge of the fixed particle on the left, $q_B$ be the charge of the fixed particle on the right, and $q_C$ be the charge of the third particle. Then the $x$ component of the net force on the third particle is

$$F^{\text{elec}}_x = k\frac{q_A q_C}{(x + a)^2} + k\frac{q_B q_C}{x^2}.$$

This is zero if

$$\frac{q_A}{(x + a)^2} = -\frac{q_B}{x^2}.$$

The solution for $x$ is

$$x = \frac{\sqrt{q_B}}{\sqrt{-q_A} - \sqrt{q_B}} a = \frac{\sqrt{2.00q}}{\sqrt{5.00q} - \sqrt{2.00q}} a = 1.72a \,.$$

## 35

The electric force of the particle with charge $Q$ is the centripetal force that keeps the particle with charge $q$ moving in uniform circular motion, so

$$k \frac{|Q| q}{r^2} = \frac{mv^2}{r} \,,$$

where $r$ is the radius of the circular path, and $v$ is the speed of the particle. The radius is $r = 20.0 \, \text{cm}$. The solution for $|Q|$ is

$$|Q| = \frac{mv^2 r}{kq} = \frac{(0.800 \times 10^{-3} \, \text{kg})(50.0 \, \text{m/s})^2 (0.200 \, \text{m})}{(8.99 \times 10^9 \, \text{N} \cdot \text{m}^2/\text{C}^2)(4.00 \times 10^{-6} \, \text{C})} = 1.11 \times 10^{-5} \, \text{C} \,.$$

The force must be one of attraction, so $Q$ and $q$ must have opposite signs. Thus $Q = -1.11 \times 10^{-5} \, \text{C}$.

## 45

The free-body diagram for the ball on the left is shown. $\vec{T}$ is the tension force of the string, $\vec{F}^{\text{elec}}$ is the electrical force of the other ball, and $\vec{F}^{\text{grav}}$ is the gravitational force of Earth. Take the positive $x$ direction to be to the right and the positive $y$ direction to be upward. Then the $x$ component of Newton's second law is $T \sin\theta - F^{\text{grav}} = 0$ and the $y$ component is $T \cos\theta - F^{\text{grav}} = 0$. Substitute $T = F^{\text{grav}}/\cos\theta$, from the second equation, into the first equation, to obtain $F^{\text{grav}} \tan\theta - F^{\text{elec}} = 0$. Now $F^{\text{grav}} = mg$, where $m$ is the mass of the ball and, according to Coulomb's law $F^{\text{elec}} = kq^2/x^2$. Thus

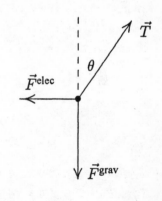

$$mg \tan\theta = k \frac{q^2}{x^2} \,.$$

In addition, $\tan\theta = (x/2)/\sqrt{L^2 - (x/2)^2}$ and if $x$ is much smaller than $x/2$ it can be approximated as $\tan\theta = x/2L$. Then

$$\frac{mgx}{2L} = k \frac{q^2}{x^2} \,.$$

The solution for $x$ is

$$x = \left( \frac{2kq^2 L}{mg} \right)^{1/3} \,.$$

**47**

Since the rod is in equilibrium the net force on it is zero and the net torque about any point is also zero. Compute the torques about the bearing. Take the positive $y$ direction to be upward and the positive $z$ direction to be out of the page. The sphere on the left exerts a force of $\vec{F}_L = k(Qq/h^2)\hat{j}$ and a torque of $\vec{\tau}_L = -k(Qq/h^2)(L/2)\hat{k}$. The hanging weight exerts a force of $\vec{F}_W = -W\hat{j}$ and a torque of $\vec{\tau}_W = -W(x - L/2)\hat{k}$. The sphere on the right exerts a force of $\vec{F}_R = k(2Qq/h^2)\hat{j}$ and a torque of $\vec{\tau}_R = k(2Qq/h^2)(L/2)\hat{k}$. The bearing exerts a force of $\vec{F}_B = F_B\hat{j}$. The $z$ component of the net torque is

$$\tau_z^{net} = -\frac{kQqL}{2h^2} - W(x - L/2) + \frac{kQqL}{h^2}.$$

Set this equal to zero and solve for $x$. The result is

$$x = \frac{L}{2}\left(1 + \frac{kQq}{h^2 W}\right).$$

The $y$ component of the net force is

$$F_y^{net} = \frac{kQq}{h^2} - W + \frac{2kQq}{h^2} + F_B.$$

Set this equal to zero and set $F_B$ equal to zero, then solve for $h$. The result is

$$h = \sqrt{\frac{3kQq}{W}}.$$

# Chapter 23

## 5

At points between the charges, the individual electric fields are in the same direction and do not cancel. Charge $q_B$ has a greater magnitude than charge $q_A$, so a point of zero field must be closer to particle A than to particle B. It must be to the right of particle A on the diagram.

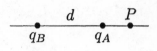

Put the origin at particle B and let $x$ be the coordinate of $P$, the point where the field vanishes. Then the net electric field at $P$ is given by

$$E = k \left[ \frac{|q_B|}{x^2} - \frac{|q_A|}{(x-d)^2} \right] .$$

If the field is to vanish,

$$\frac{|q_2|}{x^2} = \frac{|q_1|}{(x-d)^2} .$$

Take the square root of both sides to obtain $\sqrt{|q_2|}/x = \sqrt{|q_1|}/(x-d)$. The solution for $x$ is

$$x = \left( \frac{\sqrt{|q_B|}}{\sqrt{|q_B|} - \sqrt{|q_A|}} \right) d = \left( \frac{\sqrt{4.0|q_A|}}{\sqrt{4.0|q_A|} - \sqrt{|q_A|}} \right) d$$

$$= \left( \frac{2.0}{2.0 - 1.0} \right) d = 2.0d = (2.0)(50\,\text{cm}) = 100\,\text{cm} .$$

The point is 50 cm to the right of particle A.

## 9

Choose the coordinate axes as shown on the diagram to the right. At the center of the square, the electric fields produced by the particles at the lower left and upper right corners are both along the $x$ axis and each points away from the center and toward the particle that produces it. Since each particle is a distance $d = \sqrt{2}a/2 = a/\sqrt{2}$ away from the center, the net field due to these two particles is

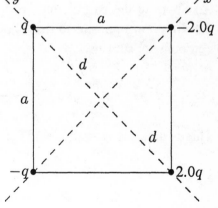

$$E_x = k \left[ \frac{2q}{a^2/2} - \frac{q}{a^2/2} \right]$$

$$= \frac{kq}{a^2/2} = \frac{(8.99 \times 10^9\,\text{N} \cdot \text{m}^2/\text{C}^2)(1.0 \times 10^{-8}\,\text{C})}{(0.050\,\text{m})^2/2} = 7.19 \times 10^4\,\text{N/C}.$$

At the center of the square, the field produced by the particles at the upper left and lower right corners are both along the $y$ axis and each points away from the charge that produces it. The net field produced at the center by these particles is

$$E_y = k\left[\frac{2q}{a^2/2} - \frac{q}{a^2/2}\right] = \frac{kq}{a^2/2} = 7.19 \times 10^4\,\text{N/C}.$$

The magnitude of the field is

$$E = \sqrt{E_x^2 + E_y^2} = \sqrt{2(7.19 \times 10^4\,\text{N/C})^2} = 1.02 \times 10^5\,\text{N/C}$$

and the angle it makes with the $x$ axis is

$$\theta = \tan^{-1}\frac{E_y}{E_x} = \tan^{-1}(1) = 45°.$$

It is upward in the diagram, from the center of the square toward the center of the upper side.

## 11

(a) Use the coordinate system shown on the right. The electric fields produced at P by the two particles have the same magnitude and point in the directions shown, parallel to the triangle sides. $\vec{E}_R$ is the field of the particle on the right and $\vec{E}_L$ is the field of the particle on the left. The $x$ components of the fields sum to zero and the $y$ components sum to

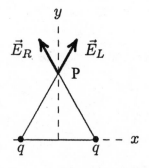

$$E_y = \frac{2kq}{L^2}\sin 60°$$

$$= \frac{2(8.99 \times 10^9\,\text{N} \cdot \text{m}^2/\text{C}^2)(12 \times 10^{-9}\,\text{C})}{(2.0\,\text{m})^2}\sin 60° = 47\,\text{N/C}.$$

This is also the magnitude of the net electric field.

(b) Suppose the particle on the right is negatively charged. Then its electric field at P points in the direction opposite to that shown in the diagram. The $y$ components sum to zero and the $x$ components sum to

$$E_x = \frac{2k|q|}{L^2}\cos 60° = \frac{2(8.99 \times 10^9\,\text{N} \cdot \text{m}^2/\text{C}^2)(12 \times 10^{-9}\,\text{C})}{(2.0\,\text{m})^2}\cos 60° = 27\,\text{N/C}.$$

This is also the magnitude of the net electric field.

## 19

Think of the quadrupole as composed of two dipoles, each with dipole moment of magnitude $p = qd$. The moments point in opposite directions and produce fields in opposite directions at points on the quadrupole axis. Consider the point P on the axis, a distance $z$ to the right of the quadrupole center and take a rightward pointing field to be positive. Then the field component produced by the right dipole of the pair is $2kqd/(z-d/2)^3$ and the field component produced by the left dipole is $-2kqd/(z+d/2)^3$, where Eq. 23–18 was used with $z-d/2$ or $z+d/2$ replacing $z$. Use the binomial expansions $(z-d/2)^{-3} \approx z^{-3} - 3z^{-4}(-d/2)$ and $(z+d/2)^{-3} \approx z^{-3} - 3z^{-4}(d/2)$ to obtain

$$E = 2kqd$$
$$\left[ \frac{1}{z^3} + \frac{3d}{2z^4} - \frac{1}{z^3} + \frac{3d}{2z^4} \right] = \frac{6kqd^2}{z^4}.$$

Let $Q = 2qd^2$. Then $E = 3kQ/z^4$.

## 21

The electric field at a point on the axis of a uniformly charged ring, a distance $z$ from the ring center, is given by

$$E_z = \frac{kqz}{(z^2 + R^2)^{3/2}},$$

where $q$ is the charge on the ring and $R$ is the radius of the ring (see Eq. 23–26). Here the $z$ axis is along the symmetry axis of the ring. For $q$ positive, the field points upward at points above the ring and downward at points below the ring. Take the positive $z$ direction to be upward. Then the vertical component of the force on an electron with coordinate $z$ on the axis is

$$F_z^{\text{elec}} = -\frac{keqz}{(z^2 + R^2)^{3/2}}.$$

For small amplitude oscillations $z$ is much less than $R$ and can be neglected in the denominator. Thus

$$F_z^{\text{elec}} = -\frac{keqz}{R^3}.$$

The force is a restoring force: it pulls the electron toward the equilibrium point $z = 0$. Furthermore, the magnitude of the force is proportional to $z$, just as if the electron were attached to a spring with spring constant $K = keq/R^3$. The electron moves in simple harmonic motion with an angular frequency given by

$$\omega = \sqrt{\frac{K}{m}} = \sqrt{\frac{keq}{mR^3}} = \sqrt{\frac{eq}{2\pi\epsilon_0 mR^3}},$$

where $m$ is the mass of the electron. The last form was obtained by replacing $k$ with $1/4\pi\epsilon_0$.

## 25

(a) The linear charge density $\lambda$ is the charge per unit length of rod. Since the charge is uniformly distributed on the rod, $\lambda = -q/L$.

(b) Position the $x$ axis along the rod with the origin at the left end of the rod, as shown in the diagram. Let $dx$ be an infinitesimal length of rod at $x$. The charge in this segment is $dq = \lambda\, dx$. The segment may be considered to be

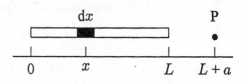

a point particle with charge $dq$. The electric field it produces at point P has only an $x$ component and this component is given by

$$dE_x = \frac{k\lambda\, dx}{(L+a-x)^2}.$$

The total electric field produced at P by the whole rod is the integral

$$E_x = k\lambda \int_0^L \frac{dx}{(L+a-x)^2} = \frac{k\lambda}{L+a-x}\bigg|_0^L = k\lambda \left[\frac{1}{a} - \frac{1}{L+a}\right] = \frac{k\lambda L}{a(L+a)}.$$

When $-q/L$ is substituted for $\lambda$ the result is $E_x = -kq/a(L+a)$. The negative sign indicates that the field is toward the rod.

(c) If $a$ is much larger than $L$, the quantity $L+a$ in the denominator can be approximated by $a$ and the expression for the electric field becomes $E_x = -kq/a^2$. This is the expression for the electric field of a point particle with charge $q$, located at the origin.

## 27

Consider an infinitesimal section of the rod of length $dx$, a distance $x$ from the left end, as shown in the diagram to the right. It contains charge $dq = \lambda\, dx$ and is a distance $r$ from $P$. The magnitude of the field it produces at $P$ is given by

$$dE = \frac{k\lambda\, dx}{r^2}.$$

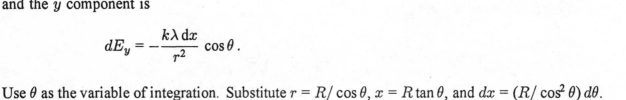

The $x$ component is

$$dE_x = -\frac{k\lambda\, dx}{r^2}\sin\theta$$

and the $y$ component is

$$dE_y = -\frac{k\lambda\, dx}{r^2}\cos\theta.$$

Use $\theta$ as the variable of integration. Substitute $r = R/\cos\theta$, $x = R\tan\theta$, and $dx = (R/\cos^2\theta)\, d\theta$. The limits of integration are 0 and $\pi/2$ rad. Thus

$$E_x = -\frac{k\lambda}{R}\int_0^{\pi/2}\sin\theta\, d\theta = \frac{k\lambda}{R}\cos\theta\bigg|_0^{\pi/2} = -\frac{k\lambda}{R}$$

and

$$E_y = -\frac{k\lambda}{R}\int_0^{\pi/2}\cos\theta\, d\theta = -\frac{k\lambda}{R}\sin\theta\bigg|_0^{\pi/2} = -\frac{k\lambda}{R}.$$

Notice that $E_x = E_y$ no matter what the value of $R$. Thus $\vec{E}$ makes an angle of 45° with the rod for all values of $R$.

**39**

(a) The magnitude of the force on the particle is given by $F^{\text{elec}} = |q|E$, where $q$ is the particle's charge and $E$ is the magnitude of the electric field at the location of the particle. Thus

$$E = \frac{F^{\text{elec}}}{|q|} = \frac{3.0 \times 10^{-6}\,\text{N}}{2.0 \times 10^{-9}\,\text{C}} = 1.5 \times 10^{3}\,\text{N/C}.$$

The force points downward and the charge is negative, so the field must be pointing upward.

(b) The magnitude of the electrostatic force on a proton is

$$F_p^{\text{elec}} = eE = (1.60 \times 10^{-19}\,\text{C})(1.5 \times 10^{3}\,\text{N/C}) = 2.4 \times 10^{-16}\,\text{N}.$$

A proton is positively charged, so the force is in the same direction as the field, upward.

(c) The magnitude of the gravitational force on the proton is

$$F_p^{\text{grav}} = mg = (1.67 \times 10^{-27}\,\text{kg})(9.8\,\text{m/s}^2) = 1.64 \times 10^{-26}\,\text{N}.$$

The force is downward.

(d) The ratio of the force magnitudes is

$$\frac{F_p^{\text{elec}}}{F_p^{\text{grav}}} = \frac{2.4 \times 10^{-16}\,\text{N}}{1.64 \times 10^{-26}\,\text{N}} = 1.5 \times 10^{10}.$$

**45**

Take the positive $x$ direction to be to the right in the figure. The $x$ component of the acceleration of the proton is $a_{px} = eE/m_p$ and the $x$ component of the acceleration of the electron is $a_{ex} = -eE/m_e$, where $E$ is the magnitude of the electric field, $m_p$ is the mass of the proton, and $m_e$ is the mass of the electron. Take the origin to be at the initial position of the proton. Then the coordinate of the proton at time $t$ is $x_p = \frac{1}{2}a_{px}t^2$ and the coordinate of the electron is $x_e = L + \frac{1}{2}a_{ex}t^2$. They pass each other when their coordinates are the same, or $\frac{1}{2}a_{px}t^2 = L + \frac{1}{2}a_{ex}t^2$. This means $t^2 = 2L/(a_{px} - a_{ex})$ and

$$x_p = \frac{a_{px}}{a_{px} - a_{ex}}L = \frac{eE/m_p}{(eE/m_p) + (eE/m_e)}L = \frac{m_e}{m_e + m_p}L$$

$$= \frac{9.11 \times 10^{-31}\,\text{kg}}{9.11 \times 10^{-31}\,\text{kg} + 1.67 \times 10^{-27}\,\text{kg}}(0.050\,\text{m}) = 2.7 \times 10^{-5}\,\text{m}.$$

**47**

The electric field is upward in the diagram and the charge is negative, so the force of the field on it is downward. The magnitude of the acceleration is $a = eE/m$, where $E$ is the magnitude of the field and $m$ is the mass of the electron. Its numerical value is

$$a = \frac{(1.60 \times 10^{-19}\,\text{C})(2.00 \times 10^{3}\,\text{N/C})}{9.11 \times 10^{-31}\,\text{kg}} = 3.51 \times 10^{14}\,\text{m/s}^2.$$

Put the origin of a coordinate system at the initial position of the electron. Take the $x$ axis to be horizontal and positive to the right; take the $y$ axis to be vertical and positive toward the top of the page. The kinematic equations are $x = v_1 t \cos\theta$, $y = v_1 t \sin\theta - \frac{1}{2}at^2$, and $v_y = v_1 \sin\theta - at$.

First find the greatest $y$ coordinate attained by the electron. If it is less than $d$, the electron does not hit the upper plate. If it is greater than $d$, it will hit the upper plate if the corresponding $x$ coordinate is less than $L$. The greatest $y$ coordinate occurs when $v_y = 0$. This means $v_1 \sin\theta - at = 0$ or $t = (v_1/a)\sin\theta$ and

$$y_{max} = \frac{v_1^2 \sin^2\theta}{a} - \frac{1}{2}a\frac{v_1^2 \sin^2\theta}{a^2} = \frac{1}{2}\frac{v_1^2 \sin^2\theta}{a}$$

$$= \frac{(6.00 \times 10^6 \text{ m/s})^2 \sin^2 45°}{2(3.51 \times 10^{14} \text{ m/s}^2)} = 2.56 \times 10^{-2} \text{ m}.$$

Since this is greater than $d$ (= 2.00 cm), the electron might hit the upper plate. Now find the $x$ coordinate of the position of the electron when $y = d$. Since

$$v_1 \sin\theta = (6.00 \times 10^6 \text{ m/s}) \sin 45° = 4.24 \times 10^6 \text{ m/s}$$

and

$$2ad = 2(3.51 \times 10^{14} \text{ m/s}^2)(0.0200 \text{ m}) = 1.40 \times 10^{13} \text{ m}^2/\text{s}^2,$$

the solution to $d = v_1 t \sin\theta - \frac{1}{2}at^2$ is

$$t = \frac{v_1 \sin\theta - \sqrt{v_1^2 \sin^2\theta - 2ad}}{a}$$

$$= \frac{4.24 \times 10^6 \text{ m/s} - \sqrt{(4.24 \times 10^6 \text{ m/s})^2 - 1.40 \times 10^{13} \text{ m}^2/\text{s}^2}}{3.51 \times 10^{14} \text{ m/s}^2} = 6.43 \times 10^{-9} \text{ s}.$$

The negative root was used because we want the *earliest* time for which $y = d$. The $x$ coordinate is

$$x = v_1 t \cos\theta$$

$$= (6.00 \times 10^6 \text{ m/s})(6.43 \times 10^{-9} \text{ s}) \cos 45° = 2.72 \times 10^{-2} \text{ m}.$$

This is less than $L$ so the electron hits the upper plate at $x = 2.72$ cm.

## 55

The magnitude of the torque acting on the dipole is given by $\tau = pE\sin\theta$, where $p$ is the magnitude of the dipole moment, $E$ is the magnitude of the electric field, and $\theta$ is the angle between the dipole moment and the field. It is a restoring torque: it always tends to rotate the dipole moment toward the direction of the electric field. If $\theta$ is positive, the torque is negative and vice versa. Write $\tau = -pE\sin\theta$. If the amplitude of the motion is small, we may replace $\sin\theta$ with $\theta$ in radians. Thus $\tau = -pE\theta$. Since the magnitude of the torque is proportional to the angle of rotation, the dipole oscillates in simple harmonic motion, just like a torsional pendulum with torsion constant $\kappa = pE$. The angular frequency $\omega$ is given by

$$\omega^2 = \frac{\kappa}{I} = \frac{pE}{I},$$

where $I$ is the rotational inertia of the dipole. The frequency of oscillation is

$$f = \frac{\omega}{2\pi} = \frac{1}{2\pi}\sqrt{\frac{pE}{I}}.$$

# Chapter 24

## 5

(a) The electric field is in the $y$ direction so we need to consider only the cube faces at $y = 0$ and $y = a$, where $a$ is the length of a cube edge. The field is zero for $y = 0$ so the flux at the left face is zero. The magnitude of the field at the right face is $E = (3.00\,\text{N/C} \cdot \text{m})(1.40\,\text{m}) = 4.20\,\text{N/C}$. The flux is $\Phi = \vec{E} \cdot \vec{A} = (4.20\,\text{N/C})\hat{j} \cdot (1.40\,\text{m}^2)\hat{j} = 8.23\,\text{N} \cdot \text{m}^2/\text{C}$.

(b) The fluxes at the left and right faces are the same as for part (a). The flux at the front face is $(E_x\,\hat{i}) \cdot (A\,\hat{i}) = E_x A$, where $E_x = -4.00\,\text{N/C}$. The flux at the back face is $(E_x\,\hat{i}) \cdot (-A\,\hat{i}) = -E_x A$. The sum of these is zero, so the net flux is the same as for part (a), $8.23\,\text{N} \cdot \text{m}^2/\text{C}$.

(c) According to Gauss' law the charge enclosed by the cube is $q^{\text{enc}} = \epsilon_0 \Phi = (8.85 \times 10^{-12}\,\text{C}^2/\text{N} \cdot \text{m}^2)(8.23\,\text{N} \cdot \text{m}^2/\text{C} = 7.29 \times 10^{-11}\,\text{C}$.

## 9

According to Gauss' law the net flux at any surface that completely surrounds a point particle with charge $q$ is $q/\epsilon_0$. If you stack identical cubes side by side and directly on top of each other, you will find that eight cubes meet at any corner. Thus one-eighth of the field lines emanating from the point particle pass through a cube with a corner at the particle and the net flux at the surface of such a cube is $q/8\epsilon_0$. Now the field lines are radial, so at each of the three cube faces that meet at the charge, the lines are parallel to the face and the flux at the face is zero. The fluxes at each of the other three faces are the same, so the flux at each of them is one-third the total. That is, the flux at each of these faces is $(1/3)(q/8\epsilon_0) = q/24\epsilon_0$.

## 13

Use Gauss' law to find an expression for the magnitude of the electric field a distance $r$ from the center of the atom. The field is radially outward and is uniform over any sphere centered at the atom's center. Take the Gaussian surface to be a sphere of radius $r$ with its center at the center of the atom. If $E$ is the magnitude of the field, then the net flux at the Gaussian sphere is $\Phi = 4\pi r^2 E$. The charge enclosed by the Gaussian surface is the positive charge at the center of the atom and that portion of the negative charge within the surface. Since the negative charge is uniformly distributed throughout a sphere of radius $R$, we can compute the charge inside the Gaussian sphere using a ratio of volumes. That is, the negative charge inside is $-Zer^3/R^3$. Thus the net charge enclosed is $Ze - Zer^3/R^3$. Gauss' law yields

$$4\pi\epsilon_0 r^2 E = Ze\left(1 - \frac{r^3}{R^3}\right).$$

Solve for $E$:

$$E = \frac{Ze}{4\pi\epsilon_0}\left(\frac{1}{r^2} - \frac{r}{R^3}\right).$$

**19**

At all points where there is an electric field, it is radially outward. For each part of the problem, use a Gaussian surface in the form of a sphere that is concentric with the charged sphere and passes through the point where the electric field is to be found. The field is uniform on the surface, so

$$\oint \mathbf{E} \cdot d\mathbf{A} = 4\pi r^2 E,$$

where $r$ is the radius of the Gaussian surface.

(a) Here $r$ is less than $a$ and the charge enclosed by the Gaussian surface is $q(r/a)^3$. Gauss' law yields

$$4\pi r^2 E = \left(\frac{q}{\epsilon_0}\right)\left(\frac{r}{a}\right)^3,$$

so

$$E = \frac{qr}{4\pi\epsilon_0 a^3}.$$

(b) Here $r$ is greater than $a$ but less than $b$. The charge enclosed by the Gaussian surface is $q$, so Gauss' law becomes

$$4\pi r^2 E = \frac{q}{\epsilon_0}$$

and

$$E = \frac{q}{4\pi\epsilon_0 r^2}.$$

(c) The shell is conducting, so the electric field inside it is zero.

(d) For $r > c$, the charge enclosed by the Gaussian surface is zero (charge $q$ is inside the shell cavity and charge $-q$ is on the shell). Gauss' law yields

$$4\pi r^2 E = 0,$$

so $E = 0$.

(e) Consider a Gaussian surface that lies completely within the conducting shell. Since the electric field is everywhere zero on the surface, $\oint \vec{E} \cdot d\vec{A} = 0$ and, according to Gauss' law, the net charge enclosed by the surface is zero. If $Q_i$ is the charge on the inner surface of the shell, then $q + Q_i = 0$ and $Q_i = -q$. Let $Q_o$ be the charge on the outer surface of the shell. Since the net charge on the shell is $-q$, $Q_i + Q_o = -q$. This means $Q_o = -q - Q_i = -q - (-q) = 0$.

**21**

Use Gauss' law to find an expression for the electric field inside the shell in terms of $A$ and the distance from the center of the shell, then select $A$ so the field does not depend on the distance. Take the Gaussian surface to be a sphere with radius $r_G$, concentric with the spherical shell and within it ($a < r_G < b$). Gauss' law will be used to find the magnitude of the electric field a distance $r_G$ from the shell center.

The charge that is both in the shell and within the Gaussian sphere is given by the integral $q_s = \int \rho \, dV$ over the portion of the shell within the Gaussian surface. Since the charge distribution has spherical symmetry, we may take $dV$ to be the volume of a spherical shell with radius $r$ and infinitesimal thickness $dr$: $dV = 4\pi r^2 \, dr$. Thus,

$$q_s = 4\pi \int_a^{r_g} \rho r^2 \, dr = 4\pi \int_a^{r_g} \frac{A}{r} r^2 \, dr = 4\pi A \int_a^{r_g} r \, dr = 2\pi A(r_G^2 - a^2).$$

The total charge inside the Gaussian surface is $q + q_s = q + 2\pi A(r_G^2 - a^2)$.

The electric field is radial, so the flux at the Gaussian surface is $\Phi = 4\pi r_G^2 E$, where $E$ is the magnitude of the field. Gauss' law yields

$$4\pi\epsilon_0 E r_G^2 = q + 2\pi A(r_G^2 - a^2).$$

Solve for $E$:

$$E = \frac{1}{4\pi\epsilon_0}\left[\frac{q}{r_G^2} + 2\pi A - \frac{2\pi A a^2}{r_G^2}\right].$$

For the field to be uniform, the first and last terms in the brackets must cancel. They do if $q - 2\pi A a^2 = 0$ or $A = q/2\pi a^2$.

## 27

Assume the charge density of both the conducting cylinder and the shell are uniform. Neglect fringing. Symmetry can be used to show that the electric field is radial, both between the cylinder and the shell and outside the shell. It is zero, of course, inside the cylinder and inside the shell.

(a) Take the Gaussian surface to be a cylinder of length $L$ and radius $r$, concentric with the conducting cylinder and shell and with its curved surface outside the shell. The field is normal to the curved portion of the surface and has uniform magnitude over it, so the flux at this portion of the surface is $\Phi = 2\pi r L E$, where $E$ is the magnitude of the field at the Gaussian surface. The flux at the ends is zero. The charge enclosed by the Gaussian surface is $q - 2q = -q$. Gauss' law yields $2\pi r \epsilon_0 L E = q$, so

$$E = \frac{q}{2\pi\epsilon_0 L r}.$$

The enclosed charge is negative, so the field points inward.

(b) Consider a Gaussian surface in the form of a cylinder of length $L$ with the curved portion of its surface completely within the shell. The electric field is zero at all points on the curved surface and is parallel to the ends, so the net electric flux at the Gaussian surface is zero and the net charge within it is zero. Since the conducting cylinder, which is inside the Gaussian cylinder, has charge $q$, the inner surface of the shell must have charge $-q$. Since the shell has total charge $-2q$ and has charge $-q$ on its inner surface, it must have charge $-q$ on its outer surface.

(c) Take the Gaussian surface to be a cylinder of length $L$ and radius $r$, concentric with the conducting cylinder and shell and with its curved surface between the conducting cylinder and the shell. As in (a), the flux at the curved portion of the surface is $\Phi = 2\pi r L E$, where $E$ is the magnitude of the field at the Gaussian surface, and the flux at the ends is zero. The charge enclosed by the Gaussian surface is only the charge $q$ on the conducting cylinder. Gauss' law yields $2\pi\epsilon_0 r L E = q$, so

$$E = \frac{q}{2\pi\epsilon_0 L r}.$$

The enclosed charge is positive, so the field points outward.

## 35

The forces acting on the ball are shown on the diagram to the right. The gravitational force has magnitude $F^{\text{grav}} = mg$, where $m$ is the mass of the ball; the electrical force has magnitude $F^{\text{elec}} = qE$, where $q$ is the charge of the ball and $E$ is the electric field at the position of the ball; and the tension in the thread is denoted by $\vec{T}$. The electric field produced by the plate is normal to the plate and points to the right. Since the ball is positively charged, the electric force on it also points to the right. The tension in the thread makes the angle $\theta$ (= 30°) with the vertical.

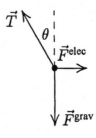

Since the ball is in equilibrium we know that the net force on it vanishes. The sum of the horizontal components yields $qE - T\sin\theta = 0$ and the sum of the vertical components yields $T\cos\theta - mg = 0$. The expression $T = qE/\sin\theta$, from the first equation, is substituted into the second to obtain $qE = mg\tan\theta$.

The electric field produced by a large uniform plane of positive charge is given by $E = \sigma/2\epsilon_0$, where $\sigma$ is the surface charge density. Thus

$$\frac{q\sigma}{2\epsilon_0} = mg\tan\theta$$

and

$$\sigma = \frac{2\epsilon_0 mg\tan\theta}{q}$$

$$= \frac{2(8.85 \times 10^{-12}\,\text{C}^2/\text{N}\cdot\text{m}^2)(1.0 \times 10^{-6}\,\text{kg})(9.8\,\text{m/s}^2)\tan 30°}{2.0 \times 10^{-8}\,\text{C}}$$

$$= 5.0 \times 10^{-9}\,\text{C/m}^2\,.$$

## 37

The charge on the metal plate, which is negative, exerts a force of repulsion on the electron and stops it. First find an expression for the acceleration of the electron, then use kinematics to find the stopping distance. Take the positive $x$ direction to be the direction of motion of the electron. Then the electric field is in the same direction and its $x$ component is given by $E_x = |\sigma|/\epsilon_0$, where $\sigma$ is the surface charge density on the plate. The $x$ component of the force on the electron is $F_x = -eE = -e|\sigma|/\epsilon_0$ and, according to Newton's second law, the $x$ component of the acceleration is

$$a_x = \frac{F}{m} = -\frac{e|\sigma|}{\epsilon_0 m},$$

where $m$ is the mass of the electron.

The force is constant, so we use constant acceleration kinematics. If $v_1$ is the initial speed of the electron, $v_2$ is its final speed, and $x$ is the distance traveled between the initial and final positions, then $v_2^2 - v_1^2 = 2a_x x$. Set $v_2 = 0$ and replace $a_x$ with $-e|\sigma|/\epsilon_0 m$, then solve for $x$. You should get

$$x = -\frac{v_1^2}{2a_x} = \frac{\epsilon_0 m v_1^2}{2e|\sigma|}.$$

Now $\frac{1}{2}mv_1^2$ is the initial kinetic energy $K_1$, so

$$x = \frac{\epsilon_0 K_1}{e|\sigma|}.$$

You must convert the given value of $K_1$ to joules. Since $1.00\,\text{eV} = 1.60 \times 10^{-19}\,\text{J}$, $100\,\text{eV} = 1.60 \times 10^{-17}\,\text{J}$. Thus,

$$x = \frac{(8.85 \times 10^{-12}\,\text{C}^2/\text{N}\cdot\text{m}^2)(1.60 \times 10^{-17}\,\text{J})}{(1.60 \times 10^{-19}\,\text{C})(2.0 \times 10^{-6}\,\text{C/m}^2)} = 4.4 \times 10^{-4}\,\text{m}.$$

## 49

(a) The number of electric field lines per unit area passing though a surface element is proportional to the magnitude of the field and to the cosine of the angle between the field and the normal to the surface element. The density of lines is greater at B than at A on two counts. First, the magnitude of the field is greater at B than at A because B is closer to the cube center, the location of the charged particle. Second, the cosine of the angle is nearly 1 over region B and is significantly less than 1 over region A.

(b) The two regions have the same area so the number of field lines through B is greater than the number through A because the density of lines is greater at B than at A.

(c) The flux at a surface is proportional to the number of field lines through the surface and so is greater at B than at A.

## 51

(a) (i.) This statement is true. The surface must be closed for Gauss' law to be valid.

(ii.) This statement is not true. The surface need not contain all the charged particles in the problem. Some may be outside the Gaussian surface.

(iii.) This statement is not true. The surface may have any shape whatsoever. Symmetry is important only when Gauss' law is used to calculate the electric field but the law is nevertheless valid no matter how much or how little symmetry is present.

(iv.) This statement is not true. The surface can be any surface whatsoever and need not be a conductor.

(v.) This statement is not true. The surface may be a physical surface or it may be purely imaginary. It does not need to be imaginary for the law to be valid.

(vi.) This statement is not true. The normals to the surface are determined by the shape of the surface. The direction of the electric field is determined by the positions and charges of the particles in the problem. The directions of the normal can be changed arbitrarily by changing the shape of the surface. This does not affect the electric field.

(b) (i.) This statement is not true. The charge $q_A$ is the net charge within the Gaussian surface.

(ii.) This statement is true: $q_A$ is the algebraic sum of the charges of all the particles within the Gaussian surface.

(iii.) This statement is not true. There may be charged particles outside the Gaussian surface.

(iv.) This statement is not true. Gauss' law is valid if the particles are stationary and if they are moving. The Gaussian surface has no influence on the motion of the particles.

(v.) This statement is true in the sense that the electric field in the integral might be the field due to charged particles within the Gaussian surface but it does not have to be. It might also be the net field of all particles in the problem. Both these choice yield the same result for the integral. If we write $\vec{E} = \vec{E}_i + \vec{E}_o$, where $\vec{E}_i$ is the field of particles within the Gaussian surface and $\vec{E}_o$ is the field of particles outside, then $\oint (\vec{E}_i + \vec{E}_o) \cdot d\vec{A} = \oint \vec{E}_i \cdot d\vec{A}$ because the net flux due to particles outside is zero.

(vi.) This statement is true in the sense that the electric field in the integral might be the net field of all particles in the problem, but it might also be the net field of only those particles within the Gaussian surface.

# Chapter 25

<u>7</u>

The potential difference between the wire and cylinder is given, not the linear charge density on the wire. Use Gauss' law to find an expression for the electric field a distance $r$ from the center of the wire, between the wire and the cylinder, in terms of the linear charge density. Then integrate with respect to $r$ to find an expression for the potential difference between the wire and cylinder in terms of the linear charge density. Use this result to obtain an expression for the linear charge density in terms of the potential difference and substitute the result into the equation for the electric field. This will give the electric field in terms of the potential difference and will allow you to compute numerical values for the field at the wire and at the cylinder.

For the Gaussian surface use a cylinder of radius $r$ and length $\ell$, concentric with the wire and cylinder. The electric field is normal to the rounded portion of the cylinder's surface and its magnitude is uniform over that surface. This means the electric flux at the Gaussian surface is given by $2\pi r \ell E$, where $E$ is the magnitude of the electric field. The charge enclosed by the Gaussian surface is $q = \lambda \ell$, where $\lambda$ is the linear charge density on the wire. Gauss' law yields $2\pi\epsilon_0 r \ell E = \lambda \ell$. Thus

$$E = \frac{\lambda}{2\pi\epsilon_0 r}.$$

Since the field is radial, the difference in the potential $V_c$ of the cylinder and the potential $V_w$ of the wire is

$$\Delta V = V_w - V_c = -\int_{r_c}^{r_w} E \, dr = \int_{r_w}^{r_c} \frac{\lambda}{2\pi\epsilon_0 r} \, dr = \frac{\lambda}{2\pi\epsilon_0} \ln \frac{r_c}{r_w},$$

where $r_w$ is the radius of the wire and $r_c$ is the radius of the cylinder. This means that

$$\lambda = \frac{2\pi\epsilon_0 \, \Delta V}{\ln(r_c/r_w)}$$

and

$$E = \frac{\lambda}{2\pi\epsilon_0 r} = \frac{\Delta V}{r \ln(r_c/r_w)}.$$

(a) Substitute $r_w$ for $r$ to obtain the field at the surface of the wire:

$$E = \frac{\Delta V}{r_w \ln(r_c/r_w)} = \frac{850 \, \text{V}}{(0.65 \times 10^{-6} \, \text{m}) \ln\left[(1.0 \times 10^{-2} \, \text{m})/(0.65 \times 10^{-6} \, \text{m})\right]}$$
$$= 1.36 \times 10^8 \, \text{V/m}.$$

(b) Substitute $r_c$ for $r$ to find the field at the surface of the cylinder:

$$E = \frac{\Delta V}{r_c \ln(r_c/r_w)} = \frac{850 \, \text{V}}{(1.0 \times 10^{-2} \, \text{m}) \ln\left[(1.0 \times 10^{-2} \, \text{m})/(0.65 \times 10^{-6} \, \text{m})\right]}$$
$$= 8.82 \times 10^3 \, \text{V/m}.$$

**9**

(a) Use Gauss' law to find expressions for the electric field inside and outside the spherical charge distribution. Since the field is radial the electric potential can be written as an integral of the field along a sphere radius, extended to infinity. Since different expressions for the field apply in different regions the integral must be split into two parts, one from infinity to the surface of the distribution and one from the surface to a point inside.

Outside the charge distribution the magnitude of the field is $E = kq/r^2$ and the potential is $V = kq/r$, where $r$ is the distance from the center of the distribution. This is the same as the field and potential of a charged point particle at the center of the spherical distribution.

To find an expression for the magnitude of the field inside the charge distribution use a Gaussian surface in the form of a sphere with radius $r$, concentric with the distribution. The field is normal to the Gaussian surface and its magnitude is uniform over it, so the electric flux at the surface is $4\pi r^2 E$. The charge enclosed is $qr^3/R^3$. Gauss' law becomes

$$4\pi\epsilon_0 r^2 E = \frac{qr^3}{R^3},$$

so

$$E = \frac{qr}{4\pi\epsilon_0 R^3} = \frac{kqr}{R^3}.$$

If $V_s$ is the potential at the surface of the distribution ($r = R$) then the potential at a point inside, a distance $r$ from the center, is

$$V = V_s - \int_R^r E\, dr = V_s - \frac{kq}{R^3}\int_R^r r\, dr = V_s - \frac{kqr^2}{2R^3} + \frac{kq}{2R}.$$

The potential at the surface can be found by replacing $r$ with $R$ in the expression for the potential at points outside the distribution. It is $V_s = kq/R$. Thus

$$V = kq\left[\frac{1}{R} - \frac{r^2}{2R^3} + \frac{1}{2R}\right] = \frac{kq}{2R^3}(3R^2 - r^2).$$

(b) In Problem 8 the electric potential was taken to be zero at the center of the sphere. In this problem it taken to be zero at infinity. According to the expression derived in part (a) the potential at the center of the sphere is $V_c = 3kq/2R$. Thus $V - V_c = -kqr^2/2R^3$. This is the result of Problem 8.

(c) The potential difference is

$$\Delta V = V_s - V_c = \frac{2kq}{2R} - \frac{3kq}{2R} = -\frac{kq}{2R}.$$

This is the same as the expression obtained in Problem 8.

(d) Only potential differences have physical significance, not the value of the potential at any particular point. The same value can be added to the potential at every point without changing the electric field, for example. Changing the reference point from the center of the distribution to infinity changes the value of the potential at every point but it does not change any potential differences.

## 11

(a) For $r > r_2$ the electric field is like that of a point particle with charge $Q$ and the electric potential is $V = kQ/r$, where the zero of potential was taken to be at infinity.

(b) To find the potential in the region $r_1 < r < r_2$, first use Gauss's law to find an expression for the electric field, then integrate along a radial path from $r_2$ to $r$. The Gaussian surface is a sphere of radius $r$, concentric with the shell. The field is radial and therefore normal to the surface. Its magnitude is uniform over the surface, so the flux at the surface is $\Phi = 4\pi r^2 E$. The volume of the shell is $(4\pi/3)(r_2^3 - r_1^3)$, so the charge density is

$$\rho = \frac{3Q}{4\pi(r_2^3 - r_1^3)}$$

and the charge enclosed by the Gaussian surface is

$$q = \left(\frac{4\pi}{3}\right)(r^3 - r_1^3)\rho = Q\left(\frac{r^3 - r_1^3}{r_2^3 - r_1^3}\right).$$

Gauss' law yields

$$4\pi\epsilon_0 r^2 E = Q\left(\frac{r^3 - r_1^3}{r_2^3 - r_1^3}\right)$$

and the magnitude of the electric field is

$$E = \frac{Q}{4\pi\epsilon_0}\frac{r^3 - r_1^3}{r^2(r_2^3 - r_1^3)} = \frac{kQ(r^3 - r_1^3)}{r^2(r_2^3 - r_1^3)}.$$

If $V_s$ is the electric potential at the outer surface of the shell ($r = r_2$) then the potential a distance $r$ from the center is given by

$$V = V_s - \int_{r_2}^r E\,dr = V_s - \frac{kQ}{r_2^3 - r_1^3}\int_{r_2}^r \left(r - \frac{r_1^3}{r^2}\right)dr$$

$$= V_s - \frac{kQ}{r_2^3 - r_1^3}\left(\frac{r^2}{2} - \frac{r_2^2}{2} + \frac{r_1^3}{r} - \frac{r_1^3}{r_2}\right).$$

The potential at the outer surface is found by placing $r = r_2$ in the expression found in part (a). It is $V_s = kQ/r_2$. Make this substitution and collect terms to find

$$V = \frac{kQ}{r_2^3 - r_1^3}\left(\frac{3r_2^2}{2} - \frac{r^2}{2} - \frac{r_1^3}{r}\right).$$

Since $\rho = 3Q/4\pi(r_2^3 - r_1^3)$ this can also be written

$$V = \frac{\rho}{3\epsilon_0}\left(\frac{3r_2^2}{2} - \frac{r^2}{2} - \frac{r_1^3}{r}\right).$$

(c) The electric field vanishes in the cavity, so the potential is everywhere the same inside and has the same value as at a point on the inside surface of the shell. Put $r = r_1$ in the result of part (b). After collecting terms the result is

$$V = \frac{3kQ(r_2^2 - r_1^2)}{2(r_2^3 - r_1^3)},$$

or, in terms of the charge density,

$$V = \frac{\rho}{2\epsilon_0}(r_2^2 - r_1^2).$$

(d) The solutions agree at $r = r_1$ and at $r = r_2$.

## 17

(a) The electric potential $V$ at the surface of the drop, the charge $q$ on the drop, and the radius $R$ of the drop are related by $V = kq/R$. Thus

$$R = \frac{kq}{V} = \frac{(8.99 \times 10^9 \, \text{N} \cdot \text{m}^2/\text{C}^2)(30 \times 10^{-12} \, \text{C})}{500 \, \text{V}} = 5.4 \times 10^{-4} \, \text{m}.$$

(b) After the drops combine the total volume is twice the volume of an original drop, so the radius $R'$ of the combined drop is given by $(R')^3 = 2R^3$ and $R' = 2^{1/3}R$. The charge is twice the charge of original drop: $q' = 2q$. Thus

$$V' = \frac{kq'}{R'} = \frac{2kq}{2^{1/3}R} = 2^{2/3}V = 2^{2/3}(500 \, \text{V}) = 790 \, \text{V}.$$

## 25

(a) Let $\ell$ (= 0.15 m) be the length of the rectangle and $w$ (= 0.050 m) be its width. Particle 1, with charge $q_1$, is a distance $\ell$ from point $A$ and particle 2, with charge $q_2$, is a distance $w$, so the electric potential at $A$ is

$$V_A = k\left[\frac{q_1}{\ell} + \frac{q_2}{w}\right]$$

$$= (8.99 \times 10^9 \, \text{N} \cdot \text{m}^2/\text{C}^2)\left[\frac{-5.0 \times 10^{-6} \, \text{C}}{0.15 \, \text{m}} + \frac{2.0 \times 10^{-6} \, \text{C}}{0.050 \, \text{m}}\right]$$

$$= 6.0 \times 10^4 \, \text{V}.$$

(b) Particle 1 is a distance $w$ from point $b$ and particle 2 is a distance $\ell$, so the electric potential at $B$ is

$$V_B = k\left[\frac{q_1}{w} + \frac{q_2}{\ell}\right]$$

$$= (8.99 \times 10^9 \, \text{N} \cdot \text{m}^2/\text{C}^2)\left[\frac{-5.0 \times 10^{-6} \, \text{C}}{0.050 \, \text{m}} + \frac{2.0 \times 10^{-6} \, \text{C}}{0.15 \, \text{m}}\right]$$

$$= -7.8 \times 10^5 \, \text{V}.$$

(c) Since the kinetic energy is zero at the beginning and end of the trip, the work done by an external agent equals the change in the potential energy of the system. The potential energy is the product of the charge $q_3$ and the electric potential. If $U_A$ is the potential energy when particle 3 is at $A$ and $U_B$ is the potential energy when particle 3 is at $B$, then the work done in moving the charge from $B$ to $A$ is

$$W = U_A - U_B = q_3(V_A - V_B) = (3.0 \times 10^{-6}\,\text{C})(6.0 \times 10^4\,\text{V} + 7.8 \times 10^5\,\text{V}) = 2.5\,\text{J}\,.$$

(d) The work done by the external agent is positive, so the energy of the three-charge system increases.

(e) and (f) The electrostatic force is conservative, so the work is the same no matter what the path.

## 31

(a) The potential energy is

$$U = \frac{kq^2}{d} = \frac{(8.99 \times 10^9\,\text{N} \cdot \text{m}^2/\text{C}^2)(5.0 \times 10^{-6}\,\text{C})^2}{1.00\,\text{m}} = 0.225\,\text{J}\,,$$

relative to the potential energy at infinite separation.

(b) Each sphere repels the other with a force that has magnitude

$$F = \frac{kq^2}{d^2} = \frac{(8.99 \times 10^9\,\text{N} \cdot \text{m}^2/\text{C}^2)(5.0 \times 10^{-6}\,\text{C})^2}{(1.00\,\text{m})^2} = 0.225\,\text{N}\,.$$

According to Newton's second law the acceleration of each sphere is the force divided by the mass of the sphere. Let $m_A$ and $m_B$ be the masses of the spheres. The acceleration of sphere $A$ is

$$a_A = \frac{F}{m_A} = \frac{0.225\,\text{N}}{5.0 \times 10^{-3}\,\text{kg}} = 45.0\,\text{m/s}^2$$

and the acceleration of sphere $B$ is

$$a_B = \frac{F}{m_B} = \frac{0.225\,\text{N}}{10 \times 10^{-3}\,\text{kg}} = 22.5\,\text{m/s}^2\,.$$

(c) Energy is conserved. The initial potential energy is $U = 0.225\,\text{J}$, as calculated in part (a). The initial kinetic energy is zero since the spheres start from rest. The final potential energy is zero since the spheres are then far apart. The final kinetic energy is $\frac{1}{2}m_A v_A^2 + \frac{1}{2}m_B v_B^2$, where $v_A$ and $v_B$ are the final speeds. Thus

$$U = \frac{1}{2}m_A v_A^2 + \frac{1}{2}m_B v_B^2\,.$$

Momentum is also conserved, so

$$0 = m_A v_A + m_B v_B\,.$$

Solve these equations simultaneously for $v_A$ and $v_B$.

Substitute $v_B = -(m_A/m_B)v_A$, from the momentum equation, into the energy equation and collect terms. You should obtain $U = \frac{1}{2}(m_A/m_B)(m_A + m_B)v_A^2$. Thus

$$v_A = \sqrt{\frac{2Um_B}{m_A(m_A + m_B)}}$$

$$= \sqrt{\frac{2(0.225\,\text{J})(10 \times 10^{-3}\,\text{kg})}{(5.0 \times 10^{-3}\,\text{kg})(5.0 \times 10^{-3}\,\text{kg} + 10 \times 10^{-3}\,\text{kg})}} = 7.75\,\text{m/s}\,.$$

Now calculate $v_B$:

$$v_B = -\frac{m_A}{m_B}v_A = -\left(\frac{5.0 \times 10^{-3}\,\text{kg}}{10 \times 10^{-3}\,\text{kg}}\right)(7.75\,\text{m/s}) = -3.87\,\text{m/s}\,.$$

## 37

A positively charged particle with charge $q$ is a distance $r - d$ from $P$, another positively charged particle, with the same charge, is a distance $r$ from $P$, and a negatively charged particle with charge $-q$ is a distance $r + d$ from $P$. Sum the individual electric potentials at $P$ to find the total:

$$V = kq\left[\frac{1}{r-d} + \frac{1}{r} - \frac{1}{r+d}\right]\,.$$

Use the binomial theorem to approximate $1/(r - d)$ for $r$ much larger than $d$:

$$\frac{1}{r-d} = (r-d)^{-1} \approx (r)^{-1} - (r)^{-2}(-d) = \frac{1}{r} + \frac{d}{r^2}\,.$$

Similarly,

$$\frac{1}{r+d} \approx \frac{1}{r} - \frac{d}{r^2}\,.$$

Only the first two terms of each expansion were retained. Thus

$$V \approx kq\left[\frac{1}{r} + \frac{d}{r^2} + \frac{1}{r} - \frac{1}{r} + \frac{d}{r^2}\right] = kq\left[\frac{1}{r} + \frac{2d}{r^2}\right] = \frac{kq}{r}\left[1 + \frac{2d}{r}\right]\,.$$

## 39

An infinitesimal segment of the rod of width $dx$, located at $x$ contains charge $dq = \lambda\,dx = cx\,dx$. The distance from the segment to $P_1$ is $r = x + d$, so the electric potential it produces at $P_1$ is $(kcx\,dx)/(x + d)$ and the total potential at $P_1$ is

$$V = kc\int_0^L \frac{x}{x+d}\,dx = kc\left[x - d\ln(x+d)\right]_0^L = kc\left[L - d\ln\left(1 + \frac{L}{d}\right)\right]\,.$$

## 43

(a) Every part of the ring is the same distance from a point $P$ on the axis. This distance is $r = \sqrt{z^2 + R^2}$, where $R$ is the radius of the ring and $z$ is the distance from the center of the ring to $P$. The electric potential at $P$ is

$$V = k\int \frac{dq}{r} = k\int \frac{dq}{\sqrt{z^2 + R^2}} = \frac{k}{\sqrt{z^2 + R^2}}\int dq = \frac{kq}{\sqrt{z^2 + R^2}}\,.$$

(b) The electric field is along the axis and its component is given by

$$E_z = -\frac{\partial V}{\partial z} = -kq\frac{\partial}{\partial z}(z^2 + R^2)^{-1/2}$$

$$= kq\left(\frac{1}{2}\right)(z^2 + R^2)^{-3/2}(2z) = \frac{kqz}{(z^2 + R^2)^{3/2}}.$$

This agrees with the result of Section 23–8.

## 45

(a) According to the result of Problem 39, the electric potential at a point with coordinate $x$ is given by

$$V = kc\left[L + x\ln\left(1 - \frac{L}{x}\right)\right].$$

where $d$ has been replaced by $-x$. Differentiate the potential with respect to $x$ to find the $x$ component of the electric field:

$$E_x = -\frac{\partial V}{\partial x} = -kc\left[\ln\left(1 - \frac{L}{x}\right) + \frac{L}{x - L}\right].$$

Substitute $x = -d$ to obtain

$$E_x = -kc\left[\ln\left(1 + \frac{L}{d}\right) - \frac{L}{d + L}\right].$$

(b) Consider two points an equal infinitesimal distance on either side of $P_1$, along a line that is parallel to the $y$ axis. The difference in the electric potential divided by their separation gives the $y$ component of the electric field. Since the two points are situated symmetrically with respect to the rod, their potentials are the same and the potential difference is zero. Thus the $y$ component of the electric field is zero.

## 49

(a) The electric potential is the sum of the contributions of the individual spheres. Let $q_A$ be the charge on sphere A, $q_B$ be the charge on sphere B, and $d$ be their separation. The point halfway between them is the same distance $d/2$ (= 1.0 m) from the center of each sphere, so the potential at the halfway point is

$$V = k\frac{q_A + q_B}{d/2} = \frac{(8.99 \times 10^9\,\text{N}\cdot\text{m}^2/\text{C}^2)(1.0 \times 10^{-8}\,\text{C} - 3.0 \times 10^{-8}\,\text{C})}{1.0\,\text{m}} = -1.80 \times 10^2\,\text{V}.$$

(b) The distance from the center of one sphere to the surface of the other is $d - R$, where $R$ is the radius of either sphere. The potential of either one of the spheres is due to the charge on that sphere and the charge on the other sphere. The potential at the surface of sphere A is

$$V_A = k\left[\frac{q_A}{R} + \frac{q_B}{d - R}\right]$$

$$= (8.99 \times 10^9\,\text{N}\cdot\text{m}^2/\text{C}^2)\left[\frac{1.0 \times 10^{-8}\,\text{C}}{0.030\,\text{m}} - \frac{3.0 \times 10^{-8}\,\text{C}}{2.0\,\text{m} - 0.030\,\text{m}}\right]$$

$$= 2.9 \times 10^3\,\text{V}.$$

The potential at the surface of sphere B is

$$V_B = k \left[ \frac{q_A}{d-R} + \frac{q_B}{R} \right]$$

$$= (8.99 \times 10^9 \, \text{N} \cdot \text{m}^2/\text{C}^2) \left[ \frac{1.0 \times 10^{-8} \, \text{C}}{2.0 \, \text{m} - 0.030 \, \text{m}} - \frac{3.0 \times 10^{-8} \, \text{C}}{0.030 \, \text{m}} \right]$$

$$= -8.9 \times 10^3 \, \text{V}.$$

# Chapter 26

## 3

Suppose the charge on the sphere increases by $\Delta q$ in time $\Delta t$. Then, in that time, its potential increases by $\Delta V = k\,\Delta q/r$, where $r$ is the radius of the sphere. This means $\Delta q = r\,\Delta V/k$. Now $\Delta q = (i_{in} - i_{out})\,\Delta t$, where $i_{in}$ is the current entering the sphere and $i_{out}$ is the current leaving. Thus

$$\Delta t = \frac{\Delta q}{i_{in} - i_{out}} = \frac{r\,\Delta V}{k(i_{in} - i_{out})}$$

$$= \frac{(0.10\,\text{m})(1000\,\text{V})}{(8.99 \times 10^9\,\text{N} \cdot \text{m}^2/\text{C}^2)(1.0000020\,\text{A} - 1.0000000\,\text{A})} = 5.6 \times 10^{-3}\,\text{s}.$$

## 11

Since the mass and density of the material do not change, the volume remains the same. If $L_0$ is the original length, $L$ is the new length, $A_0$ is the original cross-sectional area, and $A$ is the new cross-sectional area, then $L_0 A_0 = LA$ and $A = L_0 A_0/L = L_0 A_0/3L_0 = A_0/3$. The new resistance is

$$R = \frac{\rho L}{A} = \frac{\rho 3 L_0}{A_0/3} = 9\frac{\rho L_0}{A_0} = 9R_0,$$

where $R_0$ is the original resistance. Thus $R = 9(6.0\,\Omega) = 54\,\Omega$.

## 15

(a) Let $\Delta T$ be the change in temperature and $\beta$ be the coefficient of linear expansion for copper. Then $\Delta L = \beta L\,\Delta T$ and

$$\frac{\Delta L}{L} = \beta\,\Delta T = (1.7 \times 10^{-5}\,\text{K}^{-1})(1.0\,\text{K}) = 1.7 \times 10^{-5}.$$

This is 0.0017%.

The fractional change in area is

$$\frac{\Delta A}{A} = 2\beta\,\Delta T = 2(1.7 \times 10^{-5}\,\text{K}^{-1})(1.0\,\text{K}) = 3.4 \times 10^{-5}.$$

This is 0.0034%.

For small changes in the resistivity $\rho$, length $L$, and area $A$ of a wire, the change in the resistance is given by

$$\Delta R = \frac{\partial R}{\partial \rho}\,\Delta\rho + \frac{\partial R}{\partial L}\,\Delta L + \frac{\partial R}{\partial A}\,\Delta A.$$

Since $R = \rho L/A$, $\partial R/\partial \rho = L/A = R/\rho$, $\partial R/\partial L = \rho/A = R/L$, and $\partial R/\partial A = -\rho L/A^2 = -R/A$. Furthermore, $\Delta\rho/\rho = \alpha\,\Delta T$, where $\alpha$ is the temperature coefficient of resistivity for copper ($4.3 \times 10^{-3}\,\mathrm{K}^{-1}$, according to Table 27–1). Thus

$$\frac{\Delta R}{R} = \frac{\Delta\rho}{\rho} + \frac{\Delta L}{L} - \frac{\Delta A}{A} = (\alpha + \beta - 2\beta)\,\Delta T = (\alpha - \beta)\,\Delta T$$
$$= (4.3 \times 10^{-3}\,\mathrm{K}^{-1} - 1.7 \times 10^{-5}\,\mathrm{K}^{-1})(1.0\,\mathrm{K}) = 4.3 \times 10^{-3}.$$

This is 0.43%.

(b) The fractional change in resistivity is much larger than the fractional change in length and area. Changes in length and area affect the resistance much less than changes in resistivity.

### 23

(a) Let $P$ be the rate of energy dissipation, $i$ be the current in the heater, and $\Delta V$ be the potential difference across the heater. They are related by $P = i\,\Delta V$. Solve for $i$:

$$i = \frac{P}{\Delta V} = \frac{1250\,\mathrm{W}}{115\,\mathrm{V}} = 10.9\,\mathrm{A}.$$

(b) According to the definition of resistance $\Delta V = iR$, where $R$ is the resistance of the heater. Solve for $R$:

$$R = \frac{\Delta V}{i} = \frac{115\,\mathrm{V}}{10.9\,\mathrm{A}} = 10.6\,\Omega.$$

(c) The thermal energy $E$ produced by the heater in time $t$ ($= 1.0\,\mathrm{h} = 3600\,\mathrm{s}$) is

$$E = Pt = (1250\,\mathrm{W})(3600\,\mathrm{s}) = 4.5 \times 10^6\,\mathrm{J}.$$

### 27

(a) The charge $q$ that flows past any cross section of the beam in time $\Delta t$ is given by $q = i\,\Delta t$ and the number of electrons is $N = q/e = (i/e)\,\Delta t$. This is the number of electrons that are accelerated. Thus

$$N = \frac{(0.50\,\mathrm{A})(0.10 \times 10^{-6}\,\mathrm{s})}{1.60 \times 10^{-19}\,\mathrm{C}} = 3.1 \times 10^{11}.$$

(b) Over a long time $t$, the total charge is $Q = nqt$, where $n$ is the number of pulses per unit time and $q$ is the charge in one pulse. The average current is given by $\langle i \rangle = Q/t = nq$. Now $q = i\,\Delta t = (0.50\,\mathrm{A})(0.10 \times 10^{-6}\,\mathrm{s}) = 5.0 \times 10^{-8}\,\mathrm{C}$, so

$$\langle i \rangle = (500\,\mathrm{s}^{-1})(5.0 \times 10^{-8}\,\mathrm{C}) = 2.5 \times 10^{-5}\,\mathrm{A}.$$

(c) The accelerating potential difference is $\Delta V = K/e$, where $K$ is the final kinetic energy of an electron. Since $K = 50\,\mathrm{MeV}$, the accelerating potential is $\Delta V = 50\,\mathrm{MV} = 5.0 \times 10^7\,\mathrm{V}$. During a pulse the power output is

$$P = iV = (0.50\,\mathrm{A})(5.0 \times 10^7\,\mathrm{V}) = 2.5 \times 10^7\,\mathrm{W}.$$

This is the peak power. The average power is

$$\langle P \rangle = \langle i \rangle \, \Delta V = (2.5 \times 10^{-5}\,\text{A})(5.0 \times 10^{7}\,\text{V}) = 1.3 \times 10^{3}\,\text{W}\,.$$

## 31

(a) The magnitude of the current density is given by $J = nq|\langle \vec{v} \rangle|$, where $n$ is the number of particles per unit volume, $q$ is the charge on each particle, and $|\langle \vec{v} \rangle|$ is the magnitude of the average velocity of the particles. The particle concentration is $n = 2.0 \times 10^{8}\,\text{cm}^{-3} = 2.0 \times 10^{14}\,\text{m}^{-3}$, the charge is $q = 2e = 2(1.60 \times 10^{-19}\,\text{C}) = 3.20 \times 10^{-19}\,\text{C}$, and the magnitude of the average velocity is $1.0 \times 10^{5}\,\text{m/s}$. Thus

$$J = (2 \times 10^{14}\,\text{m}^{-3})(3.2 \times 10^{-19}\,\text{C})(1.0 \times 10^{5}\,\text{m/s}) = 6.4\,\text{A/m}^{2}\,.$$

Since the particles are positively charged, the current density is in the same direction as their motion, to the north.

(b) The current cannot be calculated unless the cross-sectional area of the beam is known. Then $i = JA$ can be used.

## 35

(a) The charge that strikes the surface in time $\Delta t$ is given by $\Delta q = i\,\Delta t$, where $i$ is the current. Since each particle carries charge $2e$, the number of particles that strike the surface is

$$N = \frac{\Delta q}{2e} = \frac{i\,\Delta t}{2e} = \frac{(0.25 \times 10^{-6}\,\text{A})(3.0\,\text{s})}{2(1.6 \times 10^{-19}\,\text{C})} = 2.3 \times 10^{12}\,.$$

(b) Now let $N$ be the number of particles in a length $L$ of the beam. They will all pass through the beam cross section at one end in time $t = L/v$, where $v$ is the particle speed. The current is the charge that moves through the cross section per unit time. That is, $i = 2eN/t = 2eNv/L$. Thus $N = iL/2ev$.

Now find the particle speed. The kinetic energy of a particle is

$$K = 20\,\text{MeV} = (20 \times 10^{6}\,\text{eV})(1.60 \times 10^{-19}\,\text{J/eV}) = 3.2 \times 10^{-12}\,\text{J}\,.$$

Since $K = \frac{1}{2}mv^{2}$, $v = \sqrt{2K/m}$. The mass of an alpha particle is four times the mass of a proton or $m = 4(1.67 \times 10^{-27}\,\text{kg}) = 6.68 \times 10^{-27}\,\text{kg}$, so

$$v = \sqrt{\frac{2(3.2 \times 10^{-12}\,\text{J})}{6.68 \times 10^{-27}\,\text{kg}}} = 3.1 \times 10^{7}\,\text{m/s}$$

and

$$N = \frac{iL}{2ev} = \frac{(0.25 \times 10^{-6}\,\text{A})(20 \times 10^{-2}\,\text{m})}{2(1.60 \times 10^{-19}\,\text{C})(3.1 \times 10^{7}\,\text{m/s})} = 5.0 \times 10^{3}\,.$$

(c) Use conservation of energy. The initial kinetic energy is zero, the final kinetic energy is $20\,\text{MeV} = 3.2 \times 10^{-12}\,\text{J}$, the initial potential energy is $q\,\Delta V = 2e\,\Delta V$, and the final potential energy is zero. Here $\Delta V$ is the electric potential difference through which the particles are accelerated. Conservation of energy leads to $K_f = U_i = 2e\,\Delta V$, so

$$\Delta V = \frac{K_f}{2e} = \frac{3.2 \times 10^{-12}\,\text{J}}{2(1.60 \times 10^{-19}\,\text{C})} = 10 \times 10^6\,\text{V}\,.$$

## 37

The magnitude of the current density is $J = en|\langle \vec{v} \rangle|$, where $n$ is the number of free electrons per unit volume and $|\langle \vec{v} \rangle|$ is the magnitude of the average electron velocity. Thus $|\langle \vec{v} \rangle| = J/en$. The current density is also $J = i/A$, where $i$ is the current and $A$ is the cross-sectional area of the wire, so $|\langle \vec{v} \rangle| = i/enA$.

The molar mass of copper is $63.54\,\text{g/mol}$, so the mass of a copper atom is

$$m = \frac{63.54 \times 10^{-3}\,\text{kg/mol}}{6.02 \times 10^{23}\,\text{atoms/mol}} = 1.06 \times 10^{-25}\,\text{kg}\,.$$

The density of copper is $8.96 \times 10^3\,\text{kg/m}^3$ so the number of copper atom per unit volume is

$$n = \frac{8.96 \times 10^3\,\text{kg/m}^3}{1.06 \times 10^{-25}\,\text{kg}} = 8.45 \times 10^{28}\,\text{m}^{-3}\,.$$

Since there is one free electron per atom, this is also the number of free electrons per unit volume. The magnitude of the average electron velocity is

$$|\langle \vec{v} \rangle| = \frac{300\,\text{A}}{(1.60 \times 10^{-19}\,\text{C})(8.45 \times 10^{28}\,\text{m}^{-3})(0.21 \times 10^{-4}\,\text{m}^2)} = 1.06 \times 10^{-3}\,\text{m/s}\,.$$

The time to travel the length of the wire is

$$\Delta t = \frac{L}{|\langle \vec{v} \rangle|} = \frac{0.85\,\text{m}}{1.06 \times 10^{-3}\,\text{m/s}} = 8.0 \times 10^2\,\text{s}\,.$$

This is about 13 min.

## 43

(a) The magnitude of the current density is $J = e(n_+ + n_-)|\langle \vec{v} \rangle|$, where $n_+$ is the number of positive ions per unit volume, $n_-$ is the number of negative ions per unit volume, and $|\langle \vec{v} \rangle|$ is the magnitude of the average velocity of the positive or negative ions, separately. The magnitude of the current density is related to the magnitude $E$ of the electric field by $J = \sigma E$, where $\sigma$ is the conductivity. Thus $e(n_+ + n_-)|\langle \vec{v} \rangle| = \sigma E$ and

$$|\langle \vec{v} \rangle| = \frac{\sigma E}{e(n_+ + n_-)} = \frac{(2.70 \times 10^{-14}\,(\Omega \cdot \text{m})^{-1})(120\,\text{V/m})}{(1.60 \times 10^{-19}\,\text{C})(620 \times 10^{-6}\,\text{m}^{-3} + 550 \times 10^{-6}\,\text{m}^{-3})} = 1.73 \times 10^{-2}\,\text{m/s}\,.$$

(b) The magnitude of the current density is

$$J = \sigma E = [2.70 \times 10^{-14}\,(\Omega \cdot \text{m})^{-1}](120\,\text{V/m}) = 3.24 \times 10^{-12}\,\text{A/m}^2\,.$$

It is in the same direction as the electric field, downward.

# Chapter 27

## 5

Let $i$ be the current in the circuit and take it to be positive if it is clockwise. The loop rule gives $\Delta V_B - iR_1 - iR_2 - iR_3 = 0$, so

$$i = \frac{\Delta V_B}{R_1 + R_2 + R_3} = \frac{15 \text{ V}}{12\,\Omega + 15\,\Omega + 25\,\Omega} = 0.29 \times 10^{-3} \text{ A}.$$

The potential difference across $R_1$ is $\Delta V_1 = iR_1 = (0.29 \text{ A})(12\,\Omega) = 3.48 \text{ V}$.

The potential difference across $R_2$ is $\Delta V_2 = iR_2 = (0.29 \text{ A})(15\,\Omega) = 4.35 \text{ V}$.

The potential difference across $R_3$ is $\Delta V_3 = iR_3 = (0.29 \text{ A})(25\,\Omega) = 7.25 \text{ V}$.

## 9

Take the current in $R_1$ to be $i_1$, positive if to the right, the current in $R_2$ to be $i_2$, positive if upward, and the current in $R_3$ to be $i_3$, positive if downward. Then the junction rule gives $i_3 = i_1 + i_2$. Let $\Delta V_{B1}$ be the potential difference across the battery on the left and $\Delta V_{B2}$ be the potential difference across the battery on the right. All the resistances are the same, which we take to be $R$. The loop rule applied to the left-hand loop gives $\Delta V_{B1} - i_1 R + i_2 R - \Delta V_{B2} = 0$ and applied to the right-hand loop gives $\Delta V_{B2} - i_2 R - i_3 R = 0$.

Use the junction equation to substitute for $i_3$ in the second loop equation and obtain $\Delta V_{B2} - 2i_2 R - i_1 R = 0$. Thus $i_1 = (\Delta V_{B2} - 2i_2 R)/R$. Use this to substitute for $i_1$ in the first loop equation and obtain $\Delta V_{B1} - 2\Delta V_{B2} + i_2 R = 0$. The solution is $i_2 = 0$.

Now $i_1 = (\Delta V_{B2})/R = (5.0 \text{ V})/(4.0\,\Omega) = 1.25 \text{ A}$ and $i_3 = i_1 + i_2 = 1.25 \text{ A}$.

## 13

The two resistors at the bottom of the circuit are connected in parallel, so their equivalent resistance is $R_{eq\,1} = (3.0R)(6.0R)/(3.0R + 6.0R) = 2.0R$. When they are replaced by their equivalent resistor, that resistor and the resistor with resistance $R$ are in series, so the equivalent resistance of the right side of the circuit between A and B is $R_R = 2.0R + R = 3.0R$.

The $4.0R$ and $2.0R$ resistors on the left side and top of the circuit are in series, so the equivalent resistance of the left side of the circuit between A and B is $R_L = 4.0R + 2.0R = 6.0R$.

Now $R_R$ and $R_L$ are connected in parallel so the equivalent resistance of the entire circuit between A and B is

$$R_{eq} = \frac{R_R R_L}{R_R + R_L} = \frac{(3.0R)(6.0R)}{3.0R + 6.0R} = 2.0R = (2.0)(10\,\Omega) = 20\,\Omega.$$

## 15

(a) Let $\Delta V_B$ be the potential difference across the battery. When the bulbs are connected in parallel, the potential difference across them is the same and is the same as the potential difference across the battery. The rate of energy dissipation by bulb 1 is $R = (\Delta V_B)^2/R_1$ and the rate of energy dissipation by bulb 2 is $P_2 = (\Delta V_B)^2/R_2$. If $R_1 = R_2$ the rates are the same, so the brightness is the same. If $R_1$ is greater than $R_2$, bulb 2 dissipates energy at a greater rate than bulb 1 and is the brighter of the two.

(b) When the bulbs are connected in series, the current in them is the same. The rate of energy dissipation by bulb 1 is now $P_1 = i^2 R_1$ and the rate of energy dissipation by bulb 2 is $P_2 = i^2 R_2$. If $R_1 = R_2$ the rates are the same, so the brightness is the same. If $R_1$ is greater than $R_2$, bulb 1 has the greater rate and is the brighter of the two.

## 19

Let $i_1$ be the current in $R_1$ and take it to be positive if it is to the right. Let $i_2$ be the current in $R_2$ and take it to be positive if it is upward. When the loop rule is applied to the lower loop, the result is

$$\Delta V_{B2} - i_1 R_1 = 0$$

and when it is applied to the upper loop, the result is

$$\Delta V_{B1} - \Delta V_{B2} - \Delta V_{B3} - i_2 R_2 = 0 \,.$$

The first equation yields

$$i_1 = \frac{\Delta V_{B2}}{R_1} = \frac{5.0\,\text{V}}{100\,\Omega} = 0.050\,\text{A} \,.$$

The second yields

$$i_2 = \frac{\Delta V_{B1} - \Delta V_{B2} - \Delta V_{B3}}{R_2} = \frac{6.0\,\text{V} - 5.0\,\text{V} - 4.0\,\text{V}}{50\,\Omega} = -0.060\,\text{A} \,.$$

The negative sign indicates that the current in $R_2$ is actually downward.

If $V_b$ is the potential at point $b$, then the potential at point $a$ is $V_a = V_b + \Delta V_{B3} + \Delta V_{B2}$, so $V_a - V_b = \Delta V_{B3} + \Delta V_{B2} = 4.0\,\text{V} + 5.0\,\text{V} = 9.0\,\text{V}$.

## 23

The potential difference across each resistor is $\Delta V = 25.0\,\text{V}$. Since the resistors are identical, the current in each is

$$i = \frac{V}{R} = \frac{25.0\,\text{V}}{18.0\,\Omega} = 1.39\,\text{A} \,.$$

The total current through the battery is $i_{\text{total}} = 4(1.39\,\text{A}) = 5.56\,\text{A}$.

You might use the idea of equivalent resistance. For four identical resistors in parallel, the equivalent resistance is determined by

$$\frac{1}{R_{\text{eq}}} = \frac{4}{R} \,,$$

so $R_{\text{eq}} = R/4$. When a potential difference of 25.0 V is applied to the equivalent resistor the current through it is the same as the total current through the four resistors in parallel. Thus

$$i_{\text{total}} = \frac{V}{R_{\text{eq}}} = \frac{4V}{R} = \frac{4(25.0\,\text{V})}{18.0\,\Omega} = 5.56\,\text{A} \,.$$

## 25

Let $r$ be the resistance of each of the thin wires. Since they are in parallel, the resistance $R$ of the composite can be determined from

$$\frac{1}{R} = \frac{9}{r},$$

or $R = r/9$. Now

$$r = \frac{4\rho\ell}{\pi d^2}$$

and

$$R = \frac{4\rho\ell}{\pi D^2},$$

where $\rho$ is the resistivity of copper. Here $\pi d^2/4$ was used for the cross-sectional area of any one of the original wires and $\pi D^2/4$ was used for the cross-sectional area of the replacement wire. Here $d$ and $D$ are diameters. Since the replacement wire is to have the same resistance as the composite,

$$\frac{4\rho\ell}{\pi D^2} = \frac{4\rho\ell}{9\pi d^2}.$$

Solve for $D$ and obtain $D = 3d$.

## 29

Divide the resistors into groups of $n$ resistors each, with all the resistors of a group connected in series. Suppose there are $m$ such groups, with the groups connected in parallel. The scheme is shown in the diagram on the right for $n = 3$ and $m = 2$. Let $R$ be the resistance of any one of the resistors. Then the resistance of a series group is $nR$ and the resistance of the total array can be determined from

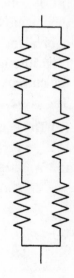

$$\frac{1}{R_{\text{total}}} = \frac{m}{nR}.$$

Since we want $R_{\text{total}} = R$, we must select $n = m$.

The current is the same in every resistor and there are $n^2$ resistors, so the maximum total rate of energy dissipation is $P_{\text{total}} = n^2 P$, where $P$ is the maximum rate of energy dissipation any one of the resistors.

You want $P_{\text{total}} > 5.0P$. Since $P = 1.0\,\text{W}$, $n^2$ must be larger than 5.0. Since $n$ must be an integer, the smallest it can be is 3. The least number of resistors is $n^2 = 9$.

## 33

(a) Let $R$ be the resistance of each of the resistors, $\Delta V_{B1}$ be the potential difference across the battery on the right side of the circuit, and $\Delta V_{B2}$ be the potential difference across the battery on the left side. The potential difference across the resistor with current $i_1$ is the same as the potential difference across the battery terminals connected to it. Thus the current is

$i_1 = (\Delta V_{B1})/R = (12.0\,\text{V})/(4.00\,\Omega) = 3.00\,\text{A}$. The positive terminal of the battery is at its upper end so the upper end of the resistor is at a higher electric potential than the lower end and the current is downward.

(b) Replace the two resistors in the upper right corner of the circuit with their equivalent resistor. Since they are in parallel the equivalent resistance is $R_{eq} = R/2 = 2.00\,\Omega$. Now take $i_2$ to be positive if it is downward and apply the loop rule to the central loop. It is $\Delta V_{B2} + i_1 R - i_2(R + R/2 + R) = 0$, so

$$i_2 = \frac{\Delta V_{B2} + i_1 R}{(5/2)R} = \frac{4.00\,\text{V} + (3.00\,\text{A})(4.00\,\Omega)}{(5/2)(4.00\,\Omega)} = 1.60\,\text{A}.$$

A positive result was obtained, so the current $i_2$ is indeed downward.

(c) The rate of energy transfer in battery 2 is $P_2 = i_2\,\Delta V_{B2} = (1.60\,\text{A})(4.00\,\text{V}) = 6.4\,\text{W}$. Since the current inside the battery is from the negative terminal to the positive terminal, the battery is supplying energy.

(d) The junction rule tells us that the current in battery 1 is $i = i_1 + i_2 = 3.00\,\text{A} + 1.60\,\text{A} = 4.60\,\text{A}$. The rate of energy transfer by the battery is $P_1 = i\,\Delta V_{B1} = (4.60\,\text{A})(12.0\,\text{V}) = 55.2\,\text{W}$. Since the current is upward, from the negative to the positive terminal, this battery is also supplying energy.

## 41

(a) The three 18-$\Omega$ resistors are connected in parallel. Since their resistances are the same, their equivalent resistance is $R_3 = (18\,\Omega)/3 = 6.0\,\Omega$. Once these resistors are replaced by the equivalent resistor the circuit becomes a single-loop circuit. If $i$ is the current in that circuit the loop rule gives $\Delta V_B - iR - iR_3 = 0$, where $R = 6.0\,\Omega$. Thus $i = (12\,\text{V})/(12\,\Omega) = 1.0\,\text{A}$. Now the current is the same in each of the 18-$\Omega$ resistors, so the junction rule yields $i_1 = i/3 = 0.33\,\text{A}$. The current is rightward.

(b) The equivalent resistance of all four resistors is $R_4 = 12\,\Omega$, so the net rate of energy dissipation is $P = i^2 R_4 = (1.0\,\text{A})(12\,\Omega) = 12\,\text{W}$. In time $\Delta t = 60\,\text{s}$ the energy dissipated is $E = P\,\Delta t = (12\,\text{W})(60\,\text{s}) = 720\,\text{J}$.

## 47

(a) Let $i$ be the current in the circuit and take it to be positive if it is to the left in $R_1$. Use Kirchhoff's loop rule: $\Delta V_{B1} - iR_2 - iR_1 - \Delta V_{B2} = 0$. Solve for $i$:

$$i = \frac{\Delta V_{B1} - \Delta V_{B2}}{R_1 + R_2} = \frac{12\,\text{V} - 6.0\,\text{V}}{4.0\,\Omega + 8.0\,\Omega} = 0.50\,\text{A}.$$

A positive value was obtained, so the current is counterclockwise around the circuit.

(b) If $i$ is the current in a resistor with resistance $R$, then the rate of energy dissipation by that resistor is given by $P = i^2 R$. For $R_1$, the rate of energy dissipation is

$$P_1 = (0.50\,\text{A})^2(4.0\,\Omega) = 1.0\,\text{W}$$

and for $R_2$, the rate of energy dissipation is

$$P_2 = (0.50\,\text{A})^2(8.0\,\Omega) = 2.0\,\text{W}.$$

(c) If $i$ is the current in a battery with potential difference $\Delta V_B$, then the battery supplies energy at the rate $P = i\,\Delta V_B$ provided the current enters the negative terminal. The battery absorbs energy at the rate $P = i\,\Delta V_B$ if the current enters the positive terminal. For battery 1 the power is

$$P_1 = (0.50\,\text{A})(12\,\text{V}) = 6.0\,\text{W}$$

and for battery 2, it is

$$P_2 = (0.50\,\text{A})(6.0\,\text{V}) = 3.0\,\text{W}.$$

The current enters battery 1 at the negative terminal, so this battery supplies energy to the circuit. The battery is discharging. The current enters battery 2 at the positive terminal, so this battery absorbs energy from the circuit. It is charging.

## 51

(a) Let $\mathcal{E}$ be the emf of the battery and $r$ be its internal resistance. When the battery is being charged with a current $i$ the potential difference across its terminals is $\Delta V_B = \mathcal{E} + ir = 12\,\text{V} + (50\,\text{A})(0.040\,\Omega) = 14\,\text{V}$.

(b) The rate of energy dissipation is $P_r = i^2 r = (50\,\text{A})^2(0.040\,\Omega) = 100\,\text{W}$.

(c) The rate of conversion to chemical energy is $P_{\mathcal{E}} = i\mathcal{E} = (50\,\text{A})(12\,\text{V}) = 600\,\text{W}$.

(d) The potential difference across the terminals is now $\Delta V_B = \mathcal{E} - ir = 12\,\text{V} - (50\,\text{A})^2(0.040\,\Omega) = 10\,\text{V}$. The rate of energy dissipation is the same, $100\,\text{W}$.

## 55

(a) and (b) The circuit is shown in the diagram on the right. The current is taken to be positive if it is clockwise. The potential difference across battery 1 is given by $V_{B1} = \mathcal{E} - ir_1$ and for this to be zero, the current must be $i = \mathcal{E}/r_1$. Kirchhoff's loop rule gives $2\mathcal{E} - ir_1 - ir_2 - iR = 0$. Substitute $i = \mathcal{E}/r_1$ and solve for $R$. You should get $R = r_1 - r_2$.

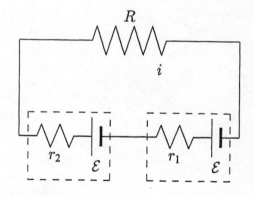

Now assume that the potential difference across battery 2 is zero and carry out the same analysis. You should find $R = r_2 - r_1$. Since $r_1 > r_2$ and $R$ must be positive, this situation is not possible. Only the potential difference across the battery with the larger internal resistance can be made to vanish with the proper choice of $R$.

## 59

(a) The batteries are identical and, because they are connected in parallel, the potential differences across them are the same. This means the currents in them are the same. Let $i$ be the current in either battery and take it to be positive if it is to the left. According to the junction rule, the current in $R$ is $2i$ and if it is positive it is to the right. The loop rule applied to either loop containing a battery and $R$ yields $\mathcal{E} - ir - 2iR = 0$, so

$$i = \frac{\mathcal{E}}{r + 2R}.$$

The rate of energy dissipation in $R$ is

$$P = (2i)^2 R = \frac{4\mathcal{E}^2 R}{(r + 2R)^2}.$$

Find the maximum by setting the derivative with respect to $R$ equal to zero. The derivative is

$$\frac{dP}{dR} = \frac{4\mathcal{E}^2}{(r + 2R)^2} - \frac{16\mathcal{E}^2 R}{(r + 2R)^3} = \frac{4\mathcal{E}^2(r - 2R)}{(r + 2R)^3}.$$

It vanishes and $P$ is a maximum if $R = r/2$.

(b) Substitute $R = r/2$ into $P = 4\mathcal{E}^2 R/(r + 2R)^2$ to obtain

$$P_{\text{max}} = \frac{4\mathcal{E}^2 r/2}{(2r)^2} = \frac{\mathcal{E}^2}{2r}.$$

# Chapter 28

## 5

(a) The capacitance of a parallel-plate capacitor is given by $C = \epsilon_0 A/d$, where $A$ is the area of each plate and $d$ is the plate separation. Since the plates are circular, the plate area is $A = \pi R^2$, where $R$ is the radius of a plate. Thus

$$C = \frac{\epsilon_0 \pi R^2}{d} = \frac{(8.85 \times 10^{-12}\,\text{F/m})\pi(8.2 \times 10^{-2}\,\text{m})^2}{1.3 \times 10^{-3}\,\text{m}} = 1.4 \times 10^{-10}\,\text{F} = 140\,\text{pF}.$$

(b) The charge on the positive plate is given by $q = C\,\Delta V$, where $\Delta V$ is the potential difference across the plates. Thus $q = (1.4 \times 10^{-10}\,\text{F})(120\,\text{V}) = 1.7 \times 10^{-8}\,\text{C} = 17\,\text{nC}$.

## 9

According to Eq. 28–17 the capacitance of a spherical capacitor is given by

$$C = 4\pi\epsilon_0 \frac{ab}{b-a},$$

where $a$ and $b$ are the radii of the spheres. If $a$ and $b$ are nearly the same then $4\pi ab$ is nearly the surface area of either sphere. Replace $4\pi ab$ with $A$ and $b - a$ with $d$ to obtain

$$C \approx \frac{\epsilon_0 A}{d}.$$

This the expression for the capacitance of a parallel-plate capacitor.

## 15

Let $x$ be the separation of the plates in the lower capacitor. Then the plate separation in the upper capacitor is $a - b - x$. The capacitance of the lower capacitor is $C_\ell = \epsilon_0 A/x$ and the capacitance of the upper capacitor is $C_u = \epsilon_0 A/(a - b - x)$, where $A$ is the plate area. Since the two capacitors are in series, the equivalent capacitance is determined from

$$\frac{1}{C_{eq}} = \frac{1}{C_\ell} + \frac{1}{C_u} = \frac{x}{\epsilon_0 A} + \frac{a - b - x}{\epsilon_0 A} = \frac{a - b}{\epsilon_0 A}.$$

Thus the equivalent capacitance is given by $C_{eq} = \epsilon_0 A/(a - b)$ and is independent of $x$.

## 17

The charge initially on the positive plate of a charged capacitor is given by $q = C_1\,|\Delta V_0|$, where $C_1$ (= 100 pF is the capacitance and $\Delta V_0$ (= 50 V) is the initial potential difference. After the battery is disconnected and the second capacitor is wired in parallel to the first, the charge on the first capacitor is $q_1 = C_1\,|\Delta V|$, where $\Delta V$ (= 35 V) is the new potential difference. Since charge is conserved in the process, the charge on the second capacitor is $q_2 = q - q_1$. Substitute $C_1|\Delta V_0|$ for $q$ and $C_1|\Delta V|$ for $q_1$ to obtain $q_2 = C_1(|\Delta V_0| - |\Delta V|)$. The potential difference across the second capacitor is also $|\Delta V|$, so the capacitance is

$$C_2 = \frac{q_2}{|\Delta V|} = \frac{|\Delta V_0| - |\Delta V|}{|\Delta V|} C_1 = \frac{50\,\text{V} - 35\,\text{V}}{35\,\text{V}}(100\,\text{pF}) = 43\,\text{pF}\,.$$

## 19

(a) After the switches are closed, the potential differences across the capacitors are the same and the two capacitors are in parallel. The potential difference from $a$ to $b$ is given by $|\Delta V_{ab}| = Q/C_{eq}$, where $Q$ is the net charge on the combination and $C_{eq}$ is the equivalent capacitance.

The equivalent capacitance is $C_{eq} = C_1 + C_2 = 4.0 \times 10^{-6}$ F. The total charge on the combination is the net charge on either pair of connected plates. The charge on capacitor 1 is

$$q_1 = C_1|\Delta V| = (1.0 \times 10^{-6}\,\text{F})(100\,\text{V}) = 1.0 \times 10^{-4}\,\text{C}$$

and the charge on capacitor 2 is

$$q_2 = C_2|\Delta V| = (3.0 \times 10^{-6}\,\text{F})(100\,\text{V}) = 3.0 \times 10^{-4}\,\text{C}\,,$$

so the net charge on the combination is $3.0 \times 10^{-4}\,\text{C} - 1.0 \times 10^{-4}\,\text{C} = 2.0 \times 10^{-4}\,\text{C}$. The potential difference is

$$V_{ab} = \frac{2.0 \times 10^{-4}\,\text{C}}{4.0 \times 10^{-6}\,\text{F}} = 50\,\text{V}\,.$$

(b) The charge on capacitor 1 is now $q_1 = C_1|\Delta V_{ab}| = (1.0 \times 10^{-6}\,\text{F})(50\,\text{V}) = 5.0 \times 10^{-5}\,\text{C}$.

(c) The charge on capacitor 2 is now $q_2 = C_2|\Delta V_{ab}| = (3.0 \times 10^{-6}\,\text{F})(50\,\text{V}) = 1.5 \times 10^{-4}\,\text{C}$.

## 29

(a) Let $q$ be the charge on the positive plate. Since the capacitance of a parallel-plate capacitor is given by $\epsilon_0 A/d$, the charge is $|q| = C|\Delta V| = \epsilon_0 A|\Delta V|/d$. After the plates are pulled apart, their separation is $2d$ and the potential difference is $\Delta V'$. Since the battery is disconnected the charge on the plates is the same. Thus $|q| = \epsilon_0 A|\Delta V'|/2d$ and

$$|\Delta V'| = \frac{2d}{\epsilon_0 A}|q| = \frac{2d}{\epsilon_0 A}\frac{\epsilon_0 A}{d}|\Delta V| = 2|\Delta V|\,.$$

(b) The initial energy stored in the capacitor is

$$U_i = \frac{1}{2}C(\Delta V)^2 = \frac{\epsilon_0 A(\Delta V)^2}{2d}$$

and the final energy stored is

$$U_f = \frac{1}{2}\frac{\epsilon_0 A}{2d}(\Delta V')^2 = \frac{1}{2}\frac{\epsilon_0 A}{2d}4(\Delta V)^2 = \frac{\epsilon_0 A(\Delta V)^2}{d}\,.$$

This is twice the initial energy.

(c) The work done to pull the plates apart is the difference in the energy: $W = U_f - U_i = \epsilon_0 A (\Delta V)^2 / 2d$.

## 31

You first need to find an expression for the energy stored in an imaginary cylinder of radius $R$ and length $L$, whose surface lies between the inner and outer cylinders of the capacitor ($a < R < b$). The energy density at any point is given by $u = \frac{1}{2}\epsilon_0 E^2$, where $E$ is the magnitude of the electric field at that point. If $q$ is the charge on the surface of the inner cylinder, then the magnitude of the electric field at a point a distance $r$ from the cylinder axis is given by

$$E = \frac{q}{2\pi\epsilon_0 L r}$$

(see Eq. 24–15 and replace $|\lambda|$ with $q/L$) and the energy density at that point is given by

$$u = \frac{1}{2}\epsilon_0 E^2 = \frac{q^2}{8\pi^2\epsilon_0 L^2 r^2}.$$

The energy in the cylinder is the volume integral

$$U_R = \int u\, dV.$$

Now $dV = 2\pi r L\, dr$, so

$$U_R = \int_a^R \frac{q^2}{8\pi^2\epsilon_0 L^2 r^2} 2\pi r L\, dr = \frac{q^2}{4\pi\epsilon_0 L}\int_a^R \frac{dr}{r} = \frac{q^2}{4\pi\epsilon_0 L}\ln\frac{R}{a}.$$

To find an expression for the total energy stored in the capacitor, replace $R$ with $b$:

$$U_b = \frac{q^2}{4\pi\epsilon_0 L}\ln\frac{b}{a}.$$

You want the ratio $U_R/U_b$ to be $1/2$, so

$$\ln\frac{R}{a} = \frac{1}{2}\ln\frac{b}{a}$$

or, since $\frac{1}{2}\ln(b/a) = \ln(\sqrt{b/a})$, $\ln(R/a) = \ln(\sqrt{b/a})$. This means $R/a = \sqrt{b/a}$ or $R = \sqrt{ab}$.

## 37

The capacitance of a cylindrical capacitor is given by

$$C = \kappa C_{\text{air}} = \frac{2\pi\kappa\epsilon_0 L}{\ln(b/a)},$$

where $C_{\text{air}}$ is the capacitance without the dielectric, $\kappa$ is the dielectric constant, $L$ is the length, $a$ is the inner radius, and $b$ is the outer radius. The capacitance per unit length of the cable is

$$\frac{C}{L} = \frac{2\pi\kappa\epsilon_0}{\ln(b/a)} = \frac{2\pi(2.6)(8.85\times 10^{-12}\,\text{F/m})}{\ln\left[(0.60\,\text{mm})/(0.10\,\text{mm})\right]} = 8.1\times 10^{-11}\,\text{F/m} = 81\,\text{pF/m}.$$

## 39

The capacitance is given by $C = \kappa C_{\text{air}} = \kappa \epsilon_0 A/d$, where $C_{\text{air}}$ is the capacitance without the dielectric, $\kappa$ is the dielectric constant, $A$ is the plate area, and $d$ is the plate separation. The electric field between the plates is given by $E = \Delta V/d$, where $\Delta V$ is the potential difference between the plates. Thus $d = |\Delta V|/E$ and $C = \kappa \epsilon_0 AE/|\Delta V|$. Solve for $A$:

$$A = \frac{C|\Delta V|}{\kappa \epsilon_0 E}.$$

For the area to be a minimum, the electric field must be the greatest it can be without breakdown occurring. That is,

$$A = \frac{(7.0 \times 10^{-8}\,\text{F})(4.0 \times 10^3\,\text{V})}{2.8(8.85 \times 10^{-12}\,\text{F/m})(18 \times 10^6\,\text{V/m})} = 0.63\,\text{m}^2.$$

## 43

(a) The magnitude of the electric field in the region between the plates is given by $E = |\Delta V|/d$, where $\Delta V$ is the potential difference between the plates and $d$ is the plate separation. The capacitance is given by $C = \kappa \epsilon_0 A/d$, where $A$ is the plate area and $\kappa$ is the dielectric constant, so $d = \kappa \epsilon_0 A/C$ and

$$E = \frac{C|\Delta V|}{\kappa \epsilon_0 A} = (100 \times 10^{-12}\,\text{F})\frac{(50\,\text{V})}{5.4(8.85 \times 10^{-12}\,\text{F/m})(100 \times 10^{-4}\,\text{m}^2)} = 1.0 \times 10^4\,\text{V/m}.$$

(b) The free charge on the positive plate is $q = C\,\Delta V = (100 \times 10^{-12}\,\text{F})(50\,\text{V}) = 5.0 \times 10^{-9}\,\text{C}$. The free charge on the negative plate is $-q = -5.0 \times 10^{-9}\,\text{C}$.

(c) The electric field is produced by both the free and induced charge. Since the field of a large uniform layer of charge is $q/2\epsilon_0 A$, the field between the plates is

$$E = \frac{q}{2\epsilon_0 A} + \frac{q}{2\epsilon_0 A} - \frac{q'}{2\epsilon_0 A} - \frac{q'}{2\epsilon_0 A},$$

where the first term is due to the positive free charge on one plate, the second is due to the negative free charge on the other plate, the third is due to the positive induced charge on one dielectric surface, and the fourth is due to the negative induced charge on the other dielectric surface. Note that the field due to the induced charge is opposite the field due to the free charge, so they tend to cancel. The induced charge is therefore

$$q' = q - \epsilon_0 AE$$
$$= 5.0 \times 10^{-9}\,\text{C} - (8.85 \times 10^{-12}\,\text{F/m})(100 \times 10^{-4}\,\text{m}^2)(1.0 \times 10^4\,\text{V/m})$$
$$= 4.1 \times 10^{-9}\,\text{C} = 4.1\,\text{nC}.$$

## 47

During charging, the charge on the positive plate of the capacitor is given by Eq. 28–39, with $RC = \tau$. That is,

$$q = C\,\Delta V_B \left[1 - e^{-t/\tau}\right],$$

where $C$ is the capacitance, $\Delta V_B$ is potential difference across the terminals of the battery used to charge the capacitor, and $\tau$ is the time constant. You want the time for which $q = 0.99 C \, \Delta V_B$, so

$$0.99 = 1 - e^{-t/\tau}.$$

Thus

$$e^{-t/\tau} = 0.01.$$

Take the natural logarithm of both sides to obtain $t/\tau = -\ln 0.01 = 4.6$ and $t = 4.6\tau$.

## 51

(a) The potential difference $\Delta V$ across the plates of a capacitor is related to the charge $q$ on the positive plate by $|\Delta V| = q/C$, where $C$ is capacitance. Since the charge on a discharging capacitor is given by $q = q_0 \, e^{-t/\tau}$, where $q_0$ is the charge at time $t = 0$ and $\tau$ is the time constant, this means

$$|\Delta V| = |\Delta V_0| \, e^{-t/\tau},$$

where $q_0/C$ was replaced by $|\Delta V_0|$, the initial potential difference. Solve for $\tau$ by dividing the equation by $|\Delta V_0|$ and taking the natural logarithm of both sides. The result is

$$\tau = -\frac{t}{\ln\left(\Delta V/\Delta V_0\right)} = -\frac{10.0\,\text{s}}{\ln\left[(1.00\,\text{V})/(100\,\text{V})\right]} = 2.17\,\text{s}.$$

(b) At $t = 17.0\,\text{s}$, $t/\tau = (17.0\,\text{s})/(2.17\,\text{s}) = 7.83$ and

$$\Delta V = \Delta V_0 \, e^{-t/\tau} = (100\,\text{V})\, e^{-7.83} = 3.96 \times 10^{-2}\,\text{V}.$$

## 53

(a) The initial energy stored in a capacitor is given by

$$U = \frac{q_0^2}{2C},$$

where $C$ is the capacitance and $q_0$ is the initial charge on one plate. Thus,

$$q_0 = \sqrt{2CU} = \sqrt{2(1.0 \times 10^{-6}\,\text{F})(0.50\,\text{J})} = 1.0 \times 10^{-3}\,\text{C}.$$

(b) The charge as a function of time is given by $q = q_0 \, e^{-t/\tau}$, where $\tau$ is the time constant. The current is the derivative with respect to time of the charge:

$$i = -\frac{dq}{dt} = \frac{q_0}{\tau} e^{-t/\tau}.$$

Here the current has been chosen to be positive when it is away from the positive plate of the capacitor. The initial current is given by this function evaluated for $t = 0$: $i_0 = q_0/\tau$. The time constant is

$$\tau = RC = (1.0 \times 10^{-6}\,\text{F})(1.0 \times 10^6\,\Omega) = 1.0\,\text{s}.$$

Thus

$$i_0 = \frac{1.0 \times 10^{-3}\,\text{C}}{1.0\,\text{s}} = 1.0 \times 10^{-3}\,\text{A}.$$

(c) Substitute $q = q_0\,e^{-t/\tau}$ into $\Delta V_C = q/C$ to obtain

$$\Delta V_C = \frac{q_0}{C}\,e^{-t/\tau} = \frac{1.0 \times 10^{-3}\,\text{C}}{1.0 \times 10^{-6}\,\text{F}}\,e^{-t/1.0\,\text{s}} = (1.0 \times 10^3\,\text{V})e^{-t/1.0\,\text{s}}.$$

Substitute $i = (q_0/\tau)\,e^{-t/\tau}$ into $V_R = iR$ to obtain

$$V_R = \frac{q_0 R}{\tau}\,e^{-t/\tau} = \frac{(1.0 \times 10^{-3}\,\text{C})(1.0 \times 10^6\,\Omega)}{1.0\,\text{s}}\,e^{-t/1.0\,\text{s}} = (1.0 \times 10^3\,\text{V})e^{-t/1.0\,\text{s}}.$$

(d) Substitute $i = (q_0/\tau)\,e^{-t/\tau}$ into $P = i^2 R$ to obtain

$$P = \frac{q_0^2 R}{\tau^2}\,e^{-2t/\tau} = \frac{(1.0 \times 10^{-3}\,\text{C})^2(1.0 \times 10^6\,\Omega)}{(1.0\,\text{s})^2}\,e^{-2t/1.0\,\text{s}} = (1.0\,\text{W})e^{-2t/1.0\,\text{s}}.$$

# Chapter 29

## 3

(a) The magnitude of the magnetic force on the proton is given by $F^{\text{mag}} = evB \sin\phi$, where $v$ is the speed of the proton, $B$ is the magnitude of the magnetic field, and $\phi$ is the angle between the particle velocity and the field when they are drawn with their tails at the same point. Thus

$$v = \frac{F^{\text{mag}}}{eB \sin\phi} = \frac{6.50 \times 10^{-17}\,\text{N}}{(1.60 \times 10^{-19}\,\text{C})(2.60 \times 10^{-3}\,\text{T}) \sin 23.0^\circ} = 4.00 \times 10^5\,\text{m/s}.$$

(b) The kinetic energy of the proton is

$$K = \frac{1}{2}mv^2 = \frac{1}{2}(1.67 \times 10^{-27}\,\text{kg})(4.00 \times 10^5\,\text{m/s})^2 = 1.34 \times 10^{-16}\,\text{J}.$$

This is $(1.34 \times 10^{-16}\,\text{J})/(1.60 \times 10^{-19}\,\text{J/eV}) = 835\,\text{eV}$.

## 7

Since the magnetic field is perpendicular to the particle velocity, the magnitude of the magnetic force is given by $F^{\text{mag}} = evB$ and the acceleration of the electron has magnitude $a = F^{\text{mag}}/m = evB/m$, where $v$ is the speed of the electron, $m$ is its mass, and $B$ is the magnitude of the magnetic field. Since the electron is traveling with uniform speed in a circle, its acceleration is $a = v^2/r$, where $r$ is the radius of the circle. Thus $evB/m = v^2/r$ and

$$B = \frac{mv}{er} = \frac{(9.11 \times 10^{-31}\,\text{kg})(1.3 \times 10^6\,\text{m/s})}{(1.60 \times 10^{-19}\,\text{C})(0.35\,\text{m})} = 2.1 \times 10^{-5}\,\text{T}.$$

## 13

Orient the magnetic field so it is perpendicular to the plane of the page. Then the electron will travel with constant speed around a circle in that plane. The magnetic force on an electron has magnitude $F^{\text{mag}} = evB$, where $v$ is the speed of the electron and $B$ is the magnitude of the magnetic field. If $r$ is the radius of the circle, the acceleration of the electron has magnitude $a = v^2/r$. Newton's second law yields $evB = mv^2/r$, so the radius of the circle is given by $r = mv/eB$. The kinetic energy of the electron is $K = \frac{1}{2}mv^2$, so $v = \sqrt{2K/m}$. Thus

$$r = \frac{m}{eB}\sqrt{\frac{2K}{m}} = \sqrt{\frac{2mK}{e^2 B^2}}.$$

This must be less than $d$, so

$$\sqrt{\frac{2mK}{e^2 B^2}} \le d$$

or

$$B \geq \sqrt{\frac{2mK}{e^2 d^2}}.$$

If the electrons are to travel as shown in Fig. 29–28, the magnetic field must be out of the page. Then the magnetic force is toward the center of the circular path.

## 17

(a) Solve the result of Touchstone Example 29–2 for $B$. You should get

$$B = \sqrt{\frac{8|\Delta V|m}{|q|x^2}}.$$

Evaluate this expression using $x = 2.00\,\text{m}$:

$$B = \sqrt{\frac{8(100 \times 10^3\,\text{V})(3.92 \times 10^{-25}\,\text{kg})}{(3.20 \times 10^{-19}\,\text{C})(2.00\,\text{m})^2}} = 0.495\,\text{T}.$$

(b) Let $N$ be the number of ions that are separated by the machine per unit time. The current is $i = qN$ and the mass that is separated per unit time is $M = mN$, where $m$ is the mass of a single ion. $M$ has the value

$$M = 100\,\text{mg/h} = \frac{100 \times 10^{-6}\,\text{kg}}{3600\,\text{s}} = 2.78 \times 10^{-8}\,\text{kg/s}.$$

Since $N = M/m$, we have

$$i = \frac{qM}{m} = \frac{(3.20 \times 10^{-19}\,\text{C})(2.78 \times 10^{-8}\,\text{kg/s})}{3.92 \times 10^{-25}\,\text{kg}} = 2.27 \times 10^{-2}\,\text{A}.$$

(c) Each ion deposits an energy of $|q\,\Delta V|$ in the cup, so the energy deposited in time $\Delta t$ is given by

$$E = N|q\,\Delta V|\,\Delta t = i|\Delta V|\,\Delta t.$$

For $\Delta t = 1.0\,\text{h}$,

$$E = (2.27 \times 10^{-2}\,\text{A})(100 \times 10^3\,\text{V})(3600\,\text{s}) = 8.17 \times 10^8\,\text{J}.$$

## 21

(a) The total force on the electron is $\vec{F} = -e(\vec{E} + \vec{v} \times \vec{B})$, where $\vec{E}$ is the electric field, $\vec{B}$ is the magnetic field, and $\vec{v}$ is the electron velocity. The magnitude of the magnetic force is $F^{\text{mag}} = evB \sin\phi$, where $\phi$ is the angle between the velocity and the field. Since the total force must vanish, $B = E/v\sin\phi$. The force is the smallest it can be when the field is perpendicular to the velocity and $\phi = 90°$. Then $B = E/v$. Use the expression $K = \frac{1}{2}mv^2$ for the kinetic energy to find the speed:

$$v = \sqrt{\frac{2K}{m}} = \sqrt{\frac{2(2.5 \times 10^3\,\text{eV})(1.60 \times 10^{-19}\,\text{J/eV})}{9.11 \times 10^{-31}\,\text{kg}}} = 2.96 \times 10^7\,\text{m/s}.$$

Thus

$$B = \frac{E}{v} = \frac{10 \times 10^3 \, \text{V/m}}{2.96 \times 10^7 \, \text{m/s}} = 3.37 \times 10^{-4} \, \text{T}.$$

The magnetic field must be perpendicular to both the electric field and the velocity of the electron. (b) A proton will pass undeflected if its velocity is the same as that of the electron. Both the electric and magnetic forces reverse direction, but they still balance each other.

## 31

The magnitude of the magnetic force on the wire is given by $F^{\text{mag}} = iLB \sin\phi$, where $i$ is the current in the wire, $L$ is the length of the wire, $B$ is the magnitude of the magnetic field, and $\phi$ is the angle between the current and the field. In this case $\phi = 70°$. Thus

$$F^{\text{mag}} = (5000 \, \text{A})(100 \, \text{m})(60.0 \times 10^{-6} \, \text{T}) \sin 70° = 28.2 \, \text{N}.$$

Apply the right-hand rule to the vector product $\vec{F}^{\text{mag}} = i\vec{L} \times \vec{B}$ to show that the force is to the west.

## 33

The magnetic force on the wire must be upward and have a magnitude equal to the magnitude of the gravitational force on the wire. Apply the right-hand rule to show that the current must be from left to right. Since the field and the current are perpendicular to each other, the magnitude of the magnetic force is given by $F^{\text{mag}} = iLB$, where $L$ is the length of the wire. Since the magnitude of the gravitational force is $F^{\text{grav}} = mg$, the condition that the tension in the supports vanish is $iLB = mg$, which yields

$$i = \frac{mg}{LB} = \frac{(0.0130 \, \text{kg})(9.8 \, \text{m/s}^2)}{(0.620 \, \text{m})(0.440 \, \text{T})} = 0.467 \, \text{A}.$$

## 35

The magnetic force must push horizontally on the rod to overcome the force of friction. But it can be oriented so it also pulls up on the rod and thereby reduces both the normal force and the maximum possible force of static friction.

Suppose the magnetic field makes the angle $\theta$ with the vertical. The diagram to the right shows the view from the end of the sliding rod. The forces are also shown: $\vec{F}^{\text{mag}}$ is the force of the magnetic field if the current is out of the page, $\vec{F}^{\text{grav}}$ is the force of gravity, $\vec{N}$ is the normal force of the stationary rails on the rod, and $\vec{f}$ is the force of friction. Notice that the magnetic force makes the angle $\theta$ with the *horizontal*. When the rod is on the verge of sliding, the net force acting on it is zero and the magnitude of the frictional force is given by $f = \mu_s N$, where $\mu_s$ is the coefficient of static friction. The magnetic field is perpendicular to the wire so the magnitude of the magnetic force is given by $F^{\text{mag}} = iLB$, where $i$ is the current in the rod and $L$ is the length of the rod.

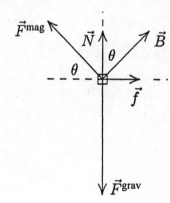

The vertical component of Newton's second law yields

$$N + iLB\sin\theta - mg = 0$$

and the horizontal component yields

$$iLB\cos\theta - \mu_s N = 0 ,$$

where $F^{\text{grav}} = mg$, where $m$ is the mass of the rod, was used. Solve the second equation for $N$ and substitute the resulting expression into the first equation, then solve for $B$. You should get

$$B = \frac{\mu_s mg}{iL(\cos\theta + \mu_s \sin\theta)} .$$

The minimum value of $B$ occurs when $\cos\theta + \mu_s \sin\theta$ is a maximum. Set the derivative of $\cos\theta + \mu_s \sin\theta$ equal to zero and solve for $\theta$. You should get $\theta = \tan^{-1}\mu_s = \tan^{-1}(0.60) = 31°$. Now evaluate the expression for the minimum value of $B$:

$$B_{\min} = \frac{0.60(1.0\,\text{kg})(9.8\,\text{m/s}^2)}{(50\,\text{A})(1.0\,\text{m})(\cos 31° + 0.60\sin 31°)} = 0.10\,\text{T} .$$

The magnetic field makes an angle of $31°$ with the vertical.

## 39

If $N$ closed loops are formed from the wire of length $L$, the circumference of each loop is $L/N$, the radius of each loop is $R = L/2\pi N$, and the area of each loop is $A = \pi R^2 = \pi(L/2\pi N)^2 = L^2/4\pi N^2$. For maximum torque, orient the plane of the loops parallel to the magnetic field, so the magnetic dipole moment is perpendicular to the field. The magnitude of the torque is then

$$\tau = NiAB = (Ni)\left(\frac{L^2}{4\pi N^2}\right)B = \frac{iL^2 B}{4\pi N} .$$

To maximize the torque, take $N$ to have the smallest possible value, 1. Then

$$\tau = \frac{iL^2 B}{4\pi} .$$

## 43

In the diagram on the right, $\vec{\mu}$ is the magnetic dipole moment of the wire loop and $\vec{B}$ is the magnetic field. Since the plane of the loop is parallel to the incline, the dipole moment is normal to the incline. The forces acting on the cylinder are the force of gravity $\vec{F}^{\text{grav}}$, acting downward from the center of mass, the normal force of the incline $\vec{N}$, acting perpendicularly to the incline through the center of mass, and the force of friction $\vec{f}$, acting up the incline at the point of contact.

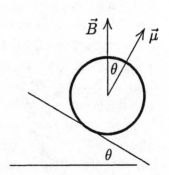

Take the positive $x$ direction to be down the incline. Then the $x$ component of Newton's second law for the center of mass yields

$$mg \sin \theta - f = ma \,,$$

where $m$ is the mass of the cylinder and $F^{\text{grav}} = mg$ was used. For purposes of calculating the torque, take the axis of the cylinder to be the axis of rotation. The magnetic field produces a torque with magnitude $\mu B \sin \theta$ and the force of friction produces a torque with magnitude $fr$, where $r$ is the radius of the cylinder. The first tends to produce a counterclockwise angular acceleration and the second tends to produce a clockwise angular acceleration. Newton's second law for rotation about the center of the cylinder, $\tau^{\text{net}} = I\alpha$, gives

$$fr - \mu B \sin \theta = I\alpha \,.$$

Since you want to find the current that holds the cylinder in place, set $a = 0$ and $\alpha = 0$, then use one equation to eliminate $f$ from the other. You should obtain $mgr = \mu B$. The loop is rectangular with two sides of length $L$ and two of length $2r$, so its area is $A = 2rL$ and the magnetic dipole moment is $\mu = NiA = 2NirL$. Thus $mgr = 2NirLB$ and

$$i = \frac{mg}{2NLB} = \frac{(0.250 \, \text{kg})(9.8 \, \text{m/s}^2)}{2(10.0)(0.100 \, \text{m})(0.500 \, \text{T})} = 2.45 \, \text{A} \,.$$

## 47

(a) The magnitude of the magnetic dipole moment of a current loop is given by $\mu = iA$, where $i$ is the current in the loop and $A$ is the area of the loop. This loop is a right triangle and its area is half the product of the sides that form the right angle. Thus

$$\mu = i\frac{1}{2}\ell_1 \ell_2 = (5.0 \, \text{A})\frac{1}{2}(0.30 \, \text{m})(0.40 \, \text{m}) = 0.30 \, \text{A} \cdot \text{m}^2 \,.$$

(b) The magnetic dipole moment is perpendicular to the plane of the loop and hence is perpendicular to the magnetic field. The magnitude of the torque is

$$\tau = \mu B = (0.30 \, \text{A} \cdot \text{m}^2)(80 \times 10^{-3} \, \text{T}) = 2.4 \times 10^{-2} \, \text{N} \cdot \text{m} \,.$$

## 51

The magnetic dipole moment is $\vec{\mu} = \mu(0.60\hat{\imath} - 0.80\hat{\jmath})$, where

$$\mu = NiA = Ni\pi r^2 = 1(0.20 \, \text{A})\pi(0.080 \, \text{m})^2 = 4.02 \times 10^{-4} \, \text{A} \cdot \text{m}^2 \,.$$

Here $i$ is the current in the loop, $N$ is the number of turns, $A$ is the area of the loop, and $r$ is its radius.

(a) The torque is

$$\begin{aligned}
\vec{\tau} = \vec{\mu} \times \vec{B} &= \mu(0.60\hat{\imath} - 0.80\hat{\jmath}) \times (0.25\hat{\imath} + 0.30\hat{k}) \\
&= \mu\left[(0.60)(0.30)(\hat{\imath} \times \hat{k}) - (0.80)(0.25)(\hat{\jmath} \times \hat{\imath}) - (0.80)(0.30)(\hat{\jmath} \times \hat{k})\right] \\
&= \mu[-0.18\hat{\jmath} + 0.20\hat{k} - 0.24\hat{\imath}] \,.
\end{aligned}$$

Units have been omitted in writing this equation and $\hat{i} \times \hat{k} = -\hat{j}$, $\hat{j} \times \hat{i} = -\hat{k}$, $\hat{j} \times \hat{k} = \hat{i}$, and $\hat{i} \times \hat{i} = 0$ were used. Substitute the value for $\mu$ to obtain

$$\vec{\tau} = [(-0.965 \times 10^{-4}\,\mathrm{N} \cdot \mathrm{m})\hat{i} - (7.23 \times 10^{-4}\,\mathrm{N} \cdot \mathrm{m})\hat{j} + (8.04 \times 10^{-4}\,\mathrm{N} \cdot \mathrm{m})\hat{k}].$$

(b) The potential energy of the dipole is given by

$$U = -\vec{\mu} \cdot \vec{B} = -\mu(0.60\,\hat{i} - 0.80\,\hat{j}) \cdot (0.25\,\hat{i} + 0.30\,\hat{k})$$
$$= -\mu(0.60)(0.25) = -0.15\mu = -6.0 \times 10^{-4}\,\mathrm{J}.$$

Here $\hat{i} \cdot \hat{i} = 1$, $\hat{i} \cdot \hat{k} = 0$, $\hat{j} \cdot \hat{i} = 0$, and $\hat{j} \cdot \hat{k} = 0$ were used.

# Chapter 30

## 3

(a) The field due to the wire, at a point 8.0 cm from the wire, must be 39 $\mu$T and must be directed toward due south. Since $B = \mu_0 i / 2\pi r$,

$$i = \frac{2\pi r B}{\mu_0} = \frac{2\pi (0.080\,\text{m})(39 \times 10^{-6}\,\text{T})}{4\pi \times 10^{-7}\,\text{T}\cdot\text{m/A}} = 16\,\text{A}.$$

(b) The current must be from west to east to produce a field to the south at points above it.

## 7

Sum the fields of the two straight wires and the circular arc. Use Eq. 30–9 for the magnitude of the magnetic field of a semi-infinite wire: $B = \mu_0 i / 4\pi R$, where $R$ is the distance from the end of the wire to the center of the arc. $R$ is the radius of the arc. The fields of both wires are out of the page at the center of the arc.

Now find an expression for the field of the arc, at its center. Divide the arc into infinitesimal segments. Each segment produces a field in the same direction. If d$s$ is the length of a segment, the magnitude of the field it produces at the arc center is $(\mu_0 i / 4\pi R^2)\,\text{d}s$. If $\theta$ is the angle subtended by the arc in radians, then $R\theta$ is the length of the arc and the total field of the arc is $\mu_0 i \theta / 4\pi R$. For the arc of the diagram, the field is into the page. The net field at the center, due to the wires and arc together, is

$$B = \frac{\mu_0 i}{4\pi R} + \frac{\mu_0 i}{4\pi R} - \frac{\mu_0 i \theta}{4\pi R} = \frac{\mu_0 i}{4\pi R}(2 - \theta).$$

For this to vanish, $\theta$ must be 2 radians.

## 11

Put the $x$ axis along the wire with the origin at the midpoint and the current in the positive $x$ direction. All segments of the wire produce magnetic fields at $P_1$ that are into the page so we simply divide the wire into infinitesimal segments and sum the magnitudes of the fields due to all the segments. The diagram shows one infinitesimal segment, with length d$x$. According to the Biot-Savart law, the magnitude of the field it produces at $P_1$ is given by

$$\text{d}B = \frac{\mu_0 i}{4\pi} \frac{\sin \theta}{r^2}\,\text{d}x.$$

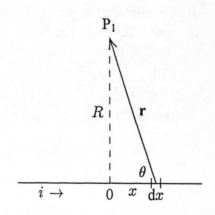

$\theta$ and $r$ are functions of $x$. Replace $r$ with $\sqrt{x^2 + R^2}$ and $\sin\theta$ with $R/r = R/\sqrt{x^2 + R^2}$, then integrate from $x = -L/2$ to $x = L/2$. The total field is

$$B = \frac{\mu_0 iR}{4\pi} \int_{-L/2}^{L/2} \frac{dx}{(x^2 + R^2)^{3/2}} = \frac{\mu_0 iR}{4\pi} \frac{1}{R^2} \frac{x}{(x^2 + R^2)^{1/2}} \bigg|_{-L/2}^{L/2} = \frac{\mu_0 i}{2\pi R} \frac{L}{\sqrt{L^2 + 4R^2}}.$$

If $L \gg R$, then $R^2$ in the denominator can be ignored and

$$B = \frac{\mu_0 i}{2\pi R}$$

is obtained. This is the field of a long straight wire. For points close to a finite wire, the field is quite similar to that of an infinitely long wire.

## 13

Follow the same steps as in the solution of Problem 11 above but change the lower limit of integration to $-L$, and the upper limit to $0$. The magnitude of the total field is

$$B = \frac{\mu_0 iR}{4\pi} \int_{-L}^{0} \frac{dx}{(x^2 + R^2)^{3/2}} = \frac{\mu_0 iR}{4\pi} \frac{1}{R^2} \frac{x}{(x^2 + R^2)^{1/2}} \bigg|_{-L}^{0} = \frac{\mu_0 i}{4\pi R} \frac{L}{\sqrt{L^2 + R^2}}.$$

## 15

The result of Problem 11 is used four times, once for each of the sides of the square loop. A point on the axis of the loop is also on a perpendicular bisector of each of the loop sides. The diagram shows the field due to one of the loop sides, the one on the left. In the expression found in Problem 11, replace $L$ with $a$ and $R$ with $\sqrt{x^2 + a^2/4} = \frac{1}{2}\sqrt{4x^2 + a^2}$. The field due to the side is therefore

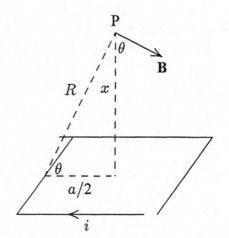

$$B = \frac{\mu_0 ia}{\pi\sqrt{4x^2 + a^2}\sqrt{4x^2 + 2a^2}}.$$

The field is in the plane of the dotted triangle shown and is perpendicular to the line from the midpoint of the loop side to the point P. Therefore it makes the angle $\theta$ with the vertical.

When the fields of the four sides are summed vectorially the horizontal components add to zero. The vertical components are all the same, so the net field is given by

$$B^{\text{net}} = 4B\cos\theta = \frac{4Ba}{2R} = \frac{4Ba}{\sqrt{4x^2 + a^2}}.$$

Thus

$$B^{\text{net}} = \frac{4\mu_0 ia^2}{\pi(4x^2 + a^2)\sqrt{4x^2 + 2a^2}}.$$

For $x = 0$, the expression reduces to

$$B^{\text{net}} = \frac{4\mu_0 i a^2}{\pi a^2 \sqrt{2}a} = \frac{2\sqrt{2}\mu_0 i}{\pi a} \,,$$

in agreement with the result of Problem 12.

## 19

Put the origin of the $x$ axis at point P and divide the strip into segments of infinitesimal width $dx$ and infinite length. Each segment produces a magnetic field that is like that of a long straight wire carrying current $di = i\,dx/w$, where $i$ is the total current and $w$ is the total width of the ribbon. The magnitude of the magnetic field produced by the segment at coordinate $x$ is $dB = (\mu_0 i/2\pi wx)\,dx$. The fields of all the segments are in the same direction, so the magnitude of the net field is the sum of the magnitudes of the individual fields. That is,

$$B = \int_d^{d+w} \frac{\mu_0 i}{2\pi wx}\,dx = \frac{\mu_0 i}{2\pi w}\ln\left(\frac{d+w}{d}\right) \,.$$

The right-hand rule shows that, since the current is into the page, the magnetic field at P points upward.

## 23

(a) If the currents are parallel, the two fields are in opposite directions in the region between the wires. Since the currents are the same, the net field is zero along the line that runs halfway between the wires. There is no possible current for which the field does not vanish.

(b) If the currents are antiparallel, the fields are in the same direction in the region between the wires. At a point halfway between, they have the same magnitude, $\mu_0 i/2\pi r$. Thus the net field at the midpoint has magnitude $B = \mu_0 i/\pi r$ and

$$i = \frac{\pi r B}{\mu_0} = \frac{\pi(0.040\,\text{m})(300 \times 10^{-6}\,\text{T})}{4\pi \times 10^{-7}\,\text{T} \cdot \text{m/A}} = 30\,\text{A} \,.$$

## 27

Each wire produces a field with magnitude given by $B = \mu_0 i/2\pi r$, where $r$ is the distance from the corner of the square to the center. According to the Pythagorean theorem, the diagonal of the square has length $\sqrt{2}a$, so $r = a/\sqrt{2}$ and $B = \mu_0 i/\sqrt{2}\pi a$. The fields due to the wires at the upper left and lower right corners both point toward the upper right corner of the square. The fields due to the wires at the upper right and lower left corners both point toward the upper left corner. The horizontal components cancel and the vertical components sum to

$$B^{\text{net}} = 4\frac{\mu_0 i}{\sqrt{2}\pi a}\cos 45° = \frac{2\mu_0 i}{\pi a}$$

$$= \frac{2(4\pi \times 10^{-7}\,\text{T} \cdot \text{m/A})(20\,\text{A})}{\pi(0.20\,\text{m})} = 8.0 \times 10^{-5}\,\text{T} \,.$$

In the calculation $\cos 45°$ was replaced with $1/\sqrt{2}$. The net field points upward.

## 33

(a) Use Ampere's law. Two of the currents are out of the page and one is into the page, so the net current enclosed by the path is $i^{\text{enc}} = 2.0\,\text{A}$, out of the page. Since the path is traversed in the clockwise sense, a current into the page is positive and a current out of the page is negative, as indicated by the right-hand rule associated with Ampere's law. Thus $i^{\text{enc}} = -2.0\,\text{A}$ and

$$\oint \mathbf{B} \cdot \mathbf{ds} = -\mu_0 (2.0\,\text{A}) = -(4\pi \times 10^{-7}\,\text{T} \cdot \text{m/A})(2.0\,\text{A}) = -2.5 \times 10^{-6}\,\text{T} \cdot \text{m}.$$

(b) The net current enclosed by the path is zero (two currents are out of the page and two are into the page), so $\oint \mathbf{B} \cdot \mathbf{ds} = \mu_0 i^{\text{enc}} = 0$.

## 39

(a) Take the magnetic field at a point within the hole to be the sum of the fields due to two current distributions. The first is the solid cylinder obtained by filling the hole and has a current density that is the same as that in the original cylinder with the hole. The second is the solid cylinder that fills the hole. It has a current density with the same magnitude as that of the original cylinder but it is in the opposite direction. Notice that if these two situations are superposed, the total current in the region of the hole is zero.

Recall that a solid cylinder carrying current $i$, uniformly distributed over a cross section, produces a magnetic field with magnitude $B = \mu_0 i r / 2\pi R^2$ a distance $r$ from its axis, inside the cylinder. Here $R$ is the radius of the cylinder.

For the cylinder of this problem, the magnitude of the current density is

$$J = \frac{i}{A} = \frac{i}{\pi(a^2 - b^2)},$$

where $A$ $(= \pi(a^2 - b^2))$ is the cross-sectional area of the cylinder with the hole. The current in the cylinder without the hole is

$$i_1 = J A_1 = \pi J a^2 = \frac{i a^2}{a^2 - b^2}$$

and the magnetic field it produces at a point inside, a distance $r_1$ from its axis, has magnitude

$$B_1 = \frac{\mu_0 i_1 r_1}{2\pi a^2} = \frac{\mu_0 i r_1 a^2}{2\pi a^2 (a^2 - b^2)} = \frac{\mu_0 i r_1}{2\pi (a^2 - b^2)}.$$

The current in the cylinder that fills the hole is

$$i_2 = \pi J b^2 = \frac{i b^2}{a^2 - b^2}$$

and the field it produces at a point inside, a distance $r_2$ from the its axis, has magnitude

$$B_2 = \frac{\mu_0 i_2 r_2}{2\pi b^2} = \frac{\mu_0 i r_2 b^2}{2\pi b^2 (a^2 - b^2)} = \frac{\mu_0 i r_2}{2\pi (a^2 - b^2)}.$$

At the center of the hole, this field is zero and the field there is exactly the same as it would be if the hole were filled. Place $r_1 = d$ in the expression for $B_1$ and obtain

$$B = \frac{\mu_0 i d}{2\pi(a^2 - b^2)}$$

for the field at the center of the hole. The field points upward in the diagram if the current is out of the page.

(b) If $b = 0$, the formula for the field becomes

$$B = \frac{\mu_0 i d}{2\pi a^2} \, .$$

This correctly gives the field of a solid cylinder carrying a uniform current $i$, at a point inside the cylinder a distance $d$ from the axis. If $d = 0$, the formula gives $B = 0$. This is correct for the field on the axis of a cylindrical shell carrying a uniform current.

(c) The diagram shows the situation in a cross-sectional plane of the cylinder. P is a point within the hole, $A$ is on the axis of the cylinder, and $C$ is on the axis of the hole. The magnetic field due to the cylinder without the hole, carrying a uniform current out of the page, is labeled $\vec{B}_1$ and the magnetic field of the cylinder that fills the hole, carrying a uniform current into the page, is labeled $\vec{B}_2$. The line from $A$ to P makes the angle $\theta_1$ with the line that joins the centers of the cylinders and the line from $C$ to P makes the angle $\theta_2$ with that line, as shown. $\vec{B}_1$ is perpendicular to the line from $A$ to P and so makes the angle $\theta_1$ with the vertical. Similarly, $\vec{B}_2$ is perpendicular to the line from $C$ to P and so makes the angle $\theta_2$ with the vertical.

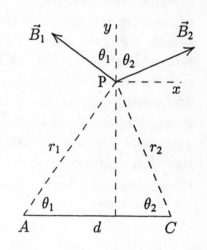

The $x$ component of the total field is

$$B_x = B_2 \sin\theta_2 - B_1 \sin\theta_1 = \frac{\mu_0 i r_2}{2\pi(a^2 - b^2)} \sin\theta_2 - \frac{\mu_0 i r_1}{2\pi(a^2 - b^2)} \sin\theta_1$$

$$= \frac{\mu_0 i}{2\pi(a^2 - b^2)} [r_2 \sin\theta_2 - r_1 \sin\theta_1] \, .$$

As the diagram shows, $r_2 \sin\theta_2 = r_1 \sin\theta_1$, so $B_x = 0$. The $y$ component is given by

$$B_y = B_2 \cos\theta_2 + B_1 \cos\theta_1 = \frac{\mu_0 i r_2}{2\pi(a^2 - b^2)} \cos\theta_2 + \frac{\mu_0 i r_1}{2\pi(a^2 - b^2)} \cos\theta_1$$

$$= \frac{\mu_0 i}{2\pi(a^2 - b^2)} [r_2 \cos\theta_2 + r_1 \cos\theta_1] \, .$$

The diagram shows that $r_2 \cos\theta_2 + r_1 \cos\theta_1 = d$, so

$$B_y = \frac{\mu_0 i d}{2\pi(a^2 - b^2)} \, .$$

This is identical to the result found in part (a) for the field on the axis of the hole. It is independent of $r_1$, $r_2$, $\theta_1$, and $\theta_2$, showing that the field is uniform in the hole.

## 41

(a) Suppose the field is not parallel to the sheet, as shown in the upper diagram. Reverse the direction of the current. According to the Biot-Savart law, the field reverses, so it will be as in the second diagram. Now rotate the sheet by $180°$ about a line that is perpendicular to the sheet. The field, of course, will rotate with it and end up in the direction shown in the third diagram. The current distribution is now exactly as it was originally, so the field must also be as it was originally. But it is not. Only if the field is parallel to the sheet will be final direction of the field be the same as the original direction. If the current is out of the page, any infinitesimal portion of the sheet in the form of a long straight wire produces a field that is to the left above the sheet and to the right below the sheet. The field must be as drawn in Fig. 30–47.

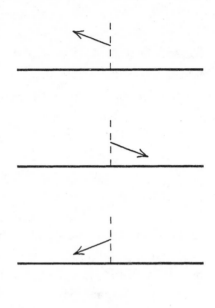

(b) Integrate the tangential component of the magnetic field around the rectangular loop shown with dotted lines. The upper and lower edges are the same distance from the current sheet and each has length $L$. This means the field has the same magnitude along these edges. It points to the left along the upper edge and to the right along the lower.

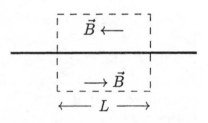

If the integration is carried out in the counterclockwise sense, the contribution of the upper edge is $BL$, the contribution of the lower edge is also $BL$, and the contribution of each of the sides is zero because the field is perpendicular to the sides. Thus $\oint \mathbf{B} \cdot d\mathbf{s} = 2BL$. The total current through the loop is $\lambda L$. Ampere's law yields $2BL = \mu_0 \lambda L$, so $B = \mu_0 \lambda/2$.

## 47

Consider a circle of radius $r$, inside the toroid and concentric with it. The current that passes through the region between the circle and the outer rim of the toroid is $Ni$, where $N$ is the number of turns and $i$ is the current. The current per unit length of circle is $\lambda = Ni/2\pi r$ and $\mu_0 \lambda$ is $\mu_0 Ni/2\pi r$, the magnitude of the magnetic field at the circle. Since the field is zero outside a toroid, this is also the change in the magnitude of the field encountered as you move from the circle to the outside.

The equality is not really surprising in light of Ampere's law. You are moving perpendicularly to the magnetic field lines. Consider an extremely narrow loop, with the narrow sides along field lines and the two long sides perpendicular to the field lines. If $B_1$ is the magnitude of the field at one end and $B_2$ is the magnitude of the field at the other end, then $\oint \mathbf{B} \cdot d\mathbf{s} = (B_2 - B_1)w$, where $w$ is the width of the loop. The current through the loop is $w\lambda$, so Ampere's law yields $(B_2 - B_1)w = \mu_0 w\lambda$ and $B_2 - B_1 = \mu_0 \lambda$.

**49**

(a) Assume that the point is inside the solenoid. The field of the solenoid at the point is parallel to the solenoid axis and the field of the wire is perpendicular to the solenoid axis. The total field makes an angle of 45° with the axis if these two fields have equal magnitudes.

The magnitude of the magnetic field produced by a solenoid at a point inside is given by $B_{\text{sol}} = \mu_0 i_{\text{sol}} n$, where $n$ is the number of turns per unit length and $i_{\text{sol}}$ is the current in the solenoid. The magnitude of the magnetic field produced by a long straight wire at a point a distance $r$ away is given by $B_{\text{wire}} = \mu_0 i_{\text{wire}}/2\pi r$, where $i_{\text{wire}}$ is the current in the wire. We want $\mu_0 n i_{\text{sol}} = \mu_0 i_{\text{wire}}/2\pi r$. The solution for $r$ is

$$r = \frac{i_{\text{wire}}}{2\pi n i_{\text{sol}}} = \frac{6.00\,\text{A}}{2\pi(10.0 \times 10^2\,\text{m}^{-1})(20.0 \times 10^{-3}\,\text{A})} = 4.77 \times 10^{-2}\,\text{m} = 4.77\,\text{cm}.$$

This distance is less than the radius of the solenoid, so the point is indeed inside as we assumed.

(b) The magnitude of the either field at the point is

$$B_{\text{sol}} = B_{\text{wire}} = \mu_0 n i_{\text{sol}} = (4\pi \times 10^{-7}\,\text{T}\cdot\text{m/A})(10.0 \times 10^2\,\text{m}^{-1})(20.0 \times 10^{-3}\,\text{A}) = 2.51 \times 10^{-5}\,\text{T}.$$

Each of the two fields is a component of the total field, so the magnitude of the total field is the square root of the sum of the squares of the individual fields: $B = \sqrt{2(2.51 \times 10^{-5}\,\text{T})^2} = 3.55 \times 10^{-5}\,\text{T}$.

**57**

(a) The magnitude of the magnetic field on the axis of a circular loop, a distance $z$ from the loop center, is given by Eq. 30–28:

$$B = \frac{N\mu_0 i R^2}{2(R^2 + z^2)^{3/2}},$$

where $R$ is the radius of the loop, $N$ is the number of turns, and $i$ is the current. Both of the loops in the problem have the same radius, the same number of turns, and carry the same current. The currents are in the same sense and the fields they produce are in the same direction in the region between them. Place the origin at the center of the left-hand loop and let $x$ be the coordinate of a point on the axis between the loops. To calculate the field of the left-hand loop, set $z = x$ in the equation above. The chosen point on the axis is a distance $s - x$ from the center of the right-hand loop. To calculate the field it produces, put $z = s - x$ in the equation above. The total field at the point is therefore

$$B = \frac{N\mu_0 i R^2}{2}\left[\frac{1}{(R^2 + x^2)^{3/2}} + \frac{1}{(R^2 + x^2 - 2sx + s^2)^{3/2}}\right].$$

Its derivative with respect to $x$ is

$$\frac{dB}{dx} = -\frac{3N\mu_0 i R^2}{2}\left[\frac{x}{(R^2 + x^2)^{5/2}} + \frac{(x - s)}{(R^2 + x^2 - 2sx + s^2)^{5/2}}\right].$$

When this is evaluated for $x = s/2$ (the midpoint between the loops), the result is

$$\frac{dB}{dx}\bigg|_{s/2} = -\frac{3N\mu_0 i R^2}{2}\left[\frac{s/2}{(R^2 + s^2/4)^{5/2}} - \frac{s/2}{(R^2 + s^2/4 - s^2 + s^2)^{5/2}}\right] = 0,$$

independently of the value of $s$.

(b) The second derivative is

$$\frac{d^2 B}{dx^2} = \frac{N\mu_0 i R^2}{2} \left[ -\frac{3}{(R^2 + x^2)^{5/2}} + \frac{15x^2}{(R^2 + x^2)^{7/2}} \right.$$

$$\left. -\frac{3}{(R^2 + x^2 - 2sx + s^2)^{5/2}} + \frac{15(x - s)^2}{(R^2 + x^2 - 2sx + s^2)^{7/2}} \right].$$

At $x = s/2$,

$$\frac{d^2 B}{dx^2}\bigg|_{s/2} = \frac{N\mu_0 i R^2}{2} \left[ -\frac{6}{(R^2 + s^2/4)^{5/2}} + \frac{30s^2/4}{(R^2 + s^2/4)^{7/2}} \right]$$

$$= \frac{N\mu_0 i R^2}{2} \frac{-6(R^2 + s^2/4) + 30s^2/4}{(R^2 + s^2/4)^{7/2}} = 3N\mu_0 i R^2 \frac{s^2 - R^2}{(R^2 + s^2/4)^{7/2}}.$$

Clearly, this is zero if $s = R$.

# Chapter 31

### 3

(a) According to Faraday's law the magnitude of the emf is rate of change of the magnetic flux. The derivative of the flux with respect to time is

$$\frac{d\Phi^{\text{mag}}}{dt} = 2(6.0 \times 10^{-3}\,\text{Wb/s}^2)t + (3.7 \times 10^{-3}\,\text{Wb/s})$$

and at $t = 2.0\,\text{s}$ this is

$$\frac{d\Phi^{\text{mag}}}{dt} = 2(6.0 \times 10^{-3}\,\text{Wb/s}^2)(2.0\,\text{s}) + (3.7 \times 10^{-3}\,\text{Wb/s}) = 2.77 \times 10^{-2}\,\text{Wb/s}.$$

The magnitude of the emf is $2.77 \times 10^{-2}\,\text{V}$.

(b) The external magnetic field is out of the page and increasing, so according to Lenz's law the magnetic field of the induced current must be into the page in the interior of the loop. This means the current is clockwise around the loop and from right to left in the resistor.

### 11

(a) In the region of the smaller loop, the magnetic field produced by the larger loop may be taken to be uniform and equal to its value at the center of the smaller loop, on the axis of the larger loop. Eq. 30–29, with $z = x$ and much greater than $R$, gives

$$B = \frac{\mu_0 i R^2}{2x^3}$$

for the magnitude. Here $\mu$. the magnitude of the dipole moment has been replaced by $i\pi R^2$. The field is upward in the diagram. The magnetic flux at the smaller loop is the product of this field and the area ($\pi r^2$) of the smaller loop:

$$\Phi^{\text{mag}} = \frac{\pi \mu_0 i r^2 R^2}{2x^3}.$$

(b) The emf is given by Faraday's law:

$$\mathcal{E} = -\frac{d\Phi^{\text{mag}}}{dt} = -\left(\frac{\pi \mu_0 i r^2 R^2}{2}\right)\frac{d}{dt}\left(\frac{1}{x^3}\right) = -\left(\frac{\pi \mu_0 i r^2 R^2}{2}\right)\left(-\frac{3}{x^4}\frac{dx}{dt}\right) = \frac{3\pi \mu_0 i r^2 R^2 v}{2x^4}.$$

(c) The field of the larger loop is upward and decreases with distance away from the loop. As the smaller loop moves away, the flux at its position decreases. The induced current is directed so as to produce a magnetic field that is upward through the smaller loop, in the same direction as the field of the larger loop. It will be counterclockwise as viewed from above, in the same direction as the current in the larger loop.

**13**

(a) The emf induced around the loop is given by Faraday's law: $\mathcal{E} = -d\Phi^{mag}/dt$ and the current in the loop is given by $i = \mathcal{E}/R = -(1/R)(d\Phi^{mag}/dt)$. The charge that passes through the resistor from time zero to time $t$ is given by the integral

$$q = \int_0^t i\,dt = -\frac{1}{R}\int_0^t \frac{d\Phi^{mag}}{dt}\,dt = -\frac{1}{R}\int_{\Phi^{mag}(0)}^{\Phi^{mag}(t)} d\Phi^{mag} = \frac{1}{R}\left[\Phi^{mag}(0) - \Phi^{mag}(t)\right].$$

All that matters is the change in the flux, not how it was changed.

(b) If $\Phi^{mag}(t) = \Phi^{mag}(0)$, then $q = 0$. This does not mean that the current was zero for any extended time during the interval. If $\Phi^{mag}$ increases and then decreases back to its original value, there is current in the resistor while $\Phi^{mag}$ is changing. It is in one direction at first, then in the opposite direction. When equal charge has passed through the resistor in opposite directions, the net charge is zero.

**17**

(a) Let $L$ be the length of a side of the square circuit. Then the magnetic flux through the circuit is $\Phi^{mag} = L^2 B/2$ and the induced emf is

$$\mathcal{E}_i = -\frac{d\Phi^{mag}}{dt} = -\frac{L^2}{2}\frac{dB}{dt}.$$

Now $B = (0.042\,\text{T}) - (0.870\,\text{T/s})t$ and $dB/dt = -0.870\,\text{T/s}$. Thus

$$\mathcal{E}_i = \frac{(2.00\,\text{m})^2}{2}(0.870\,\text{T/s}) = 1.74\,\text{V}.$$

The magnetic field is out of the page and decreasing so the induced emf is counterclockwise around the circuit, in the same direction as the emf of the battery. The total emf is $\mathcal{E} + \mathcal{E}_i = 20.0\,\text{V} + 1.74\,\text{V} = 21.7\,\text{V}$.

(b) The current is in the sense of the total emf, counterclockwise.

**23**

Use Faraday's law to find an expression for the emf induced by the changing magnetic field. First, find an expression for the magnetic flux through the loop. Since the field depends on $y$ but not on $x$, divide the area into strips of length $L$ and width $dy$, parallel to the $x$ axis. Here $L$ is the length of one side of the square. At time $t$, the flux through a strip with coordinate $y$ is $d\Phi^{mag} = BL\,dy = (4.0\,\text{T/m}\cdot\text{s}^2)Lt^2 y\,dy$ and the total flux through the square is

$$\Phi^{mag} = \int_0^L (4.0\,\text{T/m}\cdot\text{s}^2)Lt^2 y\,dy = (2.0\,\text{T/m}\cdot\text{s}^2)L^3 t^2.$$

According to Faraday's law, the magnitude of the emf around the square is

$$|\mathcal{E}| = \frac{d\Phi^{mag}}{dt} = \frac{d}{dt}\left[(2.0\,\text{T/m}\cdot\text{s}^2)L^3 t^2\right] = (4.0\,\text{T/m}\cdot\text{s}^2)L^3 t.$$

At $t = 2.5\,\text{s}$, this is $(4.0\,\text{T/m}\cdot\text{s}^2)(0.020\,\text{m})^3(2.5\,\text{s}) = 8.0 \times 10^{-5}\,\text{V}$.

The externally-produced magnetic field is out of the page and is increasing with time. The induced current produces a field that is into the page, so it must be clockwise. The induced emf is also clockwise.

## 25

(a) Suppose each wire has radius $R$ and the distance between their axes is $a$. Consider a single wire and calculate the flux through a rectangular area with the axis of the wire along one side. Take this side to have length $L$ and the other dimension of the rectangle to be $a$. The magnetic field is everywhere perpendicular to the rectangle. First, consider the part of the rectangle that is inside the wire. The magnitude of the field a distance $r$ from the axis is given by $B_{\text{in}} = \mu_0 i r / 2\pi R^2$ and the flux through the strip of length $L$ and width $dr$ at that distance is $(\mu_0 i r / 2\pi R^2) L\, dr$. Thus the flux through the area inside the wire is

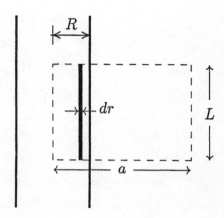

$$\Phi_{\text{in}}^{\text{mag}} = \int_0^R \frac{\mu_0 i L}{2\pi R^2}\, r\, dr = \frac{\mu_0 i L}{4\pi}.$$

Now consider the region outside the wire. There the field is given by $B_{\text{out}} = \mu_0 i / 2\pi r$ and the flux through an infinitesimally thin strip is $(\mu_0 i / 2\pi r) L\, dr$. The flux through the whole region is

$$\Phi_{\text{out}}^{\text{mag}} = \int_R^a \frac{\mu_0 i L}{2\pi} \frac{dr}{r} = \frac{\mu_0 i L}{2\pi} \ln\left(\frac{a}{R}\right).$$

The total flux through the area bounded by the dashed lines is the sum of the two contributions:

$$\Phi^{\text{mag}} = \frac{\mu_0 i L}{4\pi} \left[1 + 2\ln\left(\frac{a}{R}\right)\right].$$

Now include the contribution of the other wire. Since the currents are in the same direction, the two contributions have the same sign. They also have the same magnitude, so

$$\Phi_{\text{total}}^{\text{mag}} = \frac{\mu_0 i L}{2\pi} \left[1 + 2\ln\left(\frac{a}{R}\right)\right].$$

The total flux per unit length is

$$\frac{\Phi_{\text{total}}^{\text{mag}}}{L} = \frac{\mu_0 i}{2\pi} \left[1 + 2\ln\left(\frac{a}{R}\right)\right] = \frac{(4\pi \times 10^{-7}\,\text{T}\cdot\text{m/A})(10\,\text{A})}{2\pi} \left[1 + 2\ln\left(\frac{20\,\text{mm}}{1.25\,\text{mm}}\right)\right]$$
$$= 1.31 \times 10^{-5}\,\text{Wb/m}.$$

(b) Again consider the flux of a single wire. The flux inside the wire itself is again $\Phi_{\text{in}}^{\text{mag}} = \mu_0 i L / 4\pi$. The flux inside the region due to the other wire is

$$\Phi_{\text{out}}^{\text{mag}} = \int_{a-R}^a \frac{\mu_0 i L}{2\pi} \frac{dr}{r} = \frac{\mu_0 i L}{2\pi} \ln\left(\frac{a}{a-R}\right).$$

Add $\Phi_{in}^{mag}$ and $\Phi_{out}^{mag}$, then double the result to include the flux of the other wire and divide by $L$ to obtain the flux per unit length. The total flux per unit length that is inside the wires is

$$\frac{\Phi_{wires}^{,mag}}{L} = \frac{\mu_0 i}{2\pi}\left[1 + 2\ln\left(\frac{a}{a-R}\right)\right]$$

$$= \frac{(4\pi \times 10^{-7}\,\text{T}\cdot\text{m/A})(10\,\text{A})}{2\pi}\left[1 + 2\ln\left(\frac{20\,\text{mm}}{20\,\text{mm} - 1.25\,\text{mm}}\right)\right]$$

$$= 2.26 \times 10^{-6}\,\text{Wb/m}\,.$$

The fraction of the total flux that is inside the wires is

$$\frac{2.26 \times 10^{-6}\,\text{Wb/m}}{1.31 \times 10^{-5}\,\text{Wb/m}} = 0.17\,.$$

(c) The contributions of the two wires to the total flux have the same magnitudes but now the currents are in opposite directions, so the contributions have opposite signs. This means $\Phi_{total}^{mag} = 0$.

### 31

(a) Let $x$ be the distance from the right end of the rails to the rod. The area enclosed by the rod and rails is $Lx$ and the magnetic flux through the area is $\Phi^{mag} = BLx$. The magnitude of the emf induced is $|\mathcal{E}| = d\Phi^{mag}/dt = BL\,dx/dt = BLv$, where $v$ is the speed of the rod. Thus $|\mathcal{E}| = (1.2\,\text{T})(0.10\,\text{m})(5.0\,\text{m/s}) = 0.60\,\text{V}$.

(b) If $R$ is the resistance of the rod, the current in the loop is $i = |\mathcal{E}|/R = (0.60\,\text{V})/(0.40\,\Omega) = 1.5\,\text{A}$. Since the rod moves to the left in the diagram, the flux increases. The induced current must produce a magnetic field that is into the page in the region bounded by the rod and rails. To do this, the current must be clockwise.

(c) The rate of generation of thermal energy by the resistance of the rod is $P = \mathcal{E}/R = (0.60\,\text{V})^2/(0.40\,\Omega) = 0.90\,\text{W}$.

(d) Since the rod moves with constant velocity, the net force on it must be zero. This means the force of the external agent has the same magnitude as the magnetic force but is in the opposite direction. The magnitude of the magnetic force is $F^{mag} = iLB = (1.5\,\text{A})(0.10\,\text{m})(1.2\,\text{T}) = 0.18\,\text{N}$. Since the field is out of the page and the current is upward through the rod, the magnetic force is to the right. The force of the external agent must be 0.18 N, to the left.

(e) As the rod moves an infinitesimal distance $dx$, the external agent does work $dW^{agent} = F^{agent}\,dx$, where $F^{agent}$ is the force of the agent. The force is in the direction of motion, so the work done by the agent is positive. The rate at which the agent does work is $dW^{agent}/dt = F^{agent}\,dx/dt = F^{agent}v = (0.18\,\text{N})(5.0\,\text{m/s}) = 0.90\,\text{W}$, the same as the rate at which thermal energy is generated. The energy supplied by the external agent is converted completely to thermal energy.

### 33

(a) Let $x$ be the distance from the right end of the rails to the rod and find an expression for the magnetic flux through the area enclosed by the rod and rails. The magnetic field is not uniform but varies with distance from the long straight wire. The field is normal to the area and has magnitude $B = \mu_0 i / 2\pi r$, where $r$ is the distance from the wire and $i$ is the current in the wire. Consider an infinitesimal strip of length $x$ and width $dr$, parallel to the wire and a distance $r$ from it. The area of this strip is $A = x \, dr$ and the flux through it is $d\Phi^{\text{mag}} = (\mu_0 i x / 2\pi r) \, dr$. The total flux through the area enclosed by the rod and rails is

$$\Phi^{\text{mag}} = \frac{\mu_0 i x}{2\pi} \int_a^{a+L} \frac{dr}{r} = \frac{\mu_0 i x}{2\pi} \ln \left( \frac{a+L}{a} \right) .$$

According to Faraday's law, the magnitude of the emf induced in the loop is

$$\begin{aligned}
|\mathcal{E}| &= \frac{d\Phi}{dt} = \frac{\mu_0 i}{2\pi} \frac{dx}{dt} \ln \left( \frac{a+L}{a} \right) = \frac{\mu_0 i v}{2\pi} \ln \left( \frac{a+L}{a} \right) \\
&= \frac{(4\pi \times 10^{-7} \, \text{T} \cdot \text{m/A})(100 \, \text{A})(5.00 \, \text{m/s})}{2\pi} \ln \left( \frac{1.00 \, \text{cm} + 10.0 \, \text{cm}}{1.00 \, \text{cm}} \right) \\
&= 2.40 \times 10^{-4} \, \text{V} .
\end{aligned}$$

(b) If $R$ is the resistance of the rod, then the current in the conducting loop is

$$i_\ell = \frac{|\mathcal{E}|}{R} = \frac{2.40 \times 10^{-4} \, \text{V}}{0.400 \, \Omega} = 6.00 \times 10^{-4} \, \text{A} .$$

Since the flux is increasing, the magnetic field produced by the induced current must be into the page in the region enclosed by the rod and rails. This means the current is clockwise.

(c) Thermal energy is generated at the rate

$$P = i_\ell^2 R = (6.00 \times 10^{-4} \, \text{A})^2 (0.400 \, \Omega) = 1.44 \times 10^{-7} \, \text{W} .$$

(d) Since the rod moves with constant velocity, the net force on it is zero. The force of the external agent must have the same magnitude as the magnetic force and must be in the opposite direction. The magnitude of the magnetic force on an infinitesimal segment of the rod, with length $dr$ and a distance $r$ from the long straight wire, is $dF^{\text{mag}} = i_\ell B \, dr = (\mu_0 i_\ell i / 2\pi r) \, dr$. The total magnetic force on the rod has magnitude

$$\begin{aligned}
F^{\text{mag}} &= \frac{\mu_0 i_\ell i}{2\pi} \int_a^{a+L} \frac{dr}{r} = \frac{\mu_0 i_\ell i}{2\pi} \ln \left( \frac{a+L}{a} \right) \\
&= \frac{(4\pi \times 10^{-7} \, \text{T} \cdot \text{m/A})(6.00 \times 10^{-4} \, \text{A})(100 \, \text{A})}{2\pi} \ln \left( \frac{1.00 \, \text{cm} + 10.0 \, \text{cm}}{1.00 \, \text{cm}} \right) \\
&= 2.87 \times 10^{-8} \, \text{N} .
\end{aligned}$$

Since the field is out of the page and the current in the rod is upward in the diagram, this force is toward the right. The external agent must apply a force of $F^{\text{agent}} = 2.87 \times 10^{-8} \, \text{N}$, to the left.

(e) The external agent does work at the rate

$$P = F^{\text{agent}} v = (2.87 \times 10^{-8} \, \text{N})(5.00 \, \text{m/s}) = 1.44 \times 10^{-7} \, \text{W} .$$

This is the same as the rate at which thermal energy is generated in the rod. All the energy supplied by the agent is converted to thermal energy.

## 35

(a) The field point is inside the solenoid, so Eq. 31–26 applies. The magnitude of the induced electric field is

$$E = \frac{1}{2}\frac{dB}{dt}r = \frac{1}{2}(6.5 \times 10^{-3}\,\text{T/s})(0.0220\,\text{m}) = 7.15 \times 10^{-5}\,\text{V/m}.$$

(b) Now the field point is outside the solenoid and Eq. 31–28 applies. The magnitude of the induced field is

$$E = \frac{1}{2}\frac{dB}{dt}\frac{R^2}{r} = \frac{1}{2}(6.5 \times 10^{-3}\,\text{T/s})\frac{(0.0600\,\text{m})^2}{(0.0820\,\text{m})} = 1.43 \times 10^{-4}\,\text{V/m}.$$

## 39

Consider a circle of radius $r$ (= 6.0 mm), between the plates and with its center on the axis of the capacitor. The current through this circle is zero, so the Ampere-Maxwell law becomes

$$\oint \vec{B} \cdot d\vec{s} = \mu_0\epsilon_0\frac{d\Phi^{\text{elec}}}{dt},$$

where $\vec{B}$ is the magnetic field at points on the circle and $\Phi^{\text{elec}}$ is the electric flux through the circle. The magnetic field is tangent to the circle at all points on it, so $\oint \vec{B} \cdot d\vec{s} = 2\pi r B$. The electric flux through the circle is $\Phi^{\text{elec}} = \pi R^2 E$, where $R$ (= 3.0 mm) is the radius of a capacitor plate. When these substitutions are made, the Ampere-Maxwell law becomes

$$2\pi r B = \mu_0\epsilon_0\pi R^2\frac{dE}{dt}.$$

Thus

$$\frac{dE}{dt} = \frac{2rB}{\mu_0\epsilon_0 R^2} = \frac{2(6.0 \times 10^{-3}\,\text{m})(2.0 \times 10^{-7}\,\text{T})}{(4\pi \times 10^{-7}\,\text{H/m})(8.85 \times 10^{-12}\,\text{Fm})(3.0 \times 10^{-3}\,\text{m})^2} = 2.4 \times 10^{13}\,\text{V/m} \cdot \text{s}.$$

## 49

Consider an area $A$, normal to a uniform electric field $\vec{E}$. The displacement current density is uniform and normal to the area. Its magnitude is given by $J^{\text{dis}} = i^{\text{dis}}/A$. For this situation,

$$i^{\text{dis}} = \epsilon_0 A\frac{dE}{dt},$$

so

$$J^{\text{dis}} = \frac{1}{A}\epsilon_0 A\frac{dE}{dt} = \epsilon_0\frac{dE}{dt}.$$

## 53

If the electric field is perpendicular to a surface and has uniform magnitude over the surface then the displacement current through the surface is related to the rate of change of the electric field at the surface by

$$i^{\text{dis}} = \epsilon_0 A\frac{dE}{dt},$$

where $A$ is the area of the surface. The rate of change of the electric field is the slope of the graph.

For segment a

$$\frac{dE}{dt} = \frac{6.0 \times 10^5 \, \text{V/m} - 4.0 \times 10^5 \, \text{V/m}}{4.0 \times 10^{-6} \, \text{s}} = 5.0 \times 10^{10} \, \text{V/m} \cdot \text{s}$$

and $i^{\text{dis}} = (8.85 \times 10{-}12 \, \text{F/m})(1.6 \, \text{m}^2)(5.0 \times 10^{10} \, \text{V/m} \cdot \text{s} = 0.71 \, \text{A}.$

For segment b $dE/dt = 0$ and $i^{\text{dis}} = 0$.

For segment c

$$\frac{dE}{dt} = \frac{4.0 \times 10^5 \, \text{V/m} - 0}{5.0 \times 10^{-6} \, \text{s}} = 8.0 \times 10^{10} \, \text{V/m} \cdot \text{s}$$

and $i^{\text{dis}} = (8.85 \times 10{-}12 \, \text{F/m})(1.6 \, \text{m}^2)(8.0 \times 10^{10} \, \text{V/m} \cdot \text{s} = 1.1 \, \text{A}.$

## 59

(a) Use the Maxwell law of induction $\oint \vec{B} \cdot d\vec{s} = \mu_0 i^{\text{dis}}$, where the integral is around a circular path of radius $r$, concentric with the given circular region, and $i^{\text{dis}}$ is the displacement current through the area bounded by the path. The magnetic field is tangent to the path and its magnitude is uniform over the path, so $\oint \vec{B} \cdot d\vec{s} = 2\pi r B$. The path is inside the given region, so the displacement current through the path is

$$i^{\text{dis}} = \int_0^r J^{\text{dis}} 2\pi r \, dr = 2\pi \int_0^r (4.00 \, \text{A/m}^2) \left[ 1 - \frac{r}{R} \right] r \, dr = 2\pi (4.00 \, \text{A/m}^2) \left[ \frac{r^2}{2} - \frac{r^3}{3R} \right].$$

Thus

$$B = \frac{\mu_0 (4.00 \, \text{A/m}^2)}{r} \left[ \frac{r^2}{2} - \frac{r^3}{3R} \right] = \mu_0 (4.00 \, \text{A/m}^2) r \left[ \frac{1}{2} - \frac{r}{3R} \right]$$

$$= (4\pi \times 10^{-7} \, \text{T} \cdot \text{m/A})(4.00 \, \text{A/m}^2)(2.00 \times 10^{-2} \, \text{m}) \left[ \frac{1}{2} - \frac{2.00 \times 10^{-2} \, \text{m}}{3(3.00 \times 10^{-2} \, \text{m})} \right]$$

$$= 2.79 \times 10^{-8} \, \text{T}.$$

(b) Now the path is outside the given region and the displacement current through the path is

$$i^{\text{dis}} = \int_0^R J^{\text{dis}} 2\pi r \, dr = 2\pi \int_0^r (4.00 \, \text{A/m}^2) \left[ 1 - \frac{r}{R} \right] r \, dr$$

$$= 2\pi (4.00 \, \text{A/m}^2) \left[ \frac{R^2}{2} - \frac{R^3}{3R} \right] = 2\pi (4.00 \, \text{A/m}^2) \frac{R^2}{6}.$$

Thus

$$B = \frac{\mu_0 (4.00 \, \text{A/m}^2) R^2}{6r} = \frac{(4\pi \times 10^{-7} \, \text{T} \cdot \text{m/A})(4.00 \, \text{A/m}^2)(3.00 \times 10^{-2} \, \text{m})^2}{6(5.00 \times 10^{-2} \, \text{m})}$$

$$= 1.51 \times 10^{-8} \, \text{T}.$$

## 63

Use Gauss' law for magnetism: $\oint \vec{B} \cdot d\vec{A} = 0$. Write $\oint \vec{B} \cdot d\vec{A} = \Phi_1^{\text{mag}} + \Phi_2^{\text{mag}} + \Phi_C^{\text{mag}}$, where $\Phi_1^{\text{mag}}$ is the magnetic flux through the first end mentioned, $\Phi_2^{\text{mag}}$ is the magnetic flux through the second end mentioned, and $\Phi_C^{mag}$ is the magnetic flux through the curved surface. Over the first end, the magnetic field is inward, so the flux is $\Phi_1^{mag} = -25.0\,\mu$Wb. Over the second end, the magnetic field is uniform, normal to the surface, and outward, so the flux is $\Phi_2^{mag} = AB = \pi r^2 B$, where $A$ is the area of the end and $r$ is the radius of the cylinder. Its value is

$$\Phi_2^{mag} = \pi(0.120\,\text{m})^2(1.60 \times 10^{-3}\,\text{T}) = +7.24 \times 10^{-5}\,\text{Wb} = +72.4\,\mu\text{Wb}.$$

Since the three fluxes must sum to zero,

$$\Phi_C^{mag} = -\Phi_1^{mag} - \Phi_2^{mag} = 25.0\,\mu\text{Wb} - 72.4\,\mu\text{Wb} = -47.4\,\mu\text{Wb}.$$

The minus sign indicates that the flux is inward through the curved surface.

# Chapter 32

## 3

The area of integration for the calculation of the magnetic flux is bounded by the two dashed lines and the boundaries of the wires. If the origin is taken to be on the axis of the right-hand wire and $r$ measures distance from that axis, it extends from $r = a$ to $r = d - a$. Consider the right-hand wire first. In the region of integration, the field it produces is into the page and has magnitude $B = \mu_0 i/2\pi r$. Divide the region into strips of length $\ell$ and width $dr$, as shown. The flux through the strip a distance $r$ from the axis of the wire is $d\Phi = B\ell\, dr$ and the flux through the entire region is

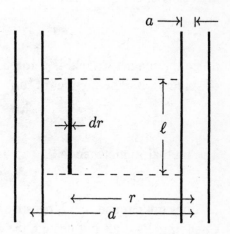

$$\Phi = \frac{\mu_0 i\ell}{2\pi} \int_a^{d-a} \frac{dr}{r} = \frac{\mu_0 i\ell}{2\pi} \ln\left(\frac{d-a}{a}\right).$$

The other wire produces the same result, so the net flux through the dotted rectangle is

$$\Phi^{\text{net}} = \frac{\mu_0 i\ell}{\pi} \ln\left(\frac{d-a}{a}\right).$$

The inductance is $\Phi^{\text{net}}$ divided by $i$:

$$L = \frac{\Phi^{\text{net}}}{i} = \frac{\mu_0 \ell}{\pi} \ln\left(\frac{d-a}{a}\right).$$

## 13

(a) The mutual inductance $M$ is given by

$$\mathcal{E}_1 = M \frac{di_2}{dt},$$

where $\mathcal{E}_1$ is the emf in coil 1 due to the changing current $i_2$ in coil 2. Thus

$$M = \frac{\mathcal{E}_1}{di_2/dt} = \frac{25.0 \times 10^{-3}\,\text{V}}{15.0\,\text{A/s}} = 1.67 \times 10^{-3}\,\text{H}.$$

(b) The flux linkage in coil 2 is

$$N_2 \Phi_{21} = M i_1 = (1.67 \times 10^{-3}\,\text{H})(3.60\,\text{A}) = 6.01 \times 10^{-3}\,\text{Wb}.$$

## 17

Assume the current in solenoid 1 is $i$ and calculate the flux linkage in solenoid 2. The mutual inductance is this flux linkage divided by $i$. The magnetic field inside solenoid 1 is parallel to the axis and has uniform magnitude $B = \mu_0 i n_1$, where $n_1$ is the number of turns per unit length of the solenoid. The cross-sectional area of the solenoid is $\pi R_1^2$ and since the field is normal to a cross section, the magnetic flux through a cross section is

$$\Phi^{\text{mag}} = AB = \pi R_1^2 \mu_0 n_1 i \,.$$

Since the magnetic field is zero outside the solenoid, this is also the flux through a cross section of solenoid 2. The number of turns in a length $l$ of solenoid 2 is $N_2 = n_2 l$ and the flux linkage is

$$N_2 \Phi^{\text{mag}} = n_2 l \pi R_1^2 \mu_0 n_1 i \,.$$

The mutual inductance is

$$M = \frac{N_2 \Phi^{\text{mag}}}{i} = \pi R_1^2 l \mu_0 n_1 n_2 \,.$$

$M$ does not depend on $R_2$ because there is no magnetic field in the region between the solenoids. Changing $R_2$ does not change the flux through solenoid 2, but changing $R_1$ does.

## 21

Starting with zero current at time $t = 0$, when the switch is closed, the current in an $RL$ series circuit at a later time $t$ is given by

$$i = \frac{\mathcal{E}}{R} \left( 1 - e^{-t/\tau_L} \right) \,,$$

where $\tau_L$ is the inductive time constant, $\mathcal{E}$ is the emf, and $R$ is the resistance. You want to calculate the time $t$ for which $i = 0.9990 \mathcal{E}/R$. This means

$$0.9990 \frac{\mathcal{E}}{R} = \frac{\mathcal{E}}{R} \left( 1 - e^{-t/\tau_L} \right) \,,$$

so

$$0.9990 = 1 - e^{-t/\tau_L}$$

or

$$e^{-t/\tau_L} = 0.0010 \,.$$

Take the natural logarithm of both sides to obtain $-(t/\tau_L) = \ln(0.0010) = -6.91$. Thus $t = 6.91\tau_L$. That is, 6.91 inductive time constants must elapse.

## 27

The current is given by $i = (\Delta V/R)(1 - e^{-Rt/L})$, where $\Delta V$ is the applied potential difference, $R$ is the resistance, and $L$ is the inductance. Its derivative with respect to time is

$$\frac{di}{dt} = \frac{\Delta V}{R} \frac{d}{dt} \left( 1 - e^{-Rt/L} \right) = \frac{\Delta V}{L} e^{-Rt/L} \,.$$

The magnitude of the exponent is $Rt/L = (180\,\Omega)(1.20 \times 10^{-3}\,\text{s})/(50.0 \times 10^{-3}\,\text{H}) = 4.32$, so

$$\frac{di}{dt} = \frac{45.0\,\text{V}}{50.0 \times 10^{-3}\,\text{H}} e^{-4.32} = 12.0\,\text{A/s}.$$

## 31

The current is given by $i = (\mathcal{E}/R)(1 - e^{-t/\tau_L})$, where $\tau_L (= L/R)$ is the inductive time constant. Solve for $t$. First, solve for the exponential and obtain $e^{-t/\tau_L} = 1 - (iR/\mathcal{E})$. Take the natural logarithm of both sides, then solve for $t$. The result is

$$t = -\tau_L \ln\left[1 - \frac{iR}{\mathcal{E}}\right].$$

(a) For $R = 1.00\,\Omega$ the time constant is $\tau_L = (18 \times 10^{-3}\,\text{H})/1.00\,\Omega) = 18 \times 10^{-3}\,\text{s}$ and

$$t = -(18 \times 10^{-3}\,\text{s}) \ln\left[1 - \frac{(2.00\,\text{A})(1.00\,\Omega)}{12\,\text{V}}\right] = 3.28 \times 10^{-3}\,\text{s}.$$

(b) For $R = 5.00\,\Omega$ the time constant is $\tau_L = (18 \times 10^{-3}\,\text{H})/5.00\,\Omega) = 3.6 \times 10^{-3}\,\text{s}$ and

$$t = -(3.6 \times 10^{-3}\,\text{s}) \ln\left[1 - \frac{(2.00\,\text{A})(5.00\,\Omega)}{12\,\text{V}}\right] = 6.45 \times 10^{-3}\,\text{s}.$$

(c) For $R = 6.00\,\Omega$ $iR/\mathcal{E} = (2.00\,\text{A}(6.00\,\Omega)/(12\,\text{V}) = 1.00$. The argument of the logarithm is zero and the logarithm is $-\infty$. Thus $t = \infty$.

(d) For $R = 6.00\,\Omega$ the given value of the current (2.00 A) is the value approached as the time becomes long. The current never actually reaches this value but it comes closer and closer as the time becomes longer. For smaller values of $R$ the given current is less than the final (maximum) value and the current reaches the value in a finite time.

(e) The smaller the value of $R$ the less time is required for the current to reach the given value. The least time is required if $R = 0$.

(f) For $R = 0$ the loop equation becomes $\mathcal{E} - L(di/dt) = 0$ and $i = (\mathcal{E}/L)t$. Thus $t = Li/\mathcal{E} = (18 \times 10^{-3}\,\text{H})(2.00\,\text{A})/(12\,\text{V}) = 3.00 \times 10^{-3}\,\text{s}$.

## 37

The amplifier is connected across the primary windings of a transformer and the resistor $R$ is connected across the secondary windings. If $i_s$ is the (rms) current in the secondary coil, then the (average) power delivered to $R$ is $P = i_s^2 R$. Now $i_s = (N_p/N_s)i_p$, where $N_p$ is the number of turns in the primary coil, $N_s$ is the number of turns in the secondary coil, and $i_p$ is the (rms) current in the primary coil. Thus

$$P = \left(\frac{i_p N_p}{N_s}\right)^2 R.$$

Now find the current in the primary circuit. It acts like a circuit consisting of a generator and two resistors in series. One resistance is the resistance $r$ of the amplifier and the other is the equivalent resistance $R_{eq}$ of the secondary circuit. Thus $i_p = \mathcal{E}/(r + R_{eq})$, where $\mathcal{E}$ is the (rms) emf of the amplifier. According to Eq. 32–34, $R_{eq} = (N_p/N_s)^2 R$, so

$$i_p = \frac{\mathcal{E}}{r + (N_p/N_s)^2 R}$$

and

$$P = \frac{\mathcal{E}^2 (N_p/N_s)^2 R}{\left[r + (N_p/N_s)^2 R\right]^2}.$$

You wish to find the value of $N_p/N_s$ so that $P$ is a maximum.
Let $x = (N_p/N_s)^2$. Then,

$$P = \frac{\mathcal{E}^2 R x}{(r + xR)^2}$$

and the derivative with respect to $x$ is

$$\frac{dP}{dx} = \frac{\mathcal{E}^2 R(r - xR)}{(r + xR)^3}.$$

This is zero for $x = r/R = (1000\,\Omega)/(10\,\Omega) = 100$. Notice that for small $x$, $P$ increases linearly with $x$ and for large $x$, it decreases in proportion to $1/x$. Thus $x = r/R$ is indeed a maximum, not a minimum.

Since $x = (N_p/N_s)^2$, maximum power is achieved for $(N_p/N_s)^2 = 100$, or $N_p/N_s = 10$.

The diagram below is a schematic of a transformer with a ten to one turns ratio. An actual transformer would have many more turns in both the primary and secondary coils.

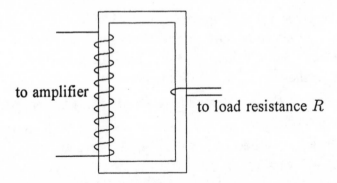

## 41

(a) The $z$ component of the orbital contribution to the magnetic dipole moment is given by $\mu_z^{orb} = -m_\ell \mu_B$, where $\mu_B$ is the Bohr magneton. Since $m_\ell = 0$, $\mu_{orb,\,z} = 0$.

(b) The potential energy associated with the orbital contribution to the magnetic dipole moment is given by $U = -\mu_z^{orb} B_z^{ext}$, where $B_z^{ext}$ is the $z$ component of the external magnetic field. Since $\mu_z^{orb} = 0$, $U = 0$.

(c) The $z$ component of the spin magnetic dipole moment is either $+\mu_B$ or $-\mu_B$, so the potential energy is either

$$U = -\mu_B B_z^{ext} = -(9.27 \times 10^{-24}\,\text{J/T})(35 \times 10^{-3}\,\text{T}) = -3.2 \times 10^{-25}\,\text{J}.$$

or $U = +3.2 \times 10^{-25}\,\text{J}$.

(d) Substitute $m_\ell = -3$ into the equations given above. The $z$ component of the orbital contribution to the magnetic dipole moment is

$$\mu_z^{\text{orb}} = -m_\ell \mu_B = -(-3)(9.27 \times 10^{-24}\,\text{J/T}) = 2.78 \times 10^{-23}\,\text{J/T}.$$

The potential energy associated with the orbital contribution to the magnetic dipole moment is

$$U = -\mu_z^{\text{orb}} B_z^{\text{ext}} = -(2.78 \times 10^{-23}\,\text{J/T})(35 \times 10^{-3}\,\text{T}) = -9.73 \times 10^{-25}\,\text{J}.$$

The potential energy associated with spin does not depend on $m_\ell$. It is $\pm 3.2 \times 10^{-25}\,\text{J}$.

## 45

The saturation magnetization corresponds to complete alignment of all atomic dipoles and its magnitude is given by $M^{\text{sat}} = \mu n$, where $n$ is the number of atoms per unit volume and $\mu$ is the magnitude of the magnetic dipole moment of each atom. The number of nickel atoms per unit volume is $n = \rho/m$, where $\rho$ is the density of nickel and $m$ is the mass of a single nickel atom, calculated using $m = M/N_A$, where $M$ is the molar mass of nickel and $N_A$ is the Avogadro constant. Thus

$$n = \frac{\rho N_A}{M} = \frac{(8.90\,\text{g/cm}^3)(6.02 \times 10^{23}\,\text{atoms/mol})}{58.71\,\text{g/mol}}$$

$$= 9.126 \times 10^{22}\,\text{atoms/cm}^3 = 9.126 \times 10^{28}\,\text{atoms/m}^3.$$

The dipole moment of a single atom of nickel is

$$\mu = \frac{M^{\text{sat}}}{n} = \frac{4.70 \times 10^5\,\text{A/m}}{9.126 \times 10^{28}\,\text{m}^3} = 5.15 \times 10^{-24}\,\text{A} \cdot \text{m}^2.$$

## 49

An electric field with circular field lines is induced as the magnetic field is turned on. Suppose the magnitude of the magnetic field increases linearly from zero to $B$ in time $t$. According to Eq. 31–26, the magnitude of the electric field at the orbit is given by

$$E = \left(\frac{r}{2}\right)\frac{dB}{dt} = \left(\frac{r}{2}\right)\frac{B}{t},$$

where $r$ is the radius of the orbit. The induced electric field is tangent to the orbit and changes the speed of the electron, the change in speed being given by

$$\Delta v = at = \frac{eE}{m}t = \left(\frac{e}{m}\right)\left(\frac{r}{2}\right)\left(\frac{B}{t}\right)t = \frac{erB}{2m}.$$

The average current associated with the circulating electron is $i = ev/2\pi r$ and the magnitude of the dipole moment is

$$\mu = Ai = \left(\pi r^2\right)\left(\frac{ev}{2\pi r}\right) = \frac{1}{2}evr.$$

The change in the dipole moment is

$$\Delta\mu = \frac{1}{2}er\,\Delta v = \frac{1}{2}er\frac{erB}{2m} = \frac{e^2r^2B}{4m}.$$

## 55

(a) A particle with charge $e$ traveling with uniform speed $v$ around a circular path of radius $r$ takes time $T = 2\pi r/v$ to complete one orbit, so the average current is

$$i = \frac{e}{T} = \frac{ev}{2\pi r}.$$

The magnitude of the dipole moment is this times the area of the orbit:

$$\mu = \frac{ev}{2\pi r}\,\pi r^2 = \frac{evr}{2}.$$

Since the magnetic force, with magnitude $evB$, is centripetal, Newton's second law yields $evB = mv^2/r$, so $r = mv/eB$. Thus

$$\mu = \frac{1}{2}(ev)\left(\frac{mv}{eB}\right) = \left(\frac{1}{B}\right)\left(\frac{1}{2}mv^2\right) = \frac{K_e}{B}.$$

The magnetic force $-e\vec{v}\times\vec{B}$ must point toward the center of the circular path. If the magnetic field is into the page, for example, the electron will travel clockwise around the circle. Since the electron is negative, the current is in the opposite direction, counterclockwise and, by the right-hand rule for dipole moments, the dipole moment is out of the page. That is, the dipole moment is directed opposite to the magnetic field vector.

(b) Notice that the charge canceled in the derivation of $\mu = K_e/B$. Thus $\mu = K_i/B$ for a positive ion. Here $K_i$ is the kinetic energy of the ion. If the magnetic field is into the page, the ion travels counterclockwise around a circular orbit and the current is in the same direction. Thus the dipole moment is again out of the page, opposite to the magnetic field.

(c) The magnetization is given by $M = \mu_e n_e + \mu_i n_i$, where $\mu_e$ is the magnitude of the dipole moment of an electron, $n_e$ is the electron concentration, $\mu_i$ is the magnitude of the dipole moment of an ion, and $n_i$ is the ion concentration. Since $n_e = n_i$, we may write $n$ for both concentrations. Substitute $\mu_e = K_e/B$ and $\mu_i = K_i/B$ to obtain

$$M = \frac{n}{B}[K_e + K_i] = \frac{5.3\times10^{21}\,\text{m}^{-3}}{1.2\,\text{T}}\left[6.2\times10^{-20}\,\text{J} + 7.6\times10^{-21}\,\text{J}\right] = 310\,\text{A/m}.$$

# Chapter 33

## 3

(a) If the battery is applied at time $t = 0$, the current is given by

$$i = \frac{\mathcal{E}}{R}\left(1 - e^{-t/\tau_L}\right),$$

where $\mathcal{E}$ is the emf of the battery, $R$ is the resistance, and $\tau_L$ is the inductive time constant. In terms of $R$ and the inductance $L$, $\tau_L = L/R$. Solve the current equation for the time constant. First obtain

$$e^{-t/\tau_L} = 1 - \frac{iR}{\mathcal{E}},$$

then take the natural logarithm of both sides to obtain

$$-\frac{t}{\tau_L} = \ln\left[1 - \frac{iR}{\mathcal{E}}\right].$$

Since

$$\ln\left[1 - \frac{iR}{\mathcal{E}}\right] = \ln\left[1 - \frac{(2.00 \times 10^{-3}\,\text{A})(10.0 \times 10^{3}\,\Omega)}{50.0\,\text{V}}\right] = -0.5108,$$

the inductive time constant is $\tau_L = t/0.5108 = (5.00 \times 10^{-3}\,\text{s})/(0.5108) = 9.79 \times 10^{-3}\,\text{s}$ and the inductance is

$$L = \tau_L R = (9.79 \times 10^{-3}\,\text{s})(10.0 \times 10^{3}\,\Omega) = 97.9\,\text{H}.$$

(b) The energy stored in the coil is

$$U^{\text{mag}} = \frac{1}{2}Li^2 = \frac{1}{2}(97.9\,\text{H})(2.00 \times 10^{-3}\,\text{A})^2 = 1.96 \times 10^{-4}\,\text{J}.$$

## 11

(a) Let $i$ be the current in the wire and $r$ be the radius of the wire. Then the magnitude of the magnetic field at the surface of the wire is

$$B = \frac{\mu_0 i}{2\pi r} = \frac{(4\pi \times 10^{-7}\,\text{T} \cdot \text{m/A})(10\,\text{A})}{2\pi(1.25 \times 10^{-3}\,\text{m})} = 1.60 \times 10^{-3}\,\text{T}.$$

The magnetic energy density at the surface of the wire is

$$u^{\text{mag}} = \frac{B^2}{2\mu_0} = \frac{(1.60 \times 10^{-3}\,\text{T})^2}{2(4\pi \times 10^{-7}\,\text{T} \cdot \text{m/A})} = 1.0\,\text{J/m}^3.$$

(b) The magnitude of the electric field is given by $E = \rho J$, where $\rho$ is the resistivity of copper and $J$ is the magnitude of the current density. The resistance $R$ of the wire is $R = \rho L/A$, where

$L$ is its length and $A$ is its cross-sectional area. Thus $\rho = AR/L$. The magnitude of the current density is $J = i/A$. Use these expressions to substitute for $\rho$ and $J$ in the equation for $E$. You should obtain

$$E = \frac{iR}{L} = (10\,\text{A})(3.3 \times 10^{-3}\,\Omega/\text{m}) = 3.3 \times 10^{-2}\,\text{V/m}.$$

Since the current density is uniform this is the magnitude of the electric field everywhere within the wire, including points on its surface. The electric energy density at the surface of the wire is

$$u^{\text{elec}} = \frac{1}{2}\epsilon_0 E^2 = \frac{1}{2}(8.85 \times 10^{-12}\,\text{F/m})(3.3 \times 10^{-2}\,\text{V/m})^2 = 4.8 \times 10^{-15}\,\text{J/m}^3.$$

## 15

(a) All the energy in the circuit resides in the capacitor when it has its maximum charge. The current is then zero. If $C$ is the capacitance and $Q$ is the maximum charge on the capacitor, then the total energy is

$$U = \frac{Q^2}{2C} = \frac{(2.90 \times 10^{-6}\,\text{C})^2}{2(3.60 \times 10^{-6}\,\text{F})} = 1.17 \times 10^{-6}\,\text{J}.$$

(b) When the capacitor is fully discharged, the current is a maximum and all the energy resides in the inductor. If $I$ is the maximum current, then $U = LI^2/2$ and

$$I = \sqrt{\frac{2U}{L}} = \sqrt{\frac{2(1.17 \times 10^{-6}\,\text{J})}{75 \times 10^{-3}\,\text{H}}} = 5.59 \times 10^{-3}\,\text{A}.$$

## 19

(a) The mass $m$ corresponds to the inductance, so $m = 1.25\,\text{kg}$.

(b) The spring constant $k$ corresponds to the reciprocal of the capacitance. Since the total energy is given by $U = Q^2/2C$, where $Q$ is the maximum charge on the capacitor and $C$ is the capacitance,

$$C = \frac{Q^2}{2U} = \frac{(175 \times 10^{-6}\,\text{C})^2}{2(5.70 \times 10^{-6}\,\text{J})} = 2.69 \times 10^{-3}\,\text{F}$$

and

$$k = \frac{1}{2.69 \times 10^{-3}\,\text{m/N}} = 372\,\text{N/m}.$$

(c) The maximum displacement $x^{\text{max}}$ corresponds to the maximum charge, so

$$x^{\text{max}} = 175 \times 10^{-6}\,\text{m}.$$

(d) The maximum speed $v^{\text{max}}$ corresponds to the maximum current. The maximum current is

$$I = Q\omega = \frac{Q}{\sqrt{LC}} = \frac{175 \times 10^{-6}\,\text{C}}{\sqrt{(1.25\,\text{H})(2.69 \times 10^{-3}\,\text{F})}} = 3.02 \times 10^{-3}\,\text{A}.$$

Thus $v^{\text{max}} = 3.02 \times 10^{-3}\,\text{m/s}$.

## 25

(a) After the switch is thrown to position $b$, the circuit is an $LC$ circuit. The angular frequency of oscillation is $\omega = 1/\sqrt{LC}$ and the frequency is

$$f = \frac{\omega}{2\pi} = \frac{1}{2\pi\sqrt{LC}} = \frac{1}{2\pi\sqrt{(54.0 \times 10^{-3}\,\text{H})(6.20 \times 10^{-6}\,\text{F})}} = 275\,\text{Hz}.$$

(b) When the switch is thrown, the capacitor is charged to $\Delta V_C = 34.0\,\text{V}$ and the current is zero. Thus the maximum charge on the capacitor is $Q = (\Delta V_C)C = (34.0\,\text{V})(6.20 \times 10^{-6}\,\text{F}) = 2.11 \times 10^{-4}\,\text{C}$. The current amplitude is

$$I = \omega Q = 2\pi f Q = 2\pi(275\,\text{Hz})(2.11 \times 10^{-4}\,\text{C}) = 0.365\,\text{A}.$$

## 31

(a) The charge on the capacitor is given by $q(t) = Q \sin \omega t$, where $Q$ is the maximum charge and $\omega$ is the angular frequency of oscillation. A sine function was chosen so that $q = 0$ at time $t = 0$. The current is

$$i(t) = \frac{dq}{dt} = \omega Q \cos \omega t$$

and at $t = 0$, it is $i = \omega Q$. This is the current amplitude $I$. Since $\omega = 1/\sqrt{LC}$,

$$Q = I\sqrt{LC} = (2.00\,\text{A})\sqrt{(3.00 \times 10^{-3}\,\text{H})(2.70 \times 10^{-6}\,\text{F})} = 1.80 \times 10^{-4}\,\text{C}.$$

(b) The energy stored in the capacitor is given by

$$U^{\text{elec}} = \frac{q^2}{2C} = \frac{Q^2 \sin^2 \omega t}{2C}$$

and its rate of change is

$$\frac{dU^{\text{elec}}}{dt} = \frac{Q^2 \omega \sin \omega t \cos \omega t}{C}.$$

Use the trigonometric identity $\cos \omega t \sin \omega t = \frac{1}{2} \sin(2\omega t)$ to write this as

$$\frac{dU^{\text{elec}}}{dt} = \frac{\omega Q^2}{2C} \sin(2\omega t).$$

The greatest rate of change occurs when $\sin(2\omega t) = 1$ or $2\omega t = \pi/2\,\text{rad}$. This means

$$t = \frac{\pi}{4\omega} = \frac{\pi T}{4(2\pi)} = \frac{T}{8},$$

where $T$ is the period of oscillation. The relationship $\omega = 2\pi/T$ was used.

(c) Substitute $\omega = 2\pi/T$ and $\sin(2\omega t) = 1$ into $dU^{\text{elec}}/dt = (\omega Q^2/2C)\sin(2\omega t)$ to obtain

$$\left(\frac{dU^{\text{elec}}}{dt}\right)^{\text{max}} = \frac{2\pi Q^2}{2TC} = \frac{\pi Q^2}{TC}.$$

Now $T = 2\pi\sqrt{LC} = 2\pi\sqrt{(3.00 \times 10^{-3}\,\text{H})(2.70 \times 10^{-6}\,\text{F})} = 5.655 \times 10^{-4}\,\text{s}$, so

$$\left(\frac{dU^{\text{elec}}}{dt}\right)^{\text{max}} = \frac{\pi(1.80 \times 10^{-4}\,\text{C})^2}{(5.655 \times 10^{-4}\,\text{s})(2.70 \times 10^{-6}\,\text{F})} = 66.7\,\text{W}.$$

Notice that this is a positive result, indicating that the energy in the capacitor is indeed increasing at $t = T/8$.

## 37

Let $T$ be the period of oscillation and approximate the angular frequency by $\omega = 1/\sqrt{LC}$. Here $L$ is the inductance and $C$ is the capacitance. The time required for 50.0 cycles is

$$t = 50.0T = (50.0)\left(\frac{2\pi}{\omega}\right) = (50.0)\left(2\pi\sqrt{LC}\right)$$

$$= (50.0)\left(2\pi\sqrt{(220 \times 10^{-3}\,\text{H})(12.0 \times 10^{-6}\,\text{F})}\right) = 0.5104\,\text{s}.$$

The maximum charge on the capacitor decays according to

$$q^{\text{max}} = Qe^{-Rt/2L},$$

where $Q$ is the charge at time $t = 0$ and $R$ is the resistance in the circuit. Divide by $Q$ and take the natural logarithm of both sides to obtain

$$\ln\left(\frac{q^{\text{max}}}{Q}\right) = -\frac{Rt}{2L}$$

and

$$R = -\frac{2L}{t}\ln\left(\frac{q^{\text{max}}}{Q}\right) = -\frac{2(220 \times 10^{-3}\,\text{H})}{0.5104\,\text{s}}\ln(0.99) = 8.66 \times 10^{-3}\,\Omega.$$

## 43

(a) The current amplitude $I$ is given by $I = \Delta V_L/X_L$, where $\Delta V_L$ is the potential difference amplitude across the inductor and $X_L$ is the inductive reactance. The reactance is given by $X_L = \omega^{\text{dr}}L = 2\pi f^{\text{dr}}L$, where $\omega^{\text{dr}}$ is the driving angular frequency, $f^{\text{dr}}$ is the driving frequency, and $L$ is the inductance. Since the circuit contains only the inductor and a sinusoidal generator, $\Delta V_L = \mathcal{E}^{\text{max}}$, where $\mathcal{E}^{\text{max}}$ is the generator emf amplitude. Thus

$$I = \frac{(\Delta V_L)}{X_L} = \frac{\mathcal{E}^{\text{max}}}{2\pi f^{\text{dr}}L} = \frac{30.0\,\text{V}}{2\pi(1.00 \times 10^3\,\text{Hz})(50.0 \times 10^{-3}\,\text{H})} = 0.0955\,\text{A}.$$

(b) The frequency is now eight times as much as in part (a), so the inductive reactance is eight times as much and the current is one-eighth as much, or $= (0.0955\,\text{A})/8 = 0.0119\,\text{A}$.

## 47

(a) The generator emf is a maximum when $\sin(\omega^{dr}t - \pi/4) = 1$ or $\omega^{dr}t - \pi/4 = (\pi/2)\pm2n\pi$, where $n$ is an integer, including zero. The first time this occurs after $t = 0$ is when $\omega^{dr}t - \pi/4 = \pi/2$ or

$$t = \frac{3\pi}{4\omega^{dr}} = \frac{3\pi}{4(350\,\text{s}^{-1})} = 6.73 \times 10^{-3}\,\text{s}.$$

(b) The current is a maximum when $\sin(\omega^{dr}t - 3\pi/4) = 1$, or $\omega^{dr}t - 3\pi/4 = \pi/2 \pm 2n\pi$. The first time this occurs after $t = 0$ is when

$$t = \frac{5\pi}{4\omega^{dr}} = \frac{5\pi}{4(350\,\text{s}^{-1})} = 1.12 \times 10^{-2}\,\text{s}.$$

(c) The current lags the inductor by $\pi/2$ rad, so the circuit element must be an inductor.

(d) The current amplitude $I$ is related to the potential difference amplitude $\Delta V_L$ by $\Delta V_L = IX_L$, where $X_L$ is the inductive reactance, given by $X_L = \omega^{dr}L$. Furthermore, since there is only one element in the circuit, the amplitude of the potential difference across the element must be the same as the amplitude of the generator emf: $\Delta V_L = \mathcal{E}^{\text{max}}$. Thus $\mathcal{E}^{\text{max}} = I\omega^{dr}L$ and

$$L = \frac{\mathcal{E}^{\text{max}}}{I\omega^{dr}} = \frac{30.0\,\text{V}}{(620 \times 10^{-3}\,\text{A})(350\,\text{rad/s})} = 0.138\,\text{H}.$$

## 55

The power factor for an alternating current RLC circuit is

$$\cos\phi = \frac{R}{Z} = \frac{R}{\sqrt{R^2 + (X_L - X_C)^2}},$$

where $\phi$ is the phase constant, $R$ is the resistance, $X_L$ is the inductive reactance, $X_C$ is the capacitive reactance, and $Z$ is the impedance. Squaring yields

$$\cos^2\phi = \frac{R^2}{R^2 + (X_L - X_C)^2}$$

and the solution for $R$ is

$$R = \frac{X_L - X_C}{\tan\phi},$$

where $\sin^2\phi = 1 - \cos^2\phi$ and $\tan\phi = \sin\phi/\cos\phi$ were used.

The angular frequency is $\omega^{dr} = 2\pi f^{dr} = 2\pi(930\,\text{Hz}) = 5.84 \times 10^3\,\text{rad/s}$. The inductive reactance is

$$X_L = \omega^{dr}L = (5.84 \times 10^3\,\text{rad/s})(88 \times 10^{-3}\,\text{H}) = 514\,\Omega.$$

The capacitive reactance is

$$X_C = \frac{1}{\omega^{dr}C} = \frac{1}{(5.84 \times 10^3\,\text{rad/s})(0.94 \times 10^{-6}\,\text{F})} = 182\,\Omega.$$

Thus the resistance is

$$R = \frac{514\,\Omega - 182\,\Omega}{\tan 75°} = 89\,\Omega.$$

## 59

At resonance the current amplitude is $I = \mathcal{E}^{max}/R$, where $\mathcal{E}^{max}$ is the amplitude of the generator emf and $R$ is the resistance. At another driving frequency the current amplitude is $\mathcal{E}^{max}/Z$, where $Z$ is the impedance. You want the driving angular frequency $\omega^{dr}$ for which $\mathcal{E}^{max}/Z = \mathcal{E}^{max}/2R$ or $Z = 2R$. The impedance is $Z = \sqrt{R^2 + (X_L - X_C)^2}$, where $X_L$ is the inductive reactance and $X_C$ is the capacitive reactance. The solution to $\sqrt{R^2 + (X_L - X_C)^2} = 2R$ is $|X_L - X_C| = \sqrt{3}R$. Now $X_L = \omega^{dr}L$ and $X_C = 1/\omega^{dr}C$. Assume that $X_L$ is greater than $X_C$. Then $(\omega^{dr}L) - (1/\omega^{dr}C) = \sqrt{3}R$. Multiplication by $\omega^{dr}C$ gives $(\omega^{dr})^2 LC - 1 = \sqrt{3}CR$. The solutions to this quadratic equation are

$$\omega^{dr} = \frac{\sqrt{3}CR \pm \sqrt{3C^2R^2 + 4LC}}{2LC}.$$

Since the driving frequency must be positive we use the plus sign.

Now assume $X_C$ is greater than $X_L$ and carry out the same analysis. You should obtain

$$\omega^{dr} = \frac{-\sqrt{3}CR + \sqrt{3C^2R^2 + 4LC}}{2LC},$$

where the negative root was rejected since it leads to a negative driving angular frequency.

The difference in the driving frequencies at half height is $\Delta\omega^{dr} = \sqrt{3}R/L$ and the fractional difference is

$$\frac{\Delta\omega^{dr}}{\omega} = \sqrt{LC}\frac{\sqrt{3}R}{L} = \sqrt{\frac{3C}{L}}R,$$

where $\omega = 1/\sqrt{LC}$ was used.

For the data of Problem 57

$$\frac{\Delta\omega^{dr}}{\omega} = \sqrt{\frac{2(20.0 \times 10^{-6}\,\text{F})}{1.00\,\text{H}}}(5.00\,\Omega) = 0.039.$$

## 69

(a) The power factor is $\cos\phi$, where $\phi$ is the phase angle when the expression for the current is written $i = I\sin(\omega^{dr}t - \phi)$. Thus $\phi = -42.0°$ and $\cos\phi = \cos(-42.0°) = 0.743$.

(b) Since $\phi < 0$, $\omega^{dr}t - \phi > \omega^{dr}t$ and the current leads the emf.

(c) The phase angle is given by $\tan\phi = (X_L - X_C)/R$, where $X_L$ is the inductive reactance, $X_C$ is the capacitive reactance, and $R$ is the resistance. Now $\tan\phi = \tan(-42.0°) = -0.900$, a negative number. This means $X_L - X_C$ is negative, or $X_C > X_L$. The circuit in the box is predominantly capacitive.

(d) If the circuit were in resonance, $X_L$ would be the same as $X_C$, $\tan\phi$ would be zero, and $\phi$ would be zero. Since $\phi$ is not zero, we conclude the circuit is not in resonance.

(e) Since $\tan\phi$ is negative and finite, neither the capacitive reactance nor the resistance are zero. This means the box must contain a capacitor and a resistor. The inductive reactance may be zero, so there need not be an inductor. If there is an inductor, its reactance must be less than that of the capacitor at the operating frequency.

(f) The average power is

$$\langle P \rangle = \frac{1}{2}\mathcal{E}^{max}I\cos\phi = \frac{1}{2}(75.0\ \text{V})(1.20\ \text{A})(0.743) = 33.4\ \text{W}\,.$$

(g) The answers above depend on the frequency only through the phase angle $\phi$, which is given. If values were given for $R$, $L$, and $C$, then the value of the frequency would also be needed to compute the power factor.

## 71

(a) The average power is given by

$$\langle P \rangle = \mathcal{E}^{rms}I^{rms}\cos\phi\,,$$

where $\mathcal{E}^{rms}$ is the root-mean-square emf of the generator, $I^{rms}$ is the root-mean-square current, and $\cos\phi$ is the power factor. Now

$$I^{rms} = \frac{I}{\sqrt{2}} = \frac{\mathcal{E}^{max}}{\sqrt{2}Z}\,,$$

where $I$ is the current amplitude, $\mathcal{E}^{max}$ is the maximum emf of the generator, and $Z$ is the impedance of the circuit. $I = \mathcal{E}^{max}/Z$ was used. In addition, $\mathcal{E}^{rms} = \mathcal{E}^{max}/\sqrt{2}$ and $\cos\phi = R/Z$, where $R$ is the resistance. Thus

$$\langle P \rangle = \frac{(\mathcal{E}^{max})^2 R}{2Z^2} = \frac{(\mathcal{E}^{max})^2 R}{2\left[R^2 + (\omega^{dr}L - 1/\omega^{dr}C)^2\right]}\,.$$

Here the expression $Z = \sqrt{R^2 + (\omega^{dr}L - 1/\omega^{dr}C)^2}$ for the impedance in terms of the angular frequency was substituted.

Considered as a function of $C$, $\langle P \rangle$ has its largest value when the factor $R^2 + (\omega^{dr}L - 1/\omega^{dr}C)^2$ has the smallest possible value. This occurs for $\omega^{dr}L = 1/\omega^{dr}C$, or

$$C = \frac{1}{(\omega^{dr})^2 L} = \frac{1}{(2\pi)^2(60.0\ \text{Hz})^2(60.0 \times 10^{-3}\ \text{H})} = 1.17 \times 10^{-4}\ \text{F}\,.$$

The circuit is then at resonance.

(b) Now you want $Z^2$ to be as large as possible. Notice that it becomes large without bound as $C$ becomes small. Thus the smallest average power occurs for $C = 0$.

(c) When $\omega^{dr}L = 1/\omega^{dr}C$, the expression for the average power becomes

$$\langle P \rangle = \frac{(\mathcal{E}^{max})^2}{2R}\,,$$

so the maximum average power is

$$\langle P \rangle = \frac{(30.0\ \text{V})^2}{2(5.00\ \Omega)} = 90.0\ \text{W}\,.$$

The minimum average power is $\langle P \rangle = 0$.

(d) At maximum power, $X_L = X_C$, where $X_L$ is the inductive reactance and $X_C$ is the capacitive reactance. The phase angle $\phi$ is

$$\tan\phi = \frac{X_L - X_C}{R} = 0,$$

so $\phi = 0$. At minimum power $X_C$ is infinite, so $\tan\phi = -\infty$ and $\phi = -90°$.

(e) At maximum power, the power factor is $\cos\phi = \cos 0° = 1$ and at minimum power, it is $\cos\phi = \cos(-90°) = 0$.

# Chapter 34

## 1

If $f$ is the frequency and $\lambda$ is the wavelength of an electromagnetic wave, then $f\lambda = c$. The frequency is the same as the frequency of oscillation of the current in the $LC$ circuit of the generator. That is, $f = 1/2\pi\sqrt{LC}$, where $C$ is the capacitance and $L$ is the inductance. Thus

$$\frac{\lambda}{2\pi\sqrt{LC}} = c.$$

The solution for $L$ is

$$L = \frac{\lambda^2}{4\pi^2 C c^2} = \frac{(550 \times 10^{-9}\,\text{m})^2}{4\pi^2(17 \times 10^{-12}\,\text{F})(3.00 \times 10^8\,\text{m/s})^2} = 5.00 \times 10^{-21}\,\text{H}.$$

This is exceedingly small.

## 13

(a) The average rate of energy flow per unit area, or intensity, is related to the electric field amplitude $E_m$ by $I = E_m^2/2\mu_0 c$, so

$$E_m = \sqrt{2\mu_0 c I} = \sqrt{2(4\pi \times 10^{-7}\,\text{H/m})(3.00 \times 10^8\,\text{m/s})(10 \times 10^{-6}\,\text{W/m}^2)}$$
$$= 8.7 \times 10^{-2}\,\text{V/m}.$$

(b) The amplitude of the magnetic field is given by

$$B_m = \frac{E_m}{c} = \frac{8.7 \times 10^{-2}\,\text{V/m}}{3.00 \times 10^8\,\text{m/s}} = 2.9 \times 10^{-10}\,\text{T}.$$

(c) At a distance $r$ from the transmitter, the intensity is $I = P/4\pi r^2$, where $P$ is the power of the transmitter. Thus

$$P = 4\pi r^2 I = 4\pi(10 \times 10^3\,\text{m})^2(10 \times 10^{-6}\,\text{W/m}^2) = 1.3 \times 10^4\,\text{W}.$$

## 15

The magnetic field component is $B = B^{\text{max}}\sin(kx - \omega t)$, where $B^{\text{max}}$ is the maximum, $k$ is the wave number, and $\omega$ is the angular frequency. Thus $\partial B/\partial t = -\omega B^{\text{max}}\cos(kx - \omega t)$ and its maximum value is $(\partial B/\partial t)^{\text{max}} = \omega B^{\text{max}}$.

The intensity is $I = (1/2c\mu_0)(E^{\text{max}})^2 = (c/2\mu_0)(B^{\text{max}})^2$, where $E^{\text{max}} = cB^{\text{max}}$ was used. Thus $B^{\text{max}} = \sqrt{2\mu_0 I/c}$. If $P_s$ is the rate of energy emission of the source, then $I = P_s/4\pi r^2$, where $r$ is the distance from the source. Thus

$$B^{\text{max}} = \sqrt{\frac{2\mu_0 P_s}{4\pi r^2 c}}.$$

The angular frequency is $\omega = 2\pi c/\lambda$, so

$$\left(\frac{\partial B}{\partial t}\right)^{\text{max}} = \frac{2\pi c}{\lambda r}\sqrt{\frac{2\mu_0 P_s}{4\pi c}} = \frac{2\pi(3.00 \times 10^8 \text{ m/s})}{(500 \times 10^{-9} \text{ m})(400 \text{ m})}\sqrt{\frac{(8\pi \times 10^{-7} \text{ H/m})(200 \text{ W})}{4\pi(3.00 \times 10^8 \text{ m/s})}}$$

$$= 3.44 \times 10^6 \text{ T/s}.$$

## 21

(a) The intensity is the rate of energy generation by the radar system divided by the area of a hemisphere, which is $A = 2\pi r^2$, where $r$ is the distance from the radar system to the plane. Thus

$$I = \frac{P_s}{2\pi r^2} = \frac{180 \times 10^3 \text{ W}}{2\pi(90 \times 10^3 \text{ m})^2} = 3.5 \times 10^{-6} \text{ W/m}^2.$$

(b) The plane reflects all energy incident on its effective cross-sectional area $A_p$. Thus the rate with which energy is reflected is $P_p = IA_p = (3.5 \times 10^{-6} \text{ W/m}^2)(0.22 \text{ m}^2) = 7.8 \times 10^{-7} \text{ W}$.

(c) Since the energy is reflected uniformly over a hemisphere, the intensity back at the source is

$$I_s = \frac{P_p}{2\pi r^2} = \frac{7.8 \times 10^{-7} \text{ W}}{2\pi(90 \times 10^3 \text{ m})^2} = 1.5 \times 10^{-17} \text{ W/m}^2.$$

(d) In terms of the maximum value of the electric field the intensity is $I_s = (E^{\text{max}})^2/2c\mu_0$, so

$$E^{\text{max}} = \sqrt{2c\mu_0 I_s} = \sqrt{2(3.00 \times 10^8 \text{ m/s})(4\pi \times 10^{-7} \text{ H/m})(1.5 \times 10^{-17} \text{ W/m}^2)}$$

$$= 1.1 \times 10^{-7} \text{ V/m}.$$

(e) The rms value of the electric field is $E^{\text{rms}} = E^{\text{max}}/\sqrt{2} = 7.5 \times 10^{-8}$ V/m and the rms value of the magnetic field is $B^{\text{rms}} = E^{\text{rms}}/c = (7.5 \times 10^{-8} \text{ V/m})/(3.00 \times 10^8 \text{ m/s}) = 2.5 \times 10^{-16}$ T.

## 27

(a) Since $c = \lambda f$, where $\lambda$ is the wavelength and $f$ is the frequency of the wave,

$$f = \frac{c}{\lambda} = \frac{3.00 \times 10^8 \text{ m/s}}{3.0 \text{ m}} = 1.0 \times 10^8 \text{ Hz}.$$

(b) The magnetic field amplitude is

$$B_m = \frac{E_m}{c} = \frac{300 \text{ V/m}}{3.00 \times 10^8 \text{ m/s}} = 1.00 \times 10^{-6} \text{ T}.$$

$\vec{B}$ must be in the positive $z$ direction when $\vec{E}$ is in the positive $y$ direction in order for $\vec{E} \times \vec{B}$ to be in the positive $x$ direction (the direction of propagation).

(c) The wave number is

$$k = \frac{2\pi}{\lambda} = \frac{2\pi}{3.0\,\text{m}} = 2.1\,\text{rad/m}.$$

The angular frequency is

$$\omega = 2\pi f = 2\pi(1.0 \times 10^8\,\text{Hz}) = 6.3 \times 10^8\,\text{rad/s}.$$

(d) The intensity of the wave is

$$I = \frac{E_m^2}{2\mu_0 c} = \frac{(300\,\text{V/m})^2}{2(4\pi \times 10^{-7}\,\text{H/m})(3.00 \times 10^8\,\text{m/s})} = 119\,\text{W/m}^2.$$

(e) Since the sheet is perfectly absorbing, the rate per unit area with which momentum is delivered to it is $I/c$, so

$$\frac{dp}{dt} = \frac{IA}{c} = \frac{(119\,\text{W/m}^2)(2.0\,\text{m}^2)}{3.00 \times 10^8\,\text{m/s}} = 8.0 \times 10^{-7}\,\text{N}.$$

The radiation pressure is

$$P^{\text{absorp}} = \frac{dp/dt}{A} = \frac{8.0 \times 10^{-7}\,\text{N}}{2.0\,\text{m}^2} = 4.0 \times 10^{-7}\,\text{Pa}.$$

## 33

(a) Let $r$ be the radius and $\rho$ be the density of the particle. Since its volume is $(4\pi/3)r^3$, its mass is $m = (4\pi/3)\rho r^3$. Let $R$ be the distance from the Sun to the particle and let $M$ be the mass of the Sun. Then the gravitational force of attraction of the Sun on the particle has magnitude

$$F^{\text{grav}} = \frac{GMm}{R^2} = \frac{4\pi GM\rho r^3}{3R^2}.$$

If $P_s$ is the power output of the Sun, then at the position of the particle, the radiation intensity is $I = P_s/4\pi R^2$ and since the particle is perfectly absorbing, the radiation pressure on it is

$$P^{\text{absorp}} = \frac{I}{c} = \frac{P_s}{4\pi R^2 c}.$$

All of the radiation that passes through a circle of radius $r$ and area $A = \pi r^2$, perpendicular to the direction of propagation, is absorbed by the particle, so the force of the radiation on the particle has magnitude

$$F^{\text{rad}} = P^{\text{absorp}}A = \frac{\pi P_s r^2}{4\pi R^2 c} = \frac{P_s r^2}{4R^2 c}.$$

The force is radially outward from the Sun. Notice that both the force of gravity and the force of the radiation are inversely proportional to $R^2$. If one of these forces is larger than the other at some distance from the Sun, then that force is larger at all distances.

The two forces depend on the particle radius $r$ differently: $F^{\text{grav}}$ is proportional to $r^3$ and $F^{\text{rad}}$ is proportional to $r^2$. We expect a small radius particle to be blown away by the radiation pressure and a large radius particle with the same density to be pulled inward toward the Sun. The critical value for the radius is the value for which the two forces are equal. Equate the expressions for $F^{\text{grav}}$ and $F^{\text{rad}}$, then solve for $r$. You should obtain

$$r = \frac{3P}{16\pi GM\rho c}.$$

(b) According to the appendix, $M = 1.99 \times 10^{30}$ kg and $P_s = 3.90 \times 10^{26}$ W. Thus

$$r = \frac{3(3.90 \times 10^{26}\,\text{W})}{16\pi(6.67 \times 10^{-11}\,\text{N} \cdot \text{m}^2/\text{kg}^2)(1.99 \times 10^{30}\,\text{kg})(1.0 \times 10^3\,\text{kg/m}^3)(3.00 \times 10^8\,\text{m/s})}$$
$$= 5.8 \times 10^{-7}\,\text{m}.$$

## 35

(a) The downward force of gravity must match the upward radiation force in magnitude. The magnitude of the gravitational force is $F^{\text{grav}} = mg$ and since the mass is $(4\pi/3)r^3\rho$, where $r$ is the radius of the sphere and $\rho$ is its density, $F^{\text{grav}} = (4\pi/3)r^3\rho g$.

The magnitude of the radiation force is $F^{\text{rad}} = P^{\text{absorp}}A$, where $P^{\text{absorp}}$ is the radiation pressure for total absorption and $A$ is the cross-sectional area of the sphere. Now $P^{\text{absorp}} = I/c$, where $I$ is the radiation intensity at the sphere. The radiation source is isotropic, so $I = P_s/4\pi R^2$, where $P_s$ is the rate of energy emission by the source and $R$ is the distance from the source to the sphere. Thus

$$F^{\text{rad}} = \frac{I}{c}A = \frac{P_s}{4\pi R^2 c}\pi r^2.$$

The magnitudes of the two forces are equal, so

$$\frac{4\pi}{3}r^3\rho g = \frac{P_s}{4\pi R^2 c}\pi r^2.$$

The solution for $P_s$ is

$$P_s = \frac{16\pi}{3}R^2 rc\rho g$$
$$= \frac{16\pi}{3}(0.500\,\text{m})^2(2.00 \times 10^{-3}\,\text{m})(3.00 \times 10^8\,\text{m/s})(19.0 \times 10^3\,\text{kg/m}^3)(9.8\,\text{m/s}^2)$$
$$= 4.68 \times 10^{11}\,\text{W}.$$

(b) If the sphere moves horizontally by any amount the two forces would not be along the same line and the sphere would continue to move away from its equilibrium position.

## 45

(a) The rotation cannot be done with a single sheet. If a sheet is placed with its polarizing direction at an angle of 90° to the direction of polarization of the incident radiation, no radiation is transmitted.

It can be done with two sheets. Place the first sheet with its polarizing direction at some angle $\theta$, between 0 and 90°, to the direction of polarization of the incident radiation. Place the second sheet with its polarizing direction at 90° to the polarization direction of the incident radiation. The transmitted radiation is then polarized at 90° to the incident polarization direction. The intensity is $I_0 \cos^2 \theta \cos^2(90° - \theta) = I_0 \cos^2 \theta \sin^2 \theta$, where $I_0$ is the incident radiation. If $\theta$ is not 0 or 90°, the transmitted intensity is not zero.

(b) Consider $n$ sheets, with the polarizing direction of the first sheet making an angle of $\theta = 90°/n$ with the direction of polarization of the incident radiation and with the polarizing direction of each successive sheet rotated $90°/n$ in the same direction from the polarizing direction of the previous sheet. The transmitted radiation is polarized with its direction of polarization making an angle of 90° with the direction of polarization of the incident radiation. The intensity is $I = I_0 \cos^{2n}(90°/n)$. You want the smallest integer value of $n$ for which this is greater than $0.60 I_0$.

Start with $n = 2$ and calculate $\cos^{2n}(90°/n)$. If the result is greater than 0.60, you have obtained the solution. If it is less, increase $n$ by 1 and try again. Repeat this process, increasing $n$ by 1 each time, until you have a value for which $\cos^{2n}(90°/n)$ is greater than 0.60. The first one will be $n = 5$.

## 53

Let $I_0$ be the intensity of the light incident on the first polarizer. Since the light is unpolarized the intensity of the transmitted light and therefore the intensity incident on the second polarizer is $I_1 = \frac{1}{2}I_0$. This light is polarized along the $y$ axis. The transmitted intensity at the second polarizer is $I_2 = I_1 \cos^2 \theta = \frac{1}{2}I_0 \cos^2 \theta$ and this light is polarized along a line that makes the angle $\theta$ with the $y$ axis. The polarizing direction of the third polarizer is along the $x$ axis, so the polarization direction of light incident on it makes the angle $90° - \theta$ with the polarizing direction and the transmitted intensity is $I_3 = I_2 \cos^2(90° - \theta) = \frac{1}{2}I_0 \cos^2 \theta \cos^2(90° - \theta)$.

Since $\cos^2(90° - \theta) = \sin^2 \theta$ and $\cos \theta \sin \theta = \frac{1}{2} \sin(2\theta)$, $I_3 = (1/8)I_0 \sin^2(2\theta)$. Thus

$$\sin(2\theta) = \sqrt{\frac{8I}{I_0}} = \sqrt{8(0.050)} = 0.632$$

and $\theta = 19.6°$. Another solution is $\theta = 70.4°$.

## 59

(a) Suppose that at time $t_1$, the moon is starting a revolution (on the verge of going behind Jupiter, say) and that at this instant, the distance between Jupiter and Earth is $\ell_1$. The time of the start of the revolution as seen on Earth is $t_1^* = t_1 + \ell_1/c$. Suppose the moon starts the next revolution at time $t_2$ and at that instant, the Earth-Jupiter distance is $\ell_2$. The start of the revolution as seen on Earth is $t_2^* = t_2 + \ell_2/c$. Now, the actual period of the moon is given by $T = t_2 - t_1$ and the period as measured on Earth is

$$T^* = t_2^* - t_1^* = t_2 - t_1 + \frac{\ell_2}{c} - \frac{\ell_1}{c} = T + \frac{\ell_2 - \ell_1}{c}.$$

The period as measured on Earth is longer than the actual period because Earth moves during a revolution and light takes a finite time to travel from Jupiter to Earth. For the situation depicted in the diagram, light emitted at the end of a revolution travels a longer distance to get to Earth than light emitted at the beginning.

Suppose the position of Earth is given by the angle $\theta$, measured from $x$. Let $R$ be the radius of Earth's orbit and $d$ be the distance from the Sun to Jupiter. Then the law of cosines, applied to the triangle with the Sun, Earth, and Jupiter at the vertices, yields $\ell^2 = d^2 + R^2 - 2dR \cos\theta$. This expression can be used to calculate $\ell_1$ and $\ell_2$. Since Earth does not move very far during one revolution of the moon, we may approximate $\ell_2 - \ell_1$ by $(d\ell/dt)T$ and $T^*$ by $T + (d\ell/dt)(T/c)$. Now

$$\frac{d\ell}{dt} = \frac{2Rd\sin\theta}{\sqrt{d^2 + R^2 - 2dR\cos\theta}} \frac{d\theta}{dt} = \frac{2vd\sin\theta}{\sqrt{d^2 + R^2 - 2dR\cos\theta}},$$

where $v = R(d\theta/dt)$ is the speed of Earth in its orbit. For $\theta = 0$, $(d\ell/dt) = 0$ and $T^* = T$. Since Earth is then moving perpendicularly to the line from the Sun to Jupiter its distance from the planet does not change much during a revolution of the moon. On the other hand, when $\theta = 90°$, $d\ell/dt = vd/\sqrt{d^2 + R^2}$ and

$$T^* = T\left(1 + \frac{vd}{c\sqrt{d^2 + R^2}}\right).$$

The Earth is now moving parallel to the line from the Sun to Jupiter and its distance from the planet changes during a revolution of the moon.

(b) Let $t$ be the actual time for the moon to make $N$ revolutions and $t^*$ the time for $N$ revolutions to be observed on Earth. Then

$$t^* = t + \frac{\ell_2 - \ell_1}{c},$$

where $\ell_1$ is the Earth-Jupiter distance at the beginning of the interval and $\ell_2$ is the Earth-Jupiter distance at the end. Suppose Earth is at $x$ at the beginning of the interval and at $y$ at the end. Then $\ell_1 = d - R$ and $\ell_2 = \sqrt{d^2 + R^2}$. Thus

$$t^* = t + \frac{\sqrt{d^2 + R^2} - (d - R)}{c}.$$

A value can be found for $t$ by measuring the observed period of revolution when Earth is at $x$ and multiplying by $N$. Notice that the observed period is the true period when Earth is at $x$. Now measure the time interval as Earth moves from $x$ to $y$. This is $t^*$. The difference is

$$t^* - t = \frac{\sqrt{d^2 + R^2} - (d - R)}{c}.$$

If the radii of the orbits of Jupiter and Earth are known, the value for $t^* - t$ can be used to compute $c$.

Since Jupiter is much further from the Sun than Earth, $\sqrt{d^2 + R^2}$ may be approximated by $d$ and $t^* - t$ may be approximated by $R/c$. In this approximation, only the radius of Earth's orbit need be known.

# Chapter 35

## 3

Note that the normal to the refracting surface is vertical in the diagram. The angle of refraction is $\theta_2 = 90°$ and the angle of incidence is given by $\tan \theta_1 = w/h$, where $h$ is the height of the tank and $w$ is its width. Thus

$$\theta_1 = \tan^{-1}\left(\frac{w}{h}\right) = \tan^{-1}\left(\frac{1.10\,\text{m}}{0.850\,\text{m}}\right) = 52.31°.$$

The law of refraction yields

$$n_1 = n_2 \frac{\sin \theta_2}{\sin \theta_1} = (1.00)\left(\frac{\sin 90°}{\sin 52.31°}\right) = 1.26,$$

where the index of refraction of air was taken to be unity.

## 5

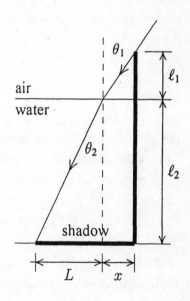

Consider a ray that grazes the top of the pole, as shown in the diagram to the right. Here $\theta_1 = 35°$, $\ell_1 = 0.50\,\text{m}$, and $\ell_2 = 1.50\,\text{m}$. The length of the shadow is $x + L$. $x$ is given by $x = \ell_1 \tan \theta_1 = (0.50\,\text{m}) \tan 35° = 0.35\,\text{m}$. According to the law of refraction, $n_2 \sin \theta_2 = n_1 \sin \theta_1$. Take $n_1 = 1$ and $n_2 = 1.33$ (from Table 34–1). Then

$$\theta_2 = \sin^{-1}\left(\frac{\sin \theta_1}{n_2}\right) = \sin^{-1}\left(\frac{\sin 35.0°}{1.33}\right) = 25.55°.$$

$L$ is given by

$$L = \ell_2 \tan \theta_2 = (1.50\,\text{m}) \tan 25.55° = 0.72\,\text{m}.$$

The length of the shadow is $0.35\,\text{m} + 0.72\,\text{m} = 1.07\,\text{m}$.

## 19

The angle of incidence $\theta_B$ for which reflected light is fully polarized is given by Eq. 35–8 of the text. If $n_1$ is the index of refraction for the medium of incidence and $n_2$ is the index of refraction for the second medium, then $\theta_B = \tan^{-1}\left(n_2/n_1\right) = \tan^{-1}\left(1.53/1.33\right) = 63.8°$.

## 23

(a) There are three images. Two are formed by single reflections from each of the mirrors and the third is formed by successive reflections from both mirrors.

(b) The positions of the images are shown on the two diagrams below. The diagram on the left below shows the image $I_1$, formed by reflections from the left-hand mirror. It is the same distance behind the mirror as the object $O$ is in front and is on the line that is perpendicular to the mirror and through the object. Image $I_2$ is formed by light that is reflected from both mirrors. You may consider $I_2$ to be the image of $I_1$ formed by the right-hand mirror, extended. $I_2$ is the same distance behind the line of the right-hand mirror as $I_1$ is in front and it is on the line that is perpendicular to the line of the mirror.

The diagram on the right below shows image $I_3$, formed by reflections from the right-hand mirror. It is the same distance behind the mirror as the object is in front and is on the line that is perpendicular to the mirror and through the object. As the diagram shows, light that is first reflected from the right-hand mirror and then from the left-hand mirror forms an image at $I_2$.

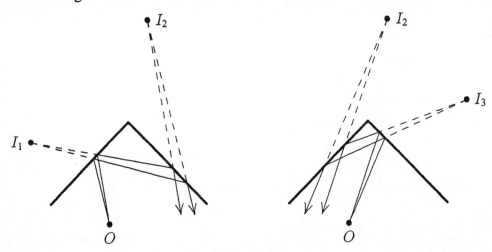

## 27

The intensity of light from a point source varies as the inverse of the square of the distance from the source. Before the mirror is in place, the intensity at the center of the screen is given by $I_0 = A/d^2$, where $A$ is a constant of proportionality. After the mirror is in place, the light that goes directly to the screen contributes intensity $I_0$, as before. Reflected light also reaches the screen. This light appears to come from the image of the source, a distance $d$ behind the mirror and a distance $3d$ from the screen. Its contribution to the intensity at the center of the screen is

$$ I_r = \frac{A}{(3d)^2} = \frac{A}{9d^2} = \frac{I_0}{9}. $$

The total intensity at the center of the screen is

$$ I = I_0 + I_r = I_0 + \frac{I_0}{9} = \frac{10}{9} I_0. $$

The ratio of the new intensity to the original intensity is $I/I_0 = 10/9$.

## 31

(a) Suppose one end of the object is a distance $o$ from the mirror and the other end is a distance $o + L$. The position $i_1$ of the image of the first end is given by

$$ \frac{1}{o} + \frac{1}{i_1} = \frac{1}{f}, $$

where $f$ is the focal length of the mirror. Thus

$$i_1 = \frac{fo}{o - f}.$$

The image of the other end is at

$$i_2 = \frac{f(o + L)}{o + L - f},$$

so the length of the image is

$$L' = i_1 - i_2 = \frac{fo}{o - f} - \frac{f(o + L)}{o + L - f} = \frac{f^2 L}{(o - f)(o + L - f)}.$$

Since the object is short compared to $o - f$, we may neglect the $L$ in the denominator and write

$$L' = L \left( \frac{f}{o - f} \right)^2.$$

(b) The lateral magnification is $m = -i/o$ and since $i = fo/(o - f)$, this can be written $m = -f/(o - f)$. The longitudinal magnification is

$$m' = \frac{L'}{L} = \left( \frac{f}{o - f} \right)^2 = m^2.$$

## 33

If $o$ is the object distance, $i$ is the image distance, and $r$ is the radius of curvature of a spherical refracting surface, then

$$\frac{n_1}{o} + \frac{n_2}{i} = \frac{n_2 - n_1}{r},$$

where $n_1$ is the index of refraction of the medium of incidence and $n_2$ is the index of refraction of the medium of refraction.

(a) In this case $o = \infty$, $n_1 = 1.00$, $i = 2r$, and $n_2 = n$, so

$$\frac{n}{2r} = \frac{n - 1.00}{r}.$$

The solution for $n$ is $n = 2.00$.

(b) Now $i = r$, so

$$\frac{n}{r} = \frac{n - 1}{r}.$$

There is no solution for $n$. An image cannot be formed at the center of the sphere.

## 39

Use the lens maker's equation, Eq. 35-19:

$$\frac{1}{f} = (n - 1) \left( \frac{1}{r_1} - \frac{1}{r_2} \right),$$

where $f$ is the focal length, $n$ is the index of refraction, $r_1$ is the radius of curvature of the first surface encountered by the light and $r_2$ is the radius of curvature of the second surface. Since one surface has twice the radius of the other and since one surface is convex to the incoming light while the other is concave, set $r_2 = -2r_1$ to obtain

$$\frac{1}{f} = (n-1)\left(\frac{1}{r_1} + \frac{1}{2r_1}\right) = \frac{3(n-1)}{2r_1}.$$

Solve for $r_1$:

$$r_1 = \frac{3(n-1)f}{2} = \frac{3(1.5-1)(60\,\text{mm})}{2} = 45\,\text{mm}.$$

The radii are 45 mm and 90 mm.

## 45

For an object in front of a thin lens, the object distance $o$ and the image distance $i$ are related by $(1/o) + (1/i) = (1/f)$, where $f$ is the focal length of the lens. For the situation described by the problem all quantities are positive, so the distance $x$ between the object and image is $x = o + i$. Substitute $i = x - o$ into the thin lens equation and solve for $x$. You should get

$$x = \frac{o^2}{p-f}.$$

To find the minimum value of $x$, set $dx/do = 0$ and solve for $o$. Since

$$\frac{dx}{do} = \frac{o(o-2f)}{(o-f)^2},$$

the result is $o = 2f$. The minimum distance is

$$x_{\text{min}} = \frac{o^2}{o-f} = \frac{(2f)^2}{2f-f} = 4f.$$

This is a minimum, rather than a maximum, since the image distance $i$ becomes large without bound as the object approaches the focal point.

## 49

Place an object far away from the composite lens and find the image distance $i$. Since the image is at a focal point, $i = f$, the effective focal length of the composite. The final image is produced by two lenses, with the image of the first lens being the object for the second. For the first lens, $(1/o_1) + (1/i_1) = (1/f_1)$, where $f_1$ is the focal length of this lens and $i_1$ is the image distance for the image it forms. Since $o_1 = \infty$, $i_1 = f_1$.

The thin lens equation, applied to the second lens, is $(1/o_2) + (1/i_2) = (1/f_2)$, where $o_2$ is the object distance, $i_2$ is the image distance, and $f_2$ is the focal length. If the thicknesses of the lenses can be ignored, the object distance for the second lens is $o_2 = -i_1$. The negative sign must be used since the image formed by the first lens is beyond the second lens if $i_1$ is positive. This means the object for the second lens is virtual and the object distance is negative. If $i_1$ is negative, the image formed by the first lens is in front of the second lens and $o_2$ is positive. In the thin lens equation, replace $o_2$ with $-f_1$ and $i_2$ with $f$ to obtain

$$-\frac{1}{f_1} + \frac{1}{f} = \frac{1}{f_2}$$

or

$$\frac{1}{f} = \frac{1}{f_1} + \frac{1}{f_2} = \frac{f_1 + f_2}{f_1 f_2}.$$

Thus

$$f = \frac{f_1 f_2}{f_1 + f_2}.$$

## 51

(a) If the object distance is $x$, then the image distance is $D - x$ and the thin lens equation becomes

$$\frac{1}{x} + \frac{1}{D - x} = \frac{1}{f}.$$

Multiply each term in the equation by $fx(D - x)$ to obtain $x^2 - Dx + Df = 0$. Solve for $x$. The two object distances for which images are formed on the screen are

$$x_1 = \frac{D - \sqrt{D(D - 4f)}}{2}$$

and

$$x_2 = \frac{D + \sqrt{D(D - 4f)}}{2}.$$

The distance between the two object positions is

$$d = x_2 - x_1 = \sqrt{D(D - 4f)}.$$

(b) The ratio of the image sizes is the same as the ratio of the lateral magnifications. If the object is at $o = x_1$, the magnitude of the lateral magnification is

$$|m_1| = \frac{i_1}{o_1} = \frac{D - x_1}{x_1}.$$

Now $x_1 = \frac{1}{2}(D - d)$, where $d = \sqrt{D(D - f)}$, so

$$|m_1| = \frac{D - (D - d)/2}{(D - d)/2} = \frac{D + d}{D - d}.$$

Similarly, when the object is at $x_2$, the magnitude of the lateral magnification is

$$|m_2| = \frac{I_2}{p_2} = \frac{D - x_2}{x_2} = \frac{D - (D + d)/2}{(D + d)/2} = \frac{D - d}{D + d}.$$

The ratio of the magnifications is

$$\frac{m_2}{m_1} = \frac{(D - d)/(D + d)}{(D + d)/(D - d)} = \left(\frac{D - d}{D + d}\right)^2.$$

## 53

(a) If $L$ is the distance between the lenses, then according to Fig. 35–28, the tube length is $s = L - f_{obj} - f_{eye} = 25.0\,\text{cm} - 4.00\,\text{cm} - 8.00\,\text{cm} = 13.0\,\text{cm}.$

(b) Solve $(1/o) + (1/i) = (1/f_{obj})$ for $o$. The image distance is $i = f_{obj} + s = 4.00\,\text{cm} + 13.0\,\text{cm} = 17.0\,\text{cm}$, so

$$o = \frac{i\,f_{obj}}{i - f_{obj}} = \frac{(17.0\,\text{cm})(4.00\,\text{cm})}{17.0\,\text{cm} - 4.00\,\text{cm}} = 5.23\,\text{cm}\,.$$

(c) The magnification of the objective is

$$m = -\frac{i}{o} = -\frac{17.0\,\text{cm}}{5.23\,\text{cm}} = -3.25\,.$$

(d) The angular magnification of the eyepiece is

$$m_\theta = \frac{25\,\text{cm}}{f_{eye}} = \frac{25\,\text{cm}}{8.00\,\text{cm}} = 3.13\,.$$

(e) The overall magnification of the microscope is

$$M = m m_\theta = (-3.25)(3.13) = -10.2\,.$$

# Chapter 36

## 7

(a) Take the phases of both waves to be zero at the front surfaces of the layers. The phase of the first wave at the back surface of the glass is given by

$$\phi_1 = k_1 L - \omega t,$$

where $k_1$ $(= 2\pi/\lambda_1)$ is the wave number and $\lambda_1$ is the wavelength in glass. Similarly, the phase of the second wave at the back surface of the plastic is given by

$$\phi_2 = k_2 L - \omega t,$$

where $k_2$ $(= 2\pi/\lambda_2)$ is the wave number and $\lambda_2$ is the wavelength in plastic. The angular frequencies are the same since the waves have the same wavelength in air and the frequency of a wave does not change when the wave enters another medium. The phase difference is

$$\phi_1 - \phi_2 = (k_1 - k_2)L = 2\pi \left( \frac{1}{\lambda_1} - \frac{1}{\lambda_2} \right) L.$$

Now $\lambda_1 = \lambda_{\text{air}}/n_1$, where $\lambda_{\text{air}}$ is the wavelength in air and $n_1$ is the index of refraction of the glass. Similarly, $\lambda_2 = \lambda_{\text{air}}/n_2$, where $n_2$ is the index of refraction of the plastic. This means that the phase difference is

$$\phi_1 - \phi_2 = \frac{2\pi}{\lambda_{\text{air}}}(n_1 - n_2)L.$$

The value of $L$ that makes this 5.65 rad is

$$L = \frac{(\phi_1 - \phi_2)\lambda_{\text{air}}}{2\pi(n_1 - n_2)} = \frac{5.65(400 \times 10^{-9}\,\text{m})}{2\pi(1.60 - 1.50)} = 3.60 \times 10^{-6}\,\text{m}.$$

(b) 5.65 rad is less than $2\pi$ rad (= 6.28 rad), the phase difference for completely constructive interference, and greater than $\pi$ rad (= 3.14 rad), the phase difference for completely destructive interference. The interference is, therefore, intermediate, neither completely constructive nor completely destructive. It is, however, closer to completely constructive than to completely destructive.

## 15

The angular positions of the maxima of a two-slit interference pattern are given by $d \sin \theta = m\lambda$, where $d$ is the slit separation, $\lambda$ is the wavelength, and $m$ is an integer. If $\theta$ is small, $\sin \theta$ may be approximated by $\theta$ in radians. Then $d\theta = m\lambda$. The angular separation of two adjacent maxima is $\Delta\theta = \lambda/d$. Let $\lambda'$ be the wavelength for which the angular separation is 10.0% greater. Then $1.10\lambda/d = \lambda'/d$ or $\lambda' = 1.10\lambda = 1.10(589\,\text{nm}) = 648\,\text{nm}$.

## 19

The maxima of a two-slit interference pattern are at angles $\theta$ given by $d \sin \theta = m\lambda$, where $d$ is the slit separation, $\lambda$ is the wavelength, and $m$ is an integer. If $\theta$ is small, $\sin \theta$ may be replaced by $\theta$ in radians. Then $d\theta = m\lambda$. The angular separation of two maxima associated with different wavelengths but the same value of $m$ is $\Delta\theta = (m/d)(\lambda_2 - \lambda_1)$ and the separation on a screen a distance $D$ away is

$$\Delta y = D \tan \Delta\theta \approx D \, \Delta\theta = \left[\frac{mD}{d}\right] (\lambda_2 - \lambda_1)$$

$$= \left[\frac{3(1.0\,\text{m})}{5.0 \times 10^{-3}\,\text{m}}\right] (600 \times 10^{-9}\,\text{m} - 480 \times 10^{-9}\,\text{m}) = 7.2 \times 10^{-5}\,\text{m}.$$

The small angle approximation $\tan \Delta\theta \approx \Delta\theta$ was made. $\Delta\theta$ must be in radians.

## 21

Consider the two waves, one from each slit, that produce the seventh bright fringe in the absence of the mica. They are in phase at the slits and travel different distances to the seventh bright fringe, where they are out of phase by $2\pi m = 14\pi$. Now a piece of mica with thickness $x$ is placed in front of one of the slits and the waves are no longer in phase at the slits. In fact, their phases at the slits differ by

$$\frac{2\pi x}{\lambda_m} - \frac{2\pi x}{\lambda} = \frac{2\pi x}{\lambda}(n-1),$$

where $\lambda_m$ is the wavelength in the mica and $n$ is the index of refraction of the mica. The relationship $\lambda_m = \lambda/n$ was used to substitute for $\lambda_m$. Since the waves are now in phase at the screen,

$$\frac{2\pi x}{\lambda}(n-1) = 14\pi$$

or

$$x = \frac{7\lambda}{n-1} = \frac{7(550 \times 10^{-9}\,\text{m})}{1.58 - 1} = 6.64 \times 10^{-6}\,\text{m}.$$

## 27

(a) To get to the detector, the wave from $S_1$ travels a distance $x$ and the wave from $S_2$ travels a distance $\sqrt{d^2 + x^2}$. The difference in phase of the two waves is

$$\Delta\phi = \frac{2\pi}{\lambda}\left[\sqrt{d^2 + x^2} - x\right],$$

where $\lambda$ is the wavelength. For a maximum in intensity, this must be a multiple of $2\pi$. Solve

$$\sqrt{d^2 + x^2} - x = m\lambda$$

for $x$. Here $m$ is an integer. Write the equation as $\sqrt{d^2 + x^2} = x + m\lambda$, then square both sides to obtain $d^2 + x^2 = x^2 + m^2\lambda^2 + 2m\lambda x$. The solution is

$$x = \frac{d^2 - m^2\lambda^2}{2m\lambda}.$$

The largest value of $m$ that produces a positive value for $x$ is $m = 3$. This corresponds to the maximum that is nearest $S_1$, at

$$x = \frac{(4.00\,\text{m})^2 - 9(1.00\,\text{m})^2}{(2)(3)(1.00\,\text{m})} = 1.17\,\text{m}\,.$$

For the next maximum, $m = 2$ and $x = 3.00\,\text{m}$. For the third maximum, $m = 1$ and $x = 7.50\,\text{m}$.

(b) Minima in intensity occur where the phase difference is $\pi$ rad; the intensity at a minimum, however, is not zero because the amplitudes of the waves are different. Although the amplitudes are the same at the sources, the waves travel different distances to get to the points of minimum intensity and each amplitude decreases in inverse proportion to the distance traveled.

## 29

Take the electric field of one wave, at the screen, to be

$$E_1 = E_0 \sin(\omega t)$$

and the electric field of the other to be

$$E_2 = 2E_0 \sin(\omega t + \Delta\phi)\,,$$

where the phase difference is given by

$$\Delta\phi = \left(\frac{2\pi d}{\lambda}\right)\sin\theta\,.$$

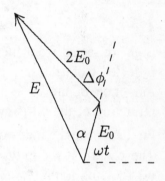

Here $d$ is the center-to-center slit separation and $\lambda$ is the wavelength. The resultant wave can be written $E = E_1 + E_2 = E\sin(\omega t + \alpha)$, where $\alpha$ is a phase constant. The phasor diagram is shown above.

The resultant amplitude $E$ is given by the trigonometric law of cosines:

$$E^2 = E_0^2 + (2E_0)^2 - 4E_0^2\cos(180° - \Delta\phi) = E_0^2(5 + 4\cos\Delta\phi)\,.$$

The intensity is given by

$$I = I_0(5 + 4\cos\Delta\phi)\,,$$

where $I_0$ is the intensity that would be produced by the first wave if the second were not present. Since $\cos\Delta\phi = 2\cos^2(\Delta\phi/2) - 1$, this may also be written

$$I = I_0\left[1 + 8\cos^2(\Delta\phi/2)\right]\,.$$

## 35

Light reflected from the front surface of the coating suffers a phase change of $\pi$ rad while light reflected from the back surface does not change phase. If $L$ is the thickness of the coating, light reflected from the back surface travels a distance $2L$ farther than light reflected from the front surface. The difference in phase of the two waves is $2L(2\pi/\lambda_c) - \pi$, where $\lambda_c$ is the wavelength in the coating. If $\lambda$ is the wavelength in vacuum, then $\lambda_c = \lambda/n$, where $n$ is the index of refraction of the coating. Thus the phase difference is $2nL(2\pi/\lambda) - \pi$. For fully constructive interference, this should be a multiple of $2\pi$. Solve

$$2nL\left(\frac{2\pi}{\lambda}\right) - \pi = 2m\pi$$

for $L$. Here $m$ is an integer. The solution is

$$L = \frac{(2m+1)\lambda}{4n}.$$

To find the smallest coating thickness, take $m = 0$. Then

$$L = \frac{\lambda}{4n} = \frac{560 \times 10^{-9}\,\text{m}}{4(2.00)} = 7.00 \times 10^{-8}\,\text{m}.$$

## 41P

Light reflected from the upper oil surface (in contact with air) changes phase by $\pi$ rad. Light reflected from the lower surface (in contact with glass) changes phase by $\pi$ rad if the index of refraction of the oil is less than that of the glass and does not change phase if the index of refraction of the oil is greater than that of the glass.

First suppose the index of refraction of the oil is greater than the index of refraction of the glass. The condition for fully destructive interference is $2n_o d = m\lambda$, where $d$ is the thickness of the oil film, $n_o$ is the index of refraction of the oil, $\lambda$ is the wavelength in vacuum, and $m$ is an integer. For the shorter wavelength $2n_o d = m_1\lambda_1$ and for the longer $2n_o d = m_2\lambda_2$. Since $\lambda_1$ is less than $\lambda_2$, $m_1$ is greater than $m_2$, and since fully destructive interference does not occur for any wavelengths between, $m_1 = m_2 + 1$. Solve $(m_2 + 1)\lambda_1 = m_2\lambda_2$ for $m_2$. The result is

$$m_2 = \frac{\lambda_1}{\lambda_2 - \lambda_1} = \frac{500\,\text{nm}}{700\,\text{nm} - 500\,\text{nm}} = 2.50.$$

Since $m_2$ must be an integer, the oil cannot have an index of refraction that is greater than that of the glass.

Now suppose the index of refraction of the oil is less than that of the glass. The condition for fully destructive interference is then $2n_o d = (2m+1)\lambda$. For the shorter wavelength $2m_o d = (2m_1+1)\lambda_1$, and for the longer $2n_o d = (2m_2 + 1)\lambda_2$. Again, $m_1 = m_2 + 1$, so $(2m_2 + 3)\lambda_1 = (2m_2 + 1)\lambda_2$. This means the value of $m_2$ is

$$m_2 = \frac{3\lambda_1 - \lambda_2}{2(\lambda_2 - \lambda_1)} = \frac{3(500\,\text{nm}) - 700\,\text{nm}}{2(700\,\text{nm} - 500\,\text{nm})} = 2.00.$$

This is an integer. Thus the index of refraction of the oil is less than that of the glass.

## 45

Assume the wedge-shaped film is in air, so the wave reflected from one surface undergoes a phase change of $\pi$ rad while the wave reflected from the other surface does not. At a place where the film thickness is $L$, the condition for fully constructive interference is $2nL = (m+\frac{1}{2})\lambda$, where $n$ is the index of refraction of the film, $\lambda$ is the wavelength in vacuum, and $m$ is an integer. The ends of the film are bright. Suppose the end where the film is narrow has thickness $L_1$ and the bright fringe there corresponds to $m = m_1$. Suppose the end where the film is thick has thickness $L_2$ and the bright fringe there corresponds to $m = m_2$. Since there are ten bright fringes, $m_2 = m_1 + 9$. Subtract $2nL_1 = (m_1 + \frac{1}{2})\lambda$ from $2nL_2 = (m_1 + 9 + \frac{1}{2})\lambda$ to obtain $2n\,\Delta L = 9\lambda$, where $\Delta L = L_2 - L_1$ is the change in the film thickness over its length. Thus

$$\Delta L = \frac{9\lambda}{2n} = \frac{9(630 \times 10^{-9}\,\text{m})}{2(1.50)} = 1.89 \times 10^{-6}\,\text{m}.$$

## 49

Consider the interference pattern formed by waves reflected from the upper and lower surfaces of the air wedge. The wave reflected from the lower surface undergoes a $\pi$-rad phase change while the wave reflected from the upper surface does not. At a place where the thickness of the wedge is $d$ the condition for a maximum in intensity is

$$2d = (m + \tfrac{1}{2})\lambda,$$

where $\lambda$ is the wavelength in air and $m$ is an integer. Thus

$$d = \frac{(2m + 1)\lambda}{4}.$$

As the geometry of Fig. 36–26a shows, $d = R - \sqrt{R^2 - r^2}$, where $R$ is the radius of curvature of the lens and $r$ is the radius of a Newton's ring. Thus

$$\frac{(2m + 1)\lambda}{4} = R - \sqrt{R^2 - r^2}.$$

Solve for $r$. First rearrange the terms so the equation becomes

$$\sqrt{R^2 - r^2} = R - \frac{(2m + 1)\lambda}{4}.$$

Now square both sides and solve for $r^2$. When you take the square root, you should get

$$r = \sqrt{\frac{(2m + 1)R\lambda}{2} - \frac{(2m + 1)^2\lambda^2}{16}}.$$

If $R$ is much larger than a wavelength, the first term dominates the second and

$$r = \sqrt{\frac{(2m + 1)R\lambda}{2}}.$$

## 57

Let $\phi_1$ be the phase difference of the waves in the two arms when the tube has air in it and let $\phi_2$ be the phase difference when the tube is evacuated. These are different because the wavelength in air is different from the wavelength in vacuum. If $\lambda$ is the wavelength in vacuum, then the wavelength in air is $\lambda/n$, where $n$ is the index of refraction of air. This means

$$\phi_1 - \phi_2 = 2L\left[\frac{2\pi n}{\lambda} - \frac{2\pi}{\lambda}\right] = \frac{4\pi(n-1)L}{\lambda},$$

where $L$ is the length of the tube. The factor 2 arises because the light traverses the tube twice, once on the way to a mirror and once after reflection from the mirror.

Each shift by one fringe corresponds to a change in phase of $2\pi$ rad, so if the interference pattern shifts by $N$ fringes as the tube is evacuated,

$$\frac{4\pi(n-1)L}{\lambda} = 2N\pi$$

and

$$n = 1 + \frac{N\lambda}{2L} = 1 + \frac{60(500 \times 10^{-9}\,\text{m})}{2(5.0 \times 10^{-2}\,\text{m})} = 1.00030\,.$$

# Chapter 37

## 5

(a) A plane wave is incident on the lens so it is brought to focus in the focal plane of the lens, a distance of 70 cm from the lens.

(b) Waves leaving the lens at an angle $\theta$ to the forward direction interfere to produce an intensity minimum if $a \sin \theta = m\lambda$, where $a$ is the slit width, $\lambda$ is the wavelength, and $m$ is an integer. The distance on the screen from the center of the pattern to the minimum is given by $y = D \tan \theta$, where $D$ is the distance from the lens to the screen. For the conditions of this problem,

$$\sin \theta = \frac{m\lambda}{a} = \frac{(1)(590 \times 10^{-9}\,\text{m})}{0.40 \times 10^{-3}\,\text{m}} = 1.475 \times 10^{-3}\,.$$

This means $\theta = 1.475 \times 10^{-3}$ rad and $y = (70 \times 10^{-2}\,\text{m}) \tan(1.475 \times 10^{-3}\,\text{rad}) = 1.03 \times 10^{-3}\,\text{m}$.

## 7

The condition for a minimum of intensity in a single-slit diffraction pattern is $a \sin \theta = m\lambda$, where $a$ is the slit width, $\lambda$ is the wavelength, and $m$ is an integer. To find the angular position of the first minimum to one side of the central maximum, set $m = 1$:

$$\theta_1 = \sin^{-1}\left(\frac{\lambda}{a}\right) = \sin^{-1}\left(\frac{589 \times 10^{-9}\,\text{m}}{1.00 \times 10^{-3}\,\text{m}}\right) = 5.89 \times 10^{-4}\,\text{rad}\,.$$

If $D$ is the distance from the slit to the screen, the distance on the screen from the center of the pattern to the minimum is $y_1 = D \tan \theta_1 = (3.00\,\text{m}) \tan(5.89 \times 10^{-4}\,\text{rad}) = 1.767 \times 10^{-3}\,\text{m}$.

To find the second minimum, set $m = 2$:

$$\theta_2 = \sin^{-1}\left(\frac{2(589 \times 10^{-9}\,\text{m})}{1.00 \times 10^{-3}\,\text{m}}\right) = 1.178 \times 10^{-3}\,\text{rad}\,.$$

The distance from the pattern center to the minimum is $y_2 = D \tan \theta_2 = (3.00\,\text{m}) \tan(1.178 \times 10^{-3}\,\text{rad}) = 3.534 \times 10^{-3}\,\text{m}$. The separation of the two minima is $\Delta y = y_2 - y_1 = 3.534\,\text{mm} - 1.767\,\text{mm} = 1.77\,\text{mm}$.

## 11

(a) The intensity for a single-slit diffraction pattern is given by

$$I = I_m \frac{\sin^2 \alpha}{\alpha^2}\,,$$

where $\alpha = (\pi a/\lambda) \sin \theta$, $a$ is the slit width and $\lambda$ is the wavelength. The angle $\theta$ is measured from the forward direction. You want $I = I_m/2$, so

$$\sin^2 \alpha = \tfrac{1}{2}\alpha^2\,.$$

(b) Evaluate $\sin^2 \alpha$ and $\alpha^2/2$ for $\alpha = 1.39\,\mathrm{rad}$ and compare the results. To be sure that $1.39\,\mathrm{rad}$ is closer to the correct value for $\alpha$ than any other value with three significant digits, you should also try $1.385\,\mathrm{rad}$ and $1.395\,\mathrm{rad}$.

(c) Since $\alpha = (\pi a/\lambda)\sin\theta$,

$$\theta = \sin^{-1}\left(\frac{\alpha\lambda}{\pi a}\right).$$

Now $\alpha/\pi = 1.39/\pi = 0.442$, so

$$\theta = \sin^{-1}\left(\frac{0.442\lambda}{a}\right).$$

The angular separation of the two points of half intensity, one on either side of the center of the diffraction pattern, is

$$\Delta\theta = 2\theta = 2\sin^{-1}\left(\frac{0.442\lambda}{a}\right).$$

(d) For $a/\lambda = 1.0$,

$$\Delta\theta = 2\sin^{-1}(0.442/1.0) = 0.916\,\mathrm{rad} = 52.5°,$$

for $a/\lambda = 5.0$,

$$\Delta\theta = 2\sin^{-1}(0.442/5.0) = 0.177\,\mathrm{rad} = 10.1°,$$

and for $a/\lambda = 10$,

$$\Delta\theta = 2\sin^{-1}(0.442/10) = 0.0884\,\mathrm{rad} = 5.06°.$$

## 13

(a) The intensity for a single-slit diffraction pattern is given by

$$I = I_m \frac{\sin^2 \alpha}{\alpha^2},$$

where $\alpha = (\pi a/\lambda)\sin\theta$. Here $a$ is the slit width and $\lambda$ is the wavelength. To find the maxima and minima, set the derivative of $I$ with respect to $\alpha$ equal to zero and solve for $\alpha$. The derivative is

$$\frac{dI}{d\alpha} = 2I_m \frac{\sin\alpha}{\alpha^3}(\alpha\cos\alpha - \sin\alpha).$$

The derivative vanishes if $\alpha \neq 0$ but $\sin\alpha = 0$. This yields $\alpha = m\pi$, where $m$ is an integer. Except for $m = 0$, these are the intensity minima: $I = 0$ for $\alpha = m\pi$.

The derivative also vanishes for $\alpha\cos\alpha - \sin\alpha = 0$. This condition can be written $\tan\alpha = \alpha$. These are the maxima.

(b) The values of $\alpha$ that satisfy $\tan \alpha = \alpha$ can be found by trial and error on a pocket calculator or computer. Each of them is slightly less than one of the values $(m + \frac{1}{2})\pi$ rad, so start with these values. The first few are 0, 4.4934, 7.7252, 10.9041, 14.0662, and 17.2207. They can also be found graphically. As in the diagram to the right, plot $y = \tan \alpha$ and $y = \alpha$ on the same graph. The intersections of the line with the $\tan \alpha$ curves are the solutions. The first solution listed above is shown on the diagram.

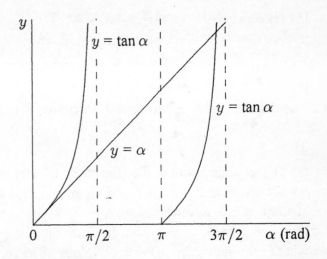

(c) Write $\alpha = (m + \frac{1}{2})\pi$ for the maxima. For the central maximum, $\alpha = 0$ and $m = -\frac{1}{2}$. For the next, $\alpha = 4.4934$ and $m = 0.930$. For the next, $\alpha = 7.7252$ and $m = 1.959$.

## 19

(a) Use the Rayleigh criteria: two objects can be resolved if their angular separation at the observer is greater than $\theta_R = 1.22\lambda/d$, where $\lambda$ is the wavelength of the light and $d$ is the diameter of the aperture (the eye or mirror). If $D$ is the distance from the observer to the objects, then the smallest separation $\ell$ they can have and still be resolvable is $\ell = D \tan \theta_R \approx D\theta_R$, where $\theta_R$ is measured in radians. The small angle approximation $\tan \theta_R \approx \theta_R$ was made. Thus

$$\ell = \frac{1.22D\lambda}{d} = \frac{1.22(8.0 \times 10^{10}\,\text{m})(550 \times 10^{-9}\,\text{m})}{5.0 \times 10^{-3}\,\text{m}} = 1.1 \times 10^7\,\text{m} = 1.1 \times 10^4\,\text{km}\,.$$

This distance is greater than the diameter of Mars. One part of the planet's surface cannot be resolved from another part.

(b) Now $d = 5.1$ m and

$$\ell = \frac{1.22(8.0 \times 10^{10}\,\text{m})(550 \times 10^{-9}\,\text{m})}{5.1\,\text{m}} = 1.1 \times 10^4\,\text{m} = 11\,\text{km}\,.$$

## 23

(a) The first minimum in the diffraction pattern is at an angular position $\theta$, measured from the center of the pattern, such that $\sin\theta = 1.22\lambda/d$, where $\lambda$ is the wavelength and $d$ is the diameter of the antenna. If $f$ is the frequency, then the wavelength is

$$\lambda = \frac{c}{f} = \frac{3.00 \times 10^8\,\text{m/s}}{220 \times 10^9\,\text{Hz}} = 1.36 \times 10^{-3}\,\text{m}\,.$$

Thus

$$\theta = \sin^{-1}\left(\frac{1.22\lambda}{d}\right) = \sin^{-1}\left(\frac{1.22(1.36 \times 10^{-3}\,\text{m})}{55.0 \times 10^{-2}\,\text{m}}\right) = 3.02 \times 10^{-3}\,\text{rad}\,.$$

The angular width of the central maximum is twice this, or $6.04 \times 10^{-3}$ rad (0.346°).

(b) Now $\lambda = 1.6\,\text{cm}$ and $d = 2.3\,\text{m}$, so

$$\theta = \sin^{-1}\left(\frac{1.22(1.6 \times 10^{-2}\,\text{m})}{2.3\,\text{m}}\right) = 8.5 \times 10^{-3}\,\text{rad}.$$

The angular width of the central maximum is $1.7 \times 10^{-2}\,\text{rad}$ ($0.97°$).

## 31

(a) The angular positions $\theta$ of the bright interference fringes are given by $d\sin\theta = m\lambda$, where $d$ is the slit separation, $\lambda$ is the wavelength, and $m$ is an integer. The first diffraction minimum occurs at the angle $\theta_1$ given by $a\sin\theta_1 = \lambda$, where $a$ is the slit width. The diffraction peak extends from $-\theta_1$ to $+\theta_1$, so you want to count the number of values of $m$ for which $-\theta_1 < \theta < +\theta_1$, or what is the same, the number of values of $m$ for which $-\sin\theta_1 < \sin\theta < +\sin\theta_1$. This means $-1/a < m/d < 1/a$ or $-d/a < m < +d/a$. Now $d/a = (0.150 \times 10^{-3}\,\text{m})/(30.0 \times 10^{-6}\,\text{m}) = 5.00$, so the values of $m$ are $m = -4, -3, -2, -1, 0, +1, +2, +3$, and $+4$. There are nine fringes.

(b) The intensity at the screen is given by

$$I = I_m \left(\cos^2\beta\right)\left(\frac{\sin\alpha}{\alpha}\right)^2,$$

where $\alpha = (\pi a/\lambda)\sin\theta$, $\beta = (\pi d/\lambda)\sin\theta$, and $I_m$ is the intensity at the center of the pattern. For the third bright interference fringe, $d\sin\theta = 3\lambda$, so $\beta = 3\pi$ rad and $\cos^2\beta = 1$. Similarly, $\alpha = 3\pi a/d = 3\pi/5.00 = 0.600\pi$ rad and

$$\left(\frac{\sin\alpha}{\alpha}\right)^2 = \left(\frac{\sin 0.600\pi}{0.600\pi}\right)^2 = 0.255.$$

The intensity ratio is $I/I_m = 0.255$.

## 37

(a) Maxima of a diffraction grating pattern occur at angles $\theta$ given by $d\sin\theta = m\lambda$, where $d$ is the slit separation, $\lambda$ is the wavelength, and $m$ is an integer. The two lines are adjacent, so their order numbers differ by unity. Let $m$ be the order number for the line with $\sin\theta = 0.2$ and $m+1$ be the order number for the line with $\sin\theta = 0.3$. Then $0.2d = m\lambda$ and $0.3d = (m+1)\lambda$. Subtract the first equation from the second to obtain $0.1d = \lambda$, or $d = \lambda/0.1 = (600 \times 10^{-9}\,\text{m})/0.1 = 6.0 \times 10^{-6}\,\text{m}$.

(b) Minima of the single-slit diffraction pattern occur at angles $\theta$ given by $a\sin\theta = m\lambda$, where $a$ is the slit width. Since the fourth-order interference maximum is missing, it must fall at one of these angles. If $a$ is the smallest slit width for which this order is missing, the angle must be given by $a\sin\theta = \lambda$. It is also given by $d\sin\theta = 4\lambda$, so $a = d/4 = (6.0 \times 10^{-6}\,\text{m})/4 = 1.5 \times 10^{-6}\,\text{m}$.

(c) First set $\theta = 90°$ and find the largest value of $m$ for which $m\lambda < d\sin\theta$. This is the highest order that is diffracted toward the screen. The condition is the same as $m < d/\lambda$ and since $d/\lambda = (6.0 \times 10^{-6}\,\text{m})/(600 \times 10^{-9}\,\text{m}) = 10.0$, the highest order seen is the $m = 9$ order. The fourth and eighth orders are missing so the observable orders are $m = 0, 1, 2, 3, 5, 6, 7$, and $9$.

## 45

Since the slit width is much less than the wavelength of the light, the central peak of the single-slit diffraction pattern is spread across the screen and the diffraction envelope can be ignored. Consider three waves, one from each slit. Since the slits are evenly spaced, the phase difference for waves from the first and second slits is the same as the phase difference for waves from the second and third slits. The electric fields of the waves at the screen can be written $E_1 = E_0 \sin(\omega t)$, $E_2 = E_0 \sin(\omega t + \phi)$, and $E_3 = E_0 \sin(\omega t + 2\phi)$, where $\phi = (2\pi d/\lambda) \sin\theta$. Here $d$ is the separation of adjacent slits and $\lambda$ is the wavelength.

The phasor diagram is shown to the right. It yields

$$E = E_0 \cos\phi + E_0 + E_0 \cos\phi = E_0(1 + 2\cos\phi)$$

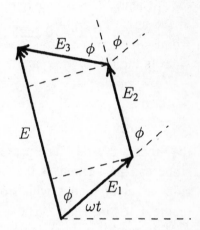

for the amplitude of the resultant wave. Since the intensity of a wave is proportional to the square of the electric field, we may write $I = AE_0^2(1 + 2\cos\phi)^2$, where $A$ is a constant of proportionality. If $I_m$ is the intensity at the center of the pattern, for which $\phi = 0$, then $I_m = 9AE_0^2$. Take $A$ to be $I_m/9E_0^2$ and obtain

$$I = \frac{I_m}{9}(1 + 2\cos\phi)^2 = \frac{I_m}{9}\left(1 + 4\cos\phi + 4\cos^2\phi\right) .$$

## 51

(a) Since the resolving power of a grating is given by $R = \lambda/\Delta\lambda$ and by $Nm$, the range of wavelengths that can just be resolved in order $m$ is $\Delta\lambda = \lambda/Nm$. Here $N$ is the number of rulings in the grating and $\lambda$ is the average wavelength. The frequency $f$ is related to the wavelength by $f\lambda = c$, where $c$ is the speed of light. This means $f\Delta\lambda + \lambda\Delta f = 0$, so $\Delta\lambda = -(\lambda/f)\Delta f = -(\lambda^2/c)\Delta f$, where $f = c/\lambda$ was used. The negative sign means that an increase in frequency corresponds to a decrease in wavelength. We may interpret $\Delta f$ as the range of frequencies that can be resolved and take it to be positive. Then $(\lambda^2/c)\Delta f = (\lambda/Nm)$ and $\Delta f = c/Nm\lambda$.

(b) The difference in travel time for waves traveling along the two extreme rays is $\Delta t = \Delta L/c$, where $\Delta L$ is the difference in path length. The waves originate at slits that are separated by $(N-1)d$, where $d$ is the slit separation and $N$ is the number of slits, so the path difference is $\Delta L = (N-1)d\sin\theta$ and the time difference is $\Delta t = (N-1)(d/c)\sin\theta$. If $N$ is large, this may be approximated by $\Delta t = (Nd/c)\sin\theta$. The lens does not affect the travel time.

(c) Substitute the expressions you derived for $\Delta t$ and $\Delta f$ to obtain

$$\Delta f \Delta t = \left(\frac{c}{Nm\lambda}\right)\left(\frac{Nd\sin\theta}{c}\right) = \frac{d\sin\theta}{m\lambda} = 1 .$$

The condition $d\sin\theta = m\lambda$ for a diffraction line was used to obtain the last result.

## 59P

(a) The sets of planes with the next five smaller interplanar spacings (after $a_0$) are shown in the diagram to the right. In terms of $a_0$, the spacings are:

(i):   $a_0/\sqrt{2} = 0.7071 a_0$

(ii):  $a_0/\sqrt{5} = 0.4472 a_0$

(iii): $a_0/\sqrt{10} = 0.3162 a_0$

(iv):  $a_0/\sqrt{13} = 0.2774 a_0$

(v):   $a_0/\sqrt{17} = 0.2425 a_0$

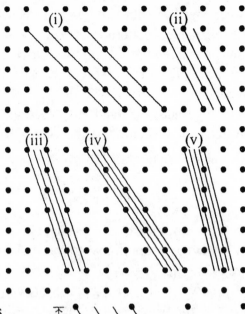

(b) Since a crystal plane passes through lattice points, its slope can be written as the ratio of two integers. Consider a set of planes with slope $m/n$, as shown in the diagram to the right. The first and last planes shown pass through adjacent lattice points along a horizontal line and there are $m - 1$ planes between. If $h$ is the separation of the first and last planes, then the interplanar spacing is $d = h/m$. If the planes make the angle $\theta$ with the horizontal, then the normal to planes (shown dotted) makes the angle $\phi = 90° - \theta$. The distance $h$ is given by $h = a_0 \cos\phi$ and the interplanar spacing is

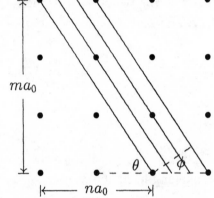

$d = h/m = (a_0/m)\cos\phi$. Since $\tan\theta = m/n$, $\tan\phi = n/m$ and $\cos\phi = 1/\sqrt{1 + \tan^2\phi} = m/\sqrt{n^2 + m^2}$. Thus

$$d = \frac{h}{m} = \frac{a_0 \cos\phi}{m} = \frac{a_0}{\sqrt{n^2 + m^2}} .$$

# Chapter 38

## 7

Assume the flash leaves the origin at time $t = 0$. If the coordinates of your clock are $x$, $y$, and $z$, then the time the flash reaches your clock is

$$t = \frac{\sqrt{x^2 + y^2 + z^2}}{c} = \frac{\sqrt{(8.0\,\text{km})^2 + (40\,\text{km})^2 + (44\,\text{km})^2}}{3.00 \times 10^5\,\text{km/s}} = 2.0 \times 10^{-4}\,\text{s}.$$

This is the time you should set on your clock.

## 9

The diagram might look like that shown on the right. Here $d = c\,\Delta\tau/2 = 7.0$ m and

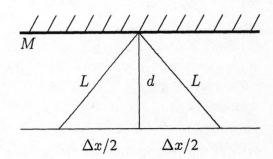

$$L = c\,\Delta t/2 = \sqrt{d^2 + (c\,\Delta x/2)^2}$$
$$= \sqrt{(7.0\,\text{m})^2 + (24\,\text{m})^2} = 25\,\text{m}.$$

Thus $\Delta t/\Delta\tau = L/d = (25\,\text{m})/(7.0\,\text{m}) = 3.6$.

## 15

(a) Your watch measures the proper time $\Delta\tau$ for your trip. The time that elapses on Earth is $\Delta t = \Delta\tau/\sqrt{1 - (v/c)^2}$, where $v$ is your speed relative to Earth. Thus

$$\frac{v}{c} = \sqrt{1 - \left(\frac{\Delta\tau}{\Delta t}\right)}.$$

Now $(\Delta\tau/\Delta t)$ is a very small number, so we use the first two terms of a binomial expansion:

$$\frac{v}{c} \approx 1 - \frac{1}{2}\left(\frac{\Delta\tau}{\Delta t}\right)^2 = 1 - \frac{1}{2}\left(\frac{1.0\,\text{y}}{1000\,\text{y}}\right)^2 = 1 - 5 \times 10^{-7} = 0.999\,999\,5.$$

(b) You age by the proper time, or 1.0 y.

(c) Special relativity is valid as long as the acceleration is small, so you need not travel in a straight line as long as you do not traverse a small-radius circle.

## 21

The mass lost is $\Delta E/c^2$, where $\Delta E$ is the energy dissipated by the bulb. The energy dissipated is $\Delta E = P\,\Delta t$, where $P$ is the rate of energy dissipation and $\Delta t$ is the time. Thus

$$m = \frac{P\,\Delta t}{c^2} = \frac{(100\,\text{W})(365.25\,d)(24\,\text{h/d})(3600\,\text{s/h})}{(3.00 \times 10^8\,\text{m/s})^2} = 3.51 \times 19^{-8}\,\text{kg}.$$

**27**

(a) The kinetic energy is given by

$$K = \frac{mc^2}{\sqrt{1 - (v/c)^2}} - mc^2,$$

and the rest energy by $mc^2$, so

$$\frac{mc^2}{\sqrt{1 - (v/c)^2}} - mc^2 = Nmc^2.$$

The solution for $v$ is

$$v = \frac{\sqrt{N(N+2)}}{N+1} c.$$

(b) The magnitude of the momentum is

$$p = \frac{mv}{\sqrt{1 - (v/c)^2}} = \frac{mc\sqrt{N(N+2)}}{(N+1)\sqrt{1 - \frac{N(N+2)}{(N+1)^2}}} = mc\sqrt{N(N+2)}.$$

**35**

(a) Use the velocity transformation equation

$$u_x = \frac{u'_x + v^{\text{rel}}}{1 + u'_x v_{\text{rel}}/c^2},$$

where $u_x$ is the $x$ component of the velocity of the speed of the flash in the laboratory frame, $u'_x$ is the $x$ component of the velocity of the flash in the rocket frame, and $v^{\text{el}}$ is the speed of the rocket frame relative to the laboratory. In this case $u_x = c\cos\phi$ and $u'_x = c\cos\phi'$. Make these substitutions, then divide by $c$ to obtain

$$\cos\phi = \frac{\cos\phi' + v^{\text{rel}}/c}{1 + (v^{\text{rel}}/c)\cos\phi'}.$$

(b) Take $\phi'$ to be $90°$. Then $\cos\phi' = 0$ and $\cos\phi = v^{\text{rel}}/c$. The faster the rocket the narrower the cone of light.

**39**

(a) In the frame of the person the distance that must be traveled is $\Delta x' = \sqrt{1 - (v^{\text{rel}}/c)^2}\,\Delta x$, where $\Delta x$ is the distance in the frame of the galaxy (23 000 ly) and $v^{\text{el}}$ is the speed of the galaxy in that frame or, what is the same, the speed of the person in the frame of the galaxy. As $v^{\text{el}}$ approaches the speed of light $\Delta x'$ approaches zero, so the person can certainly make the trip in less than a lifetime.

(b) The time to make the trip, as measured by the person, is

$$\Delta t' = \frac{\Delta x'}{v^{\mathrm{rel}}} = \frac{\sqrt{1 - (v^{\mathrm{rel}}/c)^2}}{v^{\mathrm{rel}}} \Delta x .$$

The solution for $v^{\mathrm{rel}}$ is

$$v^{\mathrm{rel}} = \frac{\Delta x}{\sqrt{(\Delta t')^2 + (\Delta x/c)^2}} .$$

Since $\Delta x/c$ is much greater than $\Delta t'$ we use the first two terms of the binomial expansion

$$\left[ (\Delta x/c)^2 + (\Delta t')^2 \right]^{-1/2} \approx \left[ (\Delta x/c)^2 \right]^{-1/2} - \frac{1}{2} \left[ (\Delta x/c)^2 \right]^{-3/2} (\Delta t')^2$$

$$= \frac{1}{(\Delta x/c)} - \frac{(\Delta t')^2}{2(\Delta x/c)^3} .$$

Thus

$$v^{\mathrm{rel}} \approx c - \frac{(\Delta t')^2 c}{(\Delta x/c)^2} .$$

Substitute $\Delta t' = 30\,\mathrm{y}$ and $\Delta x/c = (23\,000\,\mathrm{ly})/(1.0\,\mathrm{ly/y}) = 23\,000\,\mathrm{y}$ to obtain

$$v^{\mathrm{rel}} = c - \frac{(30\,\mathrm{y})^2 c}{2(23\,000\,\mathrm{y})^2} = c - 8.5 \times 10^{-7} c = 0.999\,999\,15\,c .$$

## 45

The spaceship is moving away from Earth, so the frequency received is given by

$$f = f_0 \sqrt{\frac{1 - (v^{\mathrm{rel}}/c)}{1 + (v^{\mathrm{rel}}/c)}} ,$$

where $f_0$ is the frequency in the frame of the spaceship and $v^{\mathrm{rel}}$ is the speed of the spaceship relative to Earth. See Eq. 38–33. Thus

$$f = (100\,\mathrm{MHz}) \sqrt{\frac{1 - 0.9000}{1 + 0.9000}} = 22.9\,\mathrm{MHz} .$$

## 51

The kinetic energy of the proton is given by

$$K = \frac{mc^2}{\sqrt{1 - (v/c)^2}} - mc^2 ,$$

where $m$ is its mass and $v$ is its speed. The solution for $v/c$ is

$$v/c = \sqrt{1 - \frac{(mc^2)^2}{(K + mc^2)^2}} .$$

In the Earth frame the time to cross the galaxy is $\Delta t = L/v$, where $L$ is the diameter of the galaxy. The time in the frame of the proton is the proper time for the trip and is

$$\Delta \tau = \sqrt{1 - (v/c)^2}\, \Delta t = \frac{\sqrt{1 - (v/c)^2}}{v/c}(L/c) = \frac{(mc^2)^2}{(K + mc^2)^2}(L/c) ,$$

where the equation for $v/c$ in terms of $K$ was used to make the substitution for $v/c$.

# NOTES